Lectures in PROJECTIVE GEOMETRY

A. SEIDENBERG

DOVER PUBLICATIONS, INC.
Mineola, New York

Bibliographical Note

This Dover edition, first published in 2005, is an unabridged republication of the edition published by D. Van Nostrand Company, Inc., Princeton, New Jersey, 1962.

Library of Congress Cataloging-in-Publication Data

Seidenberg, A. (Abraham), 1916–
Lectures in projective geometry / A. Seidenberg.
p. cm.
Originally published: Princeton, N.J. : Van Nostrand, 1962, in series: The university series in undergraduate mathematics.
Includes bibliographical references and index.
ISBN 0-486-44618-2 (pbk.)
1. Geometry, Projective. 2. Geometry, Analytic. I. Title.

QA471.S42 2005
516'.5—dc22

2005051967

Manufactured in the United States of America
Dover Publications, Inc., 31 East 2nd Street, Mineola, N.Y. 11501

PREFACE

The purpose of this book is to provide a text for an undergraduate course in Projective Geometry. The axiomatic approach to this subject is undoubtedly the proper one, but it seems desirable, if not necessary, to start from what the student knows, namely high school geometry, and lead him from there to new things. Accordingly, the first chapter is devoted to introducing some of the main topics in as naïve a form as possible. The subject is then begun on an axiomatic basis in the second chapter.

The axiomatic method began in geometry, and projective geometry is an especially suitable subject in which to illustrate the leading ideas of the method. In his previous studies, the student has been asked to take many things for granted. Here, however, he is engaged in a study in which all the elements are in clear view from the beginning.

Aside from the intrinsic interest of projective geometry, there is the fact that the main events of algebraic geometry, one of the principal branches of mathematics, take place in projective spaces. Therefore projective spaces are "something everyone has to know." To be sure, this basic minimum of knowledge can be conveyed efficiently by defining a point of n-space to be a class $\{(\rho x_0, \rho x_1, \ldots, \rho x_n)\}$ of $(n + 1)$-tuples, etc.; but having committed ourselves to the axiomatic method, we are led to showing how coordinates may be introduced on the basis of Pappus' Theorem (in the commutative case) and Desargues' Theorem (in the noncommutative case).

To a large extent, the book deals only with what might be called fundamental: homogeneous coordinates, higher-dimensional spaces, conics, linear transformations, quadric surfaces. Still, a number of topics are entered upon in order to achieve depth, always within the limitations set by the prerequisites for the book, namely, high school geometry and algebra. Among these are the discussion on intersection multiplicities in the case of a pair of conics, the Jordan canonical form, and the factorization of a linear transformation into polarities.

The derivation of Euclidean geometry from projective is touched upon but not entered into systematically: this derivation is undoubtedly a main aspect of the subject, but it lies beyond the goal of this book. A bibliographical note is appended which may be helpful to the student in broadening and deepening his study on the topics considered, and which also directs him to allied but more advanced subjects.

Berkeley, California A. S.
November 1961

CONTENTS

CHAPTER I. PROJECTIVE GEOMETRY AS AN EXTENSION OF HIGH SCHOOL GEOMETRY 1

1. Two approaches to projective geometry 1
2. An initial question 2
3. Projective invariants 3
4. Vanishing points 4
5. Vanishing lines 5
6. Some projective noninvariants 6
7. Betweenness 7
8. Division of a segment in a ratio 8
9. Desargues' Theorem 9
10. Perspectivity; projectivity 12
11. Harmonic tetrads; fourth harmonic 15
12. Further theorems on harmonic tetrads 20
13. The cross-ratio 21
14. Fundamental Theorem of Projective Geometry 25
15. Further remarks on the cross-ratio 27
16. Construction of the projective plane 30
17. Previous results in the constructed plane 33
18. Analytic construction of the projective plane 34
19. Elements of linear equations 38

CHAPTER II. THE AXIOMATIC FOUNDATION 42

1. Unproved propositions and undefined terms 42
2. Requirements on the axioms and undefined terms 42
3. Undefined terms and axioms for a projective plane 43
4. Initial development of the system; the Principle of Duality 43
5. Consistency of the axioms 49
6. Other models 51
7. Independence of the axioms 53

8. Isomorphism 55
9. Further axioms 56
10. Consequences of Desargues' Theorem 60
11. Free planes 65

CHAPTER III. ESTABLISHING COORDINATES IN A PLANE 68

1. Definition of a field 68
2. Consistency of the field axioms 71
3. The analytic model 72
4. Geometric description of the operations plus and times 76
5. Setting up coordinates in the projective plane 78
6. The noncommutative case 86

CHAPTER IV. RELATIONS BETWEEN THE BASIC THEOREMS 89

CHAPTER V. AXIOMATIC INTRODUCTION OF HIGHER-DIMENSIONAL SPACE 96

1. Higher-dimensional, especially 3-dimensional projective space 96
2. Desarguesian planes and higher-dimensional space 102

CHAPTER VI. CONICS 106

1. Study of the conic on the basis of high school geometry 106
2. The conic, axiomatically treated 112
3. The polar 115
4. The polar, axiomatically treated 117
5. Polarities 121

CHAPTER VII. HIGHER-DIMENSIONAL SPACES RESUMED 122

1. Theory of dependence 122
2. Application of the dependency theory to geometry 126
3. Hyperplanes 133
4. The dual space 134
5. The analytic case 137

CHAPTER VIII. COORDINATE SYSTEMS AND LINEAR TRANSFORMATIONS 141

1. Coordinate systems 141
2. Determinants 150
3. Coordinate systems resumed 152
4. Coordinate changes, alias linear transformations 153
5. A generalization from $n = 2$ to $n = 1$ 155
6. Linear transformations on a line and from one line to another 157
7. Cross-ratio 159
8. Coordinate systems and linear transformations in higher-dimensional spaces 160
9. Coordinates in affine space 162

CHAPTER IX. COORDINATE SYSTEMS ABSTRACTLY CONSIDERED 165

1. Definition of a coordinate system 165
2. Definition of a geometric object 170
3. Algebraic curves 172
4. A short cut to PNK 174
5. A result for the field of real numbers 174

CHAPTER X. CONIC SECTIONS ANALYTICALLY TREATED 177

1. Derivation of equation of conic 177
2. Uniqueness of the equation 182
3. Projective equivalence of conics 183
4. Poles and polars 184
5. Polarities and conics 190

APPENDIX TO CHAPTER X 193

A1. Factorization of linear transformations into polarities 193

CHAPTER XI. COORDINATES ON A CONIC 197

1. Coordinates on a conic 197
2. Projectivities on a conic 199

CHAPTER XII. PAIRS OF CONICS 203

1. Pencils of conics 203
2. Intersection multiplicities 204

CHAPTER XIII. QUADRIC SURFACES 208

1. Projectivities between pencils of planes 208
2. Reguli and quadric surfaces 210
3. Quadric surfaces over the complex field 213
4. Some properties of the sphere 214

CHAPTER XIV. THE JORDAN CANONICAL FORM 218

BIBLIOGRAPHICAL NOTE 226

INDEX 229

CHAPTER I

PROJECTIVE GEOMETRY AS AN EXTENSION OF HIGH SCHOOL GEOMETRY

1. Two approaches to projective geometry. There are two ways to study projective geometry: (1) as a continuation of Euclidean geometry as usually taught in high schools, and (2) as an independent discipline, with its own definitions, axioms, theorems, etc. Of these two ways, the first corresponds to the actual historical development of the subject. As Euclidean geometry developed, a large body of theorems having a common character was built up, and theorems having that character (which we will describe more explicitly below) were called *projective theorems*. But precisely because projective geometry arose from Euclidean geometry, attention began to be directed more and more to the axioms. It was for this reason that the so-called "axiomatic method," which nowadays is applied in all branches of mathematics, was first applied to projective geometry, from which subject it gradually spread to all the others.

Of the two ways of studying the subject, each has disadvantages. As to the first method, although Euclid may have been definite enough about his axioms, the high schools frequently are not, and as a consequence one is never quite sure what these axioms were. This is also partly the fault of Euclid, and today we can be somewhat more exact than he was. Another disadvantage is that projective geometry consists of theorems of a certain special type, so that Euclid's axioms, even if made precise, are not entirely suitable.

The axiomatic method of introducing the subject also has disadvantages. For one thing, one has to enter somewhat fully into what is involved in an axiomatic system, and this takes us away from geometry and other fairly familiar things to things that at first sight may seem to be bizarre and to have no bearing on the subject. Besides, the content of an axiomatic system depends on what we want to study: for algebra we would have one system of axioms; for geometry, another;

therefore if we really want to come to grips with geometric problems, we have to know, first, roughly what these problems are.

Of the two methods, we will start with the first. Since we do not say precisely what we are assuming, we shall never be able to tell precisely what we are proving. This kind of situation always gives rise to misunderstandings, especially if we insist that we are actually proving things. The first chapter will therefore not be directed toward estab-

FIG. 1.1

FIG. 1.2

lishing theorems with complete rigor, but toward supplying background material that will be helpful in the axiomatic method part of the book. At the same time, although we will not insist upon complete rigor, neither will we allow ourselves to make obvious mistakes in logic.

2. An initial question. The question toward which projective geometry first directed itself was, "*How do we see things?*" or "*What is the relation between a thing itself, say a figure, and the thing seen?*" More particularly, "Do we ever see a thing exactly as it is?" Take this rectangular table: we know it is rectangular. But a little thought shows us that we do not see it as rectangular. In fact, a little lack of thought will also show this. For suppose we try to draw the table. We draw a rectangle (Fig. 1.1). This is the top of the table. But now we try to put the legs on, and we see that we cannot. To do so, we

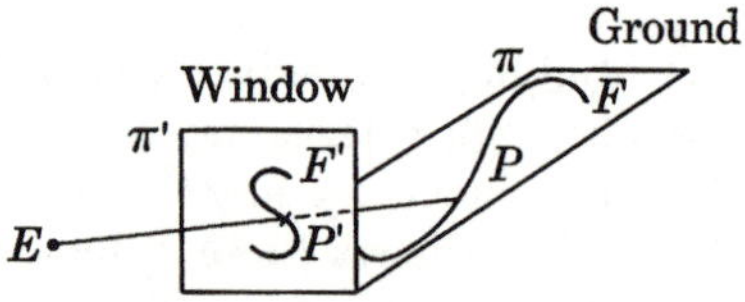

FIG. 1.3

have to draw the rectangle in a nonrectangular shape; then we can put the legs on (Fig. 1.2).

This argument will certainly convince us that angles are not always what they seem. Let us try, however, to make the argument more precise. For the following, we assume only one eye.

Let us suppose we are looking through the window onto the ground (which may be sloping), as illustrated in Fig. 1.3. On the ground is a figure, say a large letter S. The way we see this letter is by light passing from the points of S and entering the eye. We therefore have to consider lines joining the point E, representing the eye, to the various points of the figure S. These lines cut the windowpane in various points. If P is a point of S, let EP cut the windowpane in a point P', which let us call the *image of* P: the set of all the images of the points of the figure S will be called the *image of* S, or its *projection* (from E). We have two figures, then, the figure F, on the ground, and the figure F', on the windowpane. Since the rays of light emerging from F' and reaching E are the same as those emerging from F and reaching E, it is clear that as far as shape is concerned, we see the figure F' in precisely the same way we see the figure F.

Our original question becomes essentially the following: *What is the relation between a figure F, given in a plane π, and its projection in another plane π'?* (The *projection* is said to be from π to π' with *center* E.) Or "What properties of the figure F are also found in F'?" It

FIG. 1.4

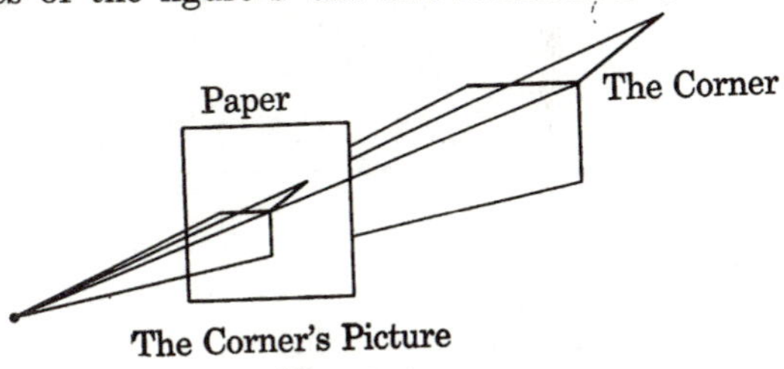

FIG. 1.5

is fairly clear that this is essentially the same question, because if we wanted to draw the figure F on π', we would draw F'.

3. Projective invariants. Let us apply these considerations to the corner of a room (Fig. 1.4). Let us *draw* this corner, as shown in Fig. 1.5. After drawing this picture, we realize that we do not always see a right angle as a right angle. In fact, there are three right angles at the corner, totaling 270°, but the angles have images totaling 360°. At least one of the angles must appear as more than 90°.

More generally, take two angles α, α'. Place α in plane π and α' in π'. Let l be the intersection of π and π', and place α and α' so that l is one of the sides of each angle (Fig. 1.6). On the other side of α take a point A; and similarly a point A' for α'. On AA' take a point E. Since, α, α' are of arbitrary size, we see that it is always possible to look at a given angle so that it appears as large as any other given angle.

If a property P of a figure F is also shared by the projection F' of F (for every possible projection), then P is called a projective invariant.

Using this terminology, we can say that *size of angle is not a projective invariant.* In particular, "*right angle*" is not a projective invariant.

Since A, A' in Fig. 1.6 could be taken arbitrarily far from O, we see that *distance is not a projective invariant.*

One may now ask for an example of a property that is a projective invariant. Roughly speaking, *a point is a projective invariant* (or the property of a figure that it be a point is a projective invariant—but we will express ourselves more simply in the above way; we will also say that a point is a *projectively invariantive* concept). A line is a projective invariant, at least roughly speaking, because if we see a plane object as a line, then it is a line. (We say "roughly speaking," because there is a slight qualification that must be made, which we shall come to shortly.)

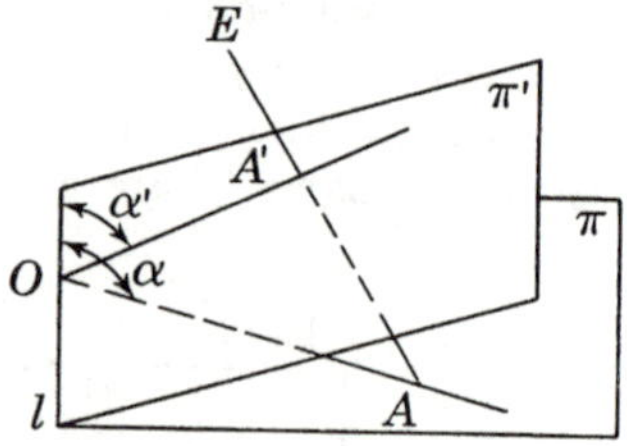

FIG. 1.6

4. Vanishing points. Let us take two planes π and π' and a projection from the center E. The center is understood not to be in π or π'. In the plane π take two lines, p, q, that intersect in a point A. Let p' and q' be the images of p and q, and let A' be the image of A. Since A is on p, A' will be on the image of p, i.e., on p'; thus A' is on p', and similarly A' is on q'. We can then say:

The intersection of the images of two lines is the image of the intersection of the two lines.

Let us call the line joining two points, the *join* of the points. Then we can say:

The join of the images of two points is the image of the join of the two points.

These two theorems will enable us to make certain simple and obvious conclusions about a figure F' that corresponds by projection to the figure F. We shall pass readily from F to F' and back to F again.

Of course, in the above situation it may well be that A has no image. This will be so if and only if EA is parallel to the plane π'. In the case that EA is parallel to π' we will say that A *disappears* (under the projection from π to π') or that it *vanishes* or is a *vanishing point.* If A vanishes, then one sees that p' and q' do not meet (for if they did, the intersection would be the image of A). Thus *intersecting lines may project into parallel lines*; and projecting back from π' to π, we see that *parallel lines may project into intersecting lines. Parallelism* is not a projective invariant. If A is a vanishing point, then the whole system of lines passing through A projects into a system of parallel lines in π'. Regarding this plane π' as the ground (as above) and π as the window-pane, we see why a system of parallel lines appears as a system of lines passing through a point.

5. Vanishing lines. Let A be a vanishing point. Then A has no image. Because of this we cannot say that a point is a projective invariant, because if a figure F consists of just one point A, then its image F' may consist of no points at all. This is a difficulty, but it is a difficulty we want to surmount. The Greeks called this *saving the phenomenon.* We want to save the idea that a point is a projective invariant. On the other hand, *size of angle* is not a projective invariant, and, moreover, we do not want to save the concept.

A line is not quite a projective invariant either. For one thing, consider all the vanishing points. These are found by taking the plane through E parallel to π' and seeing where it intersects π. It intersects it in a line, say v (Fig. 1.7). The vanishing points are just the

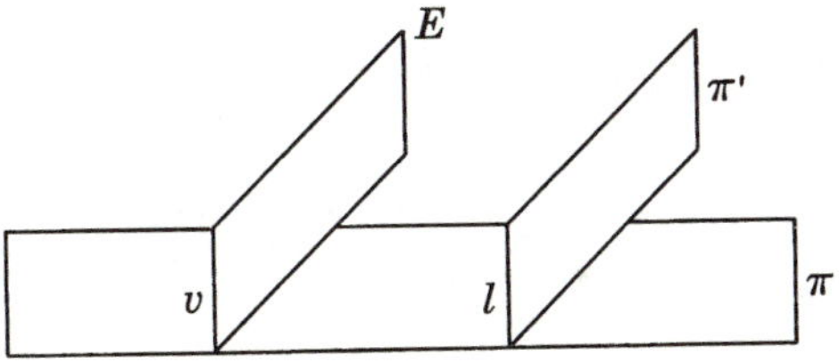

FIG. 1.7

points of v. Thus we see:

The locus of the vanishing points is a straight line.

In particular, the line v has no image, and therefore the concept of a line is not quite a projective invariant; but also in the case of a line we want to save the phenomenon.

Even if we take exception to the situation in which a figure completely vanishes, we shall still have difficulties. There is no difficulty for a point: the image of a point (if there is an image) is itself a point. But consider a line l, $l \neq v$, the vanishing line. How do we see l? Let l cross v at A; then A vanishes. No point corresponds to A on l', the line containing the images of l (= intersection of plane (E, l) with π'). For the same reason, there will be a point B on l' that, in the projection π' to π, has no image on l—namely, if B is taken so that EB is parallel to π. But then B is not the image of any point of l. Thus the image of l is not l', but l' minus the point B. Thus a line is seen as a line minus a point. One might then try *line minus a point* as a projective invariant. But a line minus a point would be seen as a line minus two points; and a line minus two points as a line minus three points, etc. The best we can get out of these considerations is:

The projection of a figure F consisting of a finite number of lines minus a finite number of points is itself a figure F' consisting of a finite number of lines minus a finite number of points.

This saves the phenomenon, but we shall have a better way.

6. Some projective noninvariants. The kind of geometrical objects we would like to study in projective geometry are the projectively invariantive objects. Thus, *distance* is out, *size of angle* is out, and so is *parallelism*. Let us then consider some of the other familiar concepts, partly to clear the way. Take a *circle*. Does it necessarily appear as a

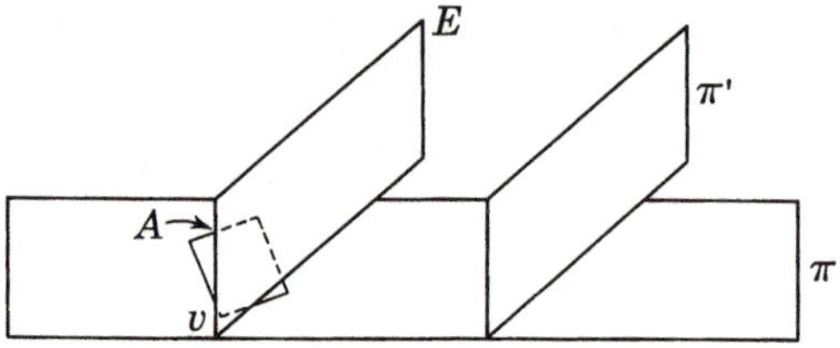

FIG. 1.8

circle? If we look at it on edge, it appears as a straight line segment. But in the situation in which we considered a projection from a plane π to a plane π', if the circle is in π, the above argument would require E to be in π, whereas we have stipulated that E should be neither in π nor in π'. Let us find another argument. We will prove that *boundedness* is not a projective invariant: by a *bounded figure* we mean one that can be enclosed within a circle. Thus a square is a bounded figure; so is a circle. A line is not bounded; neither is a parabola. We will now indicate how a bounded figure can be projected into an unbounded figure. Let us take a bounded figure, say a square for definiteness. Place this square in plane π so that it cuts the vanishing line v, say at A (Fig. 1.8). Now project. The point A disappears. The points near to A have images, but far out. This argument could be made more definite, but it should be fairly clear that the image figure is not bounded —here is an example of an argument that is far from rigorous but that we may allow temporarily.

The argument shows not only that boundedness is not a projective invariant, but that any bounded figure can be projected into an unbounded figure. In particular, a circle can be so projected, so *the concept of a circle is not a projectively invariantive concept.*

Exercises

1. If A, B are two points on a circle and P varies on the circle, then angle APB remains constant (at least if P stays on one of the two arcs into which A and B divide the circle). Use this fact (instead of the notion of boundedness) to prove that a circle is not a projective invariant.

2. Let P' be the image of a point P under a projection. We shall make use of the fact that if Q approaches P, then Q', the image of Q, approaches P'. Let l, m be lines through P, l fixed, m variable; l', m' their images. Show that if m approaches l, then m' approaches l'.

3. With the notation of the preceding exercise, let C be a curve through P and let Q be a point on C approaching P. Then the limiting position of PQ as Q approaches P is the tangent to C at P. Two curves C, D through P are called tangent at P if they have the same tangent at P. Let C, D be tangent at P. Show that under projection the curves C, D go over into tangent curves at P'. In other words, *tangency is a projective invariant.*

7. Betweenness. Let A, B be two points. We can consider the line AB: this is usually understood to mean the line extended. We also

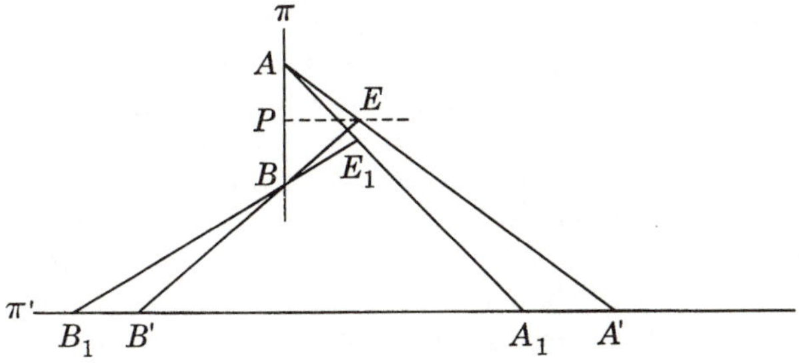

FIG. 1.9

speak of the *segment* AB. Any point P on this segment is said to be between A and B. To say that *betweenness* is invariantive is to say that if P is between A and B, and these points are projected (in any manner) into A', B', and P', then P' is between A' and B'; and to say that *betweenness* is not invariantive is to say that that conclusion is not necessarily so. In Fig. 1.9, we give a cross-section of the situation involving π, π', and E. In π we take three points A, B, P with P between A and B. First take E so that EP is parallel to plane π'. Let A', B' be the images of A, B. The point P vanishes. Now move E slightly off the line EP to a point E_1. If E has been moved very little, the image A_1 of A from the center E_1 will be near A', the image of B_1 of B near B', and the image of P will exist but be far out, hence certainly not between A_1 and B_1. Thus *betweenness is not a projective invariant.*

Exercise. A line divides the plane into two parts, which may be called the sides of the line. Show that the concept *side of a line* is not a projectively invariantive concept. (This could be done, for example, by showing that if two points lie in *different* sides of l in π, their images may well lie in the same side of l', the image of l.)

8. Division of a segment in a ratio. Let A, B be two points, C a point on AB, not necessarily between A, B (Fig. 1.10). *The point C is said to divide AB on the ratio CA/CB.* Here $C=B$ is not allowed, although $C=A$ is. One may fix a *direction* (or *orientation*) on the line, either to the right or to the left; say to the right. We then count XY positive

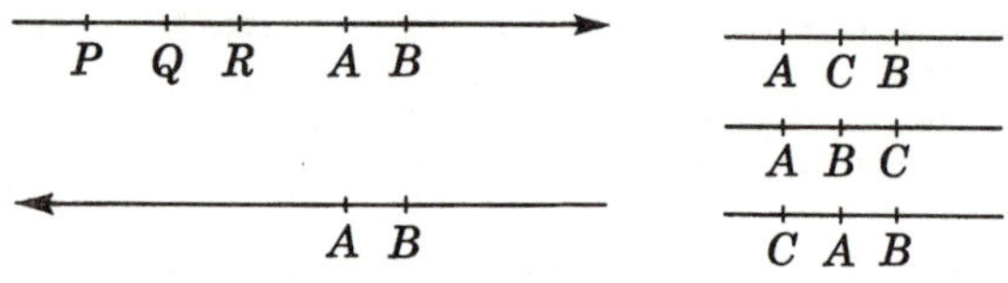

FIG. 1.10

if Y is to the right of X, negative if Y is to the left of X, and, of course, $XY=0$ if $X=Y$. With the stated orientation, in Fig. 1.10, PQ, PR, QR are positive, RP, RQ, QP negative. If the orientation were fixed in the opposite way, all these segments would have the opposite sign. Hence if C is between A and B, then CA/CB is negative, whereas if C is not between A and B, then CA/CB is positive. From this it becomes clear that *division into a given ratio is not a projective invariant,* because by projection we could change the ratio from a negative quantity to a positive. In particular, *mid-point is not a projective invariant.*

There is one other point on this subject that will prove useful. Note that no point divides AB in the ratio 1, i.e., $CA/CB \neq 1$ for every point C. Observe that if $CA/CB=1$, then C is not between A and B. In that event, however, CA and CB have different lengths, so certainly $CA/CB \neq 1$. If C is the mid-point of AB, then $CA/CB=-1$. We also want the following theorem, which applies to the so-called *real* line.

THEOREM. *Given a segment $AB(A \neq B)$ and a number r, $r \neq 1$. Then there is one and only one point C (on the line AB) such that $CA/CB=r$.*

PROOF. Fix a coordinate system on line AB, i.e., take arbitrarily a point O on that line, and to each point X on the line associate the number x giving the length OX:x positive, if X is to the right of O; x negative, if to the left. The number x is called the coordinate of the point X. For convenience, let us use a capital letter to indicate a point, and the same small letter to indicate its coordinate. Then the formula for the distance AB is simply $AB=b-a$. In algebraic terms the theorem states that for any number $r \neq 1$, there is one and only one solution c (C is the unknown) to the equation:

$$\frac{c-a}{c-b} = r.$$

To solve this equation, multiply both sides by $c-b$.

$$\begin{aligned} c-a &= rc-rb \\ c(1-r) &= a-rb \\ c &= (a-rb)/(1-r). \end{aligned}$$

In this last step we use $1-r \neq 0$ or $1 \neq r$, since division by 0 is excluded. This argument shows that there could be no solution other than $c=(a-rb)/(1-r)$. To show that it actually is a solution, we trace back from the last equality to the first. Getting $c(1-r)=a-rb$ and $c-a=rc-rb$ is easy enough. Now note that $c \neq b$, for if $c=b$, then $b-a=rb-rb=0$, i.e., $b=a$ and $B=A$, but we were given that $A \neq B$. So $c \neq b$, $c-b \neq 0$, we can divide by $c-b$ and get $\dfrac{c-a}{c-b}=r$.

9. Desargues' Theorem. In the projection from plane π to π', all the points that have no image lie on a line. This fact can frequently be used to prove that points are collinear. The general idea is this: if we have three points A, B, C that we conjecture to be collinear, we arrange a projection so that A, B disappear; we then try to prove that C also disappears, and if we can, this proves that A, B, C are collinear.

Here is an example. We say that two triangles ABC, $A'B'C'$ are *centrally perspective* if AA', BB', CC' are concurrent (Fig. 1.11a). Two triangles with sides a, b, c and a', b', c', respectively, are said to be *axially perspective* if the intersections of the corresponding sides, $a \cdot a'$, $b \cdot b'$, $c \cdot c'$, are collinear (Fig. 1.11b). (There are *degenerate* configurations that deserve some attention, i.e., configurations in which the triangles assume very special positions, say $A=A'$. What should one then mean by AA', etc.? We will not enter into these questions at this point, but will suppose that the configurations in question are not too special.)

THEOREM OF DESARGUES. *If two triangles are centrally perspective, then they are also axially perspective* (*see* Fig. 1.11c).

In Desargues' Theorem, the two triangles may be coplanar or not. For the moment we consider only the *plane* theorem.

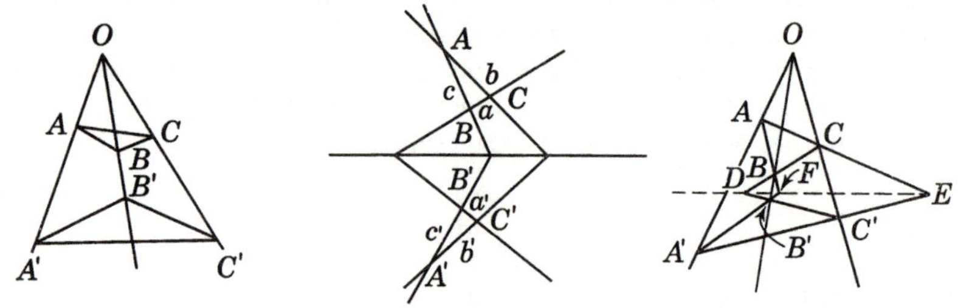

(a) Triangles Centrally Perspective

(b) Triangles Axially Perspective

(c) Figure for Desargues' Theorem

FIG. 1.11

For the proof, in Fig. 1.11, we want to prove that D, E, F, are collinear. Let, then, π be the plane of Fig. 1.11, let π' be a plane through a line parallel to DF, and let E^* be a point on a plane parallel to π' which passes through DF. This makes DF the vanishing line, so D, F disappear. Now in the projected Fig. 1.12, letting $\bar{A}$, $\bar{B}, \cdots$, $\bar{O}$ represent

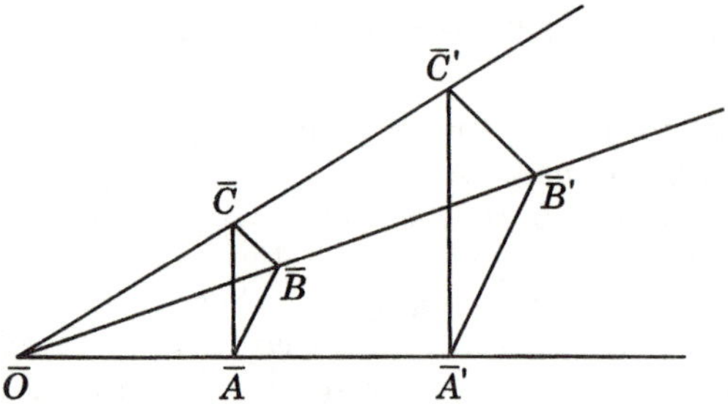

FIG. 1.12

the images, we have the following: two triangles $\bar{A}\bar{B}\bar{C}$, $\bar{A}'\bar{B}'\bar{C}'$ centrally perspective from $\bar{O}$, with $\bar{B}\bar{C}$ parallel to $\bar{B}'\bar{C}'$ (because D vanishes) and $\bar{A}\bar{B}$ parallel to $\bar{A}'\bar{B}'$ (because F vanishes); and we want to prove that $\bar{A}\bar{C}$ is parallel to $\bar{A}'\bar{C}'$ because, if so, then E also vanishes and E must lie on the vanishing line FD.

From similar triangles, $\bar{O}\bar{A}/\bar{O}\bar{A}' = \bar{O}\bar{B}/\bar{O}\bar{B}'$ and $\bar{O}\bar{B}/\bar{O}\bar{B}' = \bar{O}\bar{C}/\bar{O}\bar{C}'$. Hence, $\bar{O}\bar{A}/\bar{O}\bar{A}' = \bar{O}\bar{C}/\bar{O}\bar{C}'$; whence $\bar{A}\bar{C}$ is parallel to $\bar{A}'\bar{C}'$.

Exercise. The following theorem is called *Pappus' Theorem*. Prove it, using the above method.

PAPPUS' THEOREM. *Let l, l' be two lines, meeting in a point O. Let A, B, C be three (distinct) points on l; A', B', C', be three (distinct) points on l'; and let these six points each differ from O. Let $D = BC' \cdot B'C$, $E = CA' \cdot C'A$, $F = AB' \cdot A'B$. Then D, E, F are collinear* (Fig. 1.13).

Exercise. Let ABC be a triangle; D on BC, E on CA, F on AB; $BC \cdot EF = K$, $CA \cdot FD = L$, $AB \cdot DE = M$. Show that if K, L, M are collinear, then AD, BE, CF are concurrent.

Note that these two theorems are *projective* theorems, i.e., they are stated solely in projectively invariantive terms. The proof above of Desargues' Theorem does use nonprojective concepts—for example, the ratio of segments—but the theorem itself is projective. This is certainly a fault, because in erecting a scheme for studying projective concepts we shall want to stick entirely to projective concepts. Meanwhile we may certainly make use of anything we know to lead us to interesting or fruitful considerations.

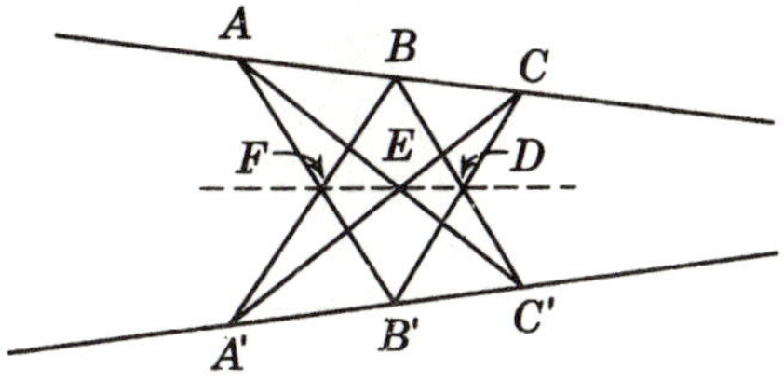

FIG. 1.13

Here is a proof of Desargues' Theorem that *is* projective. Recall the statement that the two triangles ABC, $A'B'C'$ may or may not be coplanar. Consider two cases: (1) the space case, and (2) the plane case. In the first case, the proof is particularly simple. The main point here is to show that D, E, F exist; in the plane case, this—aside from the possibility of parallelism—is obvious. In the space case, then, note that BC and $B'C'$ are coplanar, namely, they are in the plane BOC, and hence they meet—aside from the annoying possibility that they are parallel. (But in any event, there is a clear difference between two lines not meeting because they are parallel and two lines not meeting because they are skew. To overcome such annoying possibilities we shall want a situation in which *every* pair of coplanar lines meet, but non-coplanar lines will even then not intersect.) We neglect this possibility, and so at least have the point D; similarly we get the points E and F. Now we see immediately that these points are collinear, because they lie on the intersection of two planes (namely, the planes of ABC and of $A'B'C'$).

We now have the *space* case, and may use it, if we can, to establish the *plane* case. The idea of the proof is that the plane case may be regarded as a picture, or projection, of the space case. Let E^* be a point not in the plane of the figure (say your right eye). In the plane E^*OB consider another line $O\bar{B}$, not through E^*. Then from E^* we see $O\bar{B}$ as OB, or, in other words, OB is the projection from E^* of $O\bar{B}$ onto the plane of the given figure. Let $\bar{B}$, $\bar{B}'$ be on $O\bar{B}$ and such that they project from E^* onto B, B', respectively. The plane $A\bar{B}C$ does not pass through E^* (otherwise A, B, C would be collinear) and also plane $A'\bar{B}'C'$ does not pass through E^*. Now consider the space triangles $A\bar{B}C$, $A'\bar{B}'C'$. They are centrally perspective from O, so $\bar{D}, \bar{E}, \bar{F}$, where $\bar{D} = \bar{B}C \cdot \bar{B}'C'$, $\bar{E} = CA \cdot C'A'$, $\bar{F} = A\bar{B} \cdot A'\bar{B}'$, are collinear. What is the projection of $\bar{D}$ onto the given plane from E^*? It must lie on the projection of $\bar{B}C$, and this is BC; it must lie on the projection of $\bar{B}'C'$, and this is $B'C'$, so it must lie on both, i.e., the projection is $BC \cdot B'C' = D$. Similarly, the projections of $\bar{E}$, $\bar{F}$ are E, F, respectively.

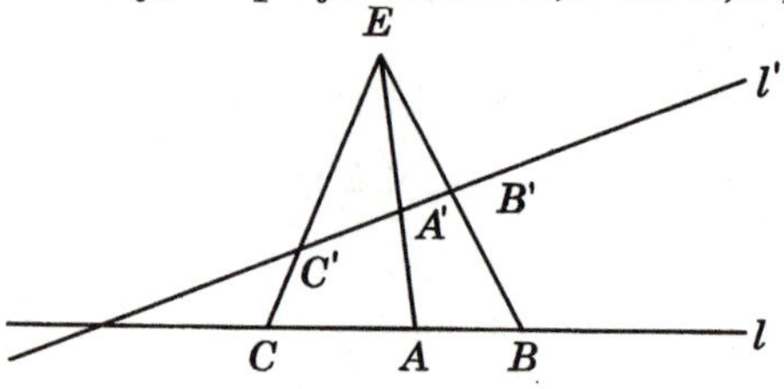

FIG. 1.14

So D, E, F are at any rate in the plane of E^* and the line $\bar{D}\bar{E}\bar{F}$; and they are in the given plane, so they are on the intersection of two planes, i.e., they are on a line. Q.E.D.

10. Perspectivity; projectivity. It would be well, for simplicity in drawing figures, if we could stick to a plane. Figure 1.14 is a cross-section of the situation just considered. The point E is the center of projection, l is a line in the plane π, l' is its image in π'. But now we will consider the whole discussion to be taking place in the plane of E, l, l'. If, then, in a plane we have lines l, l', not necessarily distinct, and a point E not on l or l', then the associating of points $A, B, C, \cdots$ on l, respectively, with the intersections $A', B', C', \cdots$ of $EA, EB, EC, \cdots$ with l' is called a *perspectivity*, and we write $(l) \overset{E}{\underset{\wedge}{-}} (l')$; also $(ABC \cdots) \overset{E}{\underset{\wedge}{-}} (A'B'C' \cdots)$: the perspectivity is said to be from l to l' and to have *center* E. Calling the perspectivity T, we write $(A)T = A'$, $(B)T = B'$, $(C)T = C'$, $\cdots$. A perspectivity is an example of a *transformation* (i.e., a method of passing from certain points A, B, $C, \cdots$ to other points A', B', $C', \cdots$; a *rotation* is a transformation; so is *reflection in a line*), and for a transformation to be perspectivity, it must be of the special character just described.

Consider a number of perspectivities carried out successively, say from l to l', then from l' to l'', then from l'' to l''', etc. Let $T: (l) \overset{E}{\underset{\wedge}{-}} (l')$; $U: (l') \overset{E'}{\underset{\wedge}{-}} (l'')$; $V: (l'') \overset{E''}{\underset{\wedge}{-}} (l''')$. In Fig. 1.15, l, l', l'', l''', E, E', E'' are fixed, given to start with, while X is variable; i.e., we consider various points X on l. We have $(X)T = X'$, $(X')U = X''$, $(X'')V = X'''$. For brevity we write $(X)TUV = X'''$. Thus TUV is also a transformation, not necessarily a perspectivity; it is a transformation carrying points of l into points of l'''. By a *product of transformations*, we mean the transformation that results by carrying out successively a number of given

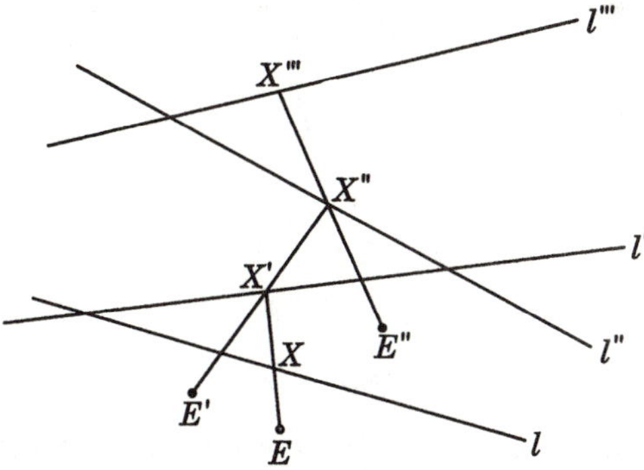

FIG. 1.15

transformations; thus the transformation just described carrying X into X''' is the product of T, U, and V (in that order).

DEFINITION. A product of perspectivities is called a *projectivity*.

NOTATION. For a projectivity we write: $(l) \overset{=}{\wedge} (l')$; also $(ABC \cdots) \overset{=}{\wedge} (A'B'C' \cdots)$, where A', B', $C', \cdots$ correspond, respectively, to $A, B, C, \cdots$.

Since, by definition, projections preserve projective invariants, and since a perspectivity is a cross-section of a projection, we see that perspectivities preserve projective invariants. A projectivity, since it is the product of perspectivities, also preserves projective invariants. Conversely, as one sees in a similar way, a property of a set of collinear points that is preserved under perspectivities is a projective invariant.

THEOREM. *Given three (distinct) collinear points A, B, C; and another three (distinct) collinear points A', B', C'. Then there exists a projectivity sending A, B, C, respectively, into A', B', C'.*

PROOF. Consider first the case that $A = A'$ and that line ABC is different from line $A'B'C'$ (Fig. 1.16a). Let P be the intersection of BB' and CC'; of course, BB' may be parallel to CC', but we will neglect this possibility for the present, on the ground that we are not seeking rigorous proofs. Supposing, then, that BB' and CC' do intersect, we have the point P, and clearly $(ABC) \overset{P}{\underset{\wedge}{-}} (AB'C')$, where A goes into $A'(=A)$, B into B', C into C', as required. Suppose now, as a second case, that A' is not on line ABC (Fig. 1.16b). On AA' take a point E

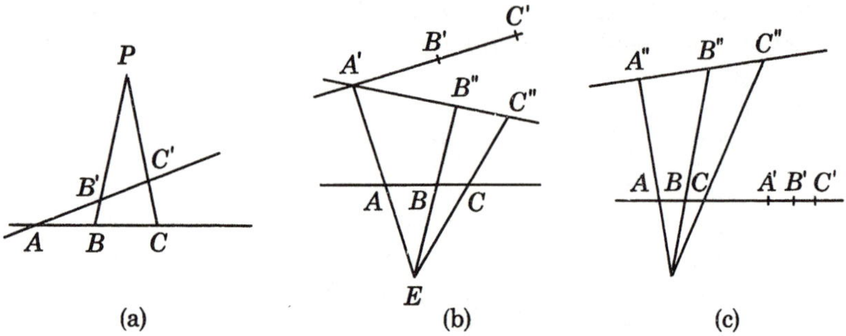

FIG. 1.16

$(E \neq A,\ E \neq A')$; and take a line l through A', but otherwise in no special position, in particular $l \neq A'B'$ and l not through E. From E project ABC into $A'B''C''$ on l. Now by the first case we can project $A'B''C''$ into $A'B'C'$. The only remaining cases to consider are those in which A, B, C, A', B', C' are in a very special relation to each other; one such possibility is that they are all collinear (Fig. 1.16c). In that event, project ABC off the given line into $A''B''C''$ on some line lying in no special position. Then by previous considerations, we can by a projectivity send $A''B''C''$, respectively, into $A'B'C'$.

Exercise. Any three noncollinear points can be sent into any other three noncollinear points by a product of projections.

We are looking for projective invariants. Any pair of points can be projected into any other pair by a projectivity, but not every triple of points can be projected into any other triple by a product of projections (in space). Any three noncollinear points can be sent into any other three noncollinear points; and any three collinear points can be sent into any other three. But *collinearity* is a projective invariant, and so *projectively* we can distinguish the one kind of triple from the other. Considering now the collinear triples, we see that there is no way projectively to distinguish one triple from another. In ordinary geometry, to make an analogy, one studies the so-called *rigid motions*, i.e., the transformations in the plane which are such that if A, B are sent into A', B' then distance AB = distance $A'B'$ (for all pairs A, B). In such geometry, one counts two figures as equivalent if one can be transformed into the other by a rigid motion: and it is clear that no general assertion can be made of the first figure that cannot be made of the second, and vice versa; the two figures are equivalent under the study in question. Similarly in projective geometry, any three collinear points are equivalent to any other three collinear points.

The next question to ask is whether any four collinear points are projectively equivalent to any other four. We shall see that the answer is *No*.

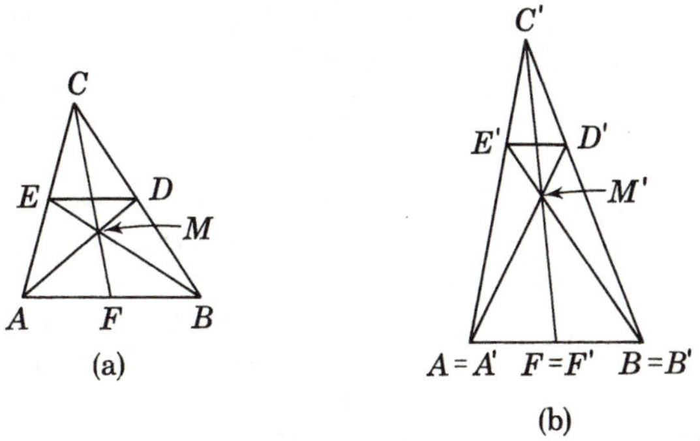

FIG. 1.17

11. Harmonic tetrads; fourth harmonic. In projecting from plane π to π', we see that certain things change and that other things do not. Let us try to use this fact to get generalizations of known theorems. For example, from plane geometry we know that the medians of a triangle meet in a point. Let us take a triangle ABC in plane π with AB lying on the intersection l of π and π'. (Fig. 1.17.) Let D be the mid-point of CB, E the mid-point of CA, F the mid-point of AB. Then by a theorem of plane geometry, DE is parallel to AB. Let AD and BE meet in M. Then CM passes through F. Now project this configuration onto π'; the points A, B, F remain as before ($A' = A$, $B' = B$, $F' = F$), the points C, D, E, M go into C', D', E', M'. The points D', E' are not necessarily the mid-points of $C'B'$ and $C'A'$, respectively; but $D'E'$ is parallel to $A'B'$. If we consider the two figures together we begin to suspect that the fact that CM bisects AB (where $M = BE$ intersected with AD—let us write this: $M = BE \cdot AD$) does not depend so much on the fact that D, E are mid-points of CB, CA, but only on the fact that ED is parallel to AB; because in the second figure $C'M'$ bisects $A'B'$, although E', D' are not mid-points of the corresponding sides. We are thus led to *conjecture* the following theorem.

THEOREM. *If a line parallel to the base AB of a triangle ABC cuts the sides CB, CA in D, E, respectively, and $M = BE \cdot AD$, then CM passes through the mid-point of AB.*

We say *conjecture*, because the above considerations do not constitute a proof. For a proof, one should start with the figure involved in the theorem itself, and try to pass by a projection into the special case of medians. The question is: Given Fig. 1.17b, with $D'E'$ parallel to $A'B'$, can one project it into Fig. 1.17a so that $A'D'$ and $B'E'$ project into medians of the image triangle? Yes, we can. We again suppose $A'B'$ to be on $l = \pi \cdot \pi'$. Now we can project E' so that in the image E is the mid-point of AC. Since ED will be parallel to AB, it will then be true that D is also the mid-point of AB, so $F(=F')$ is the mid-point of AB. Q.E.D.

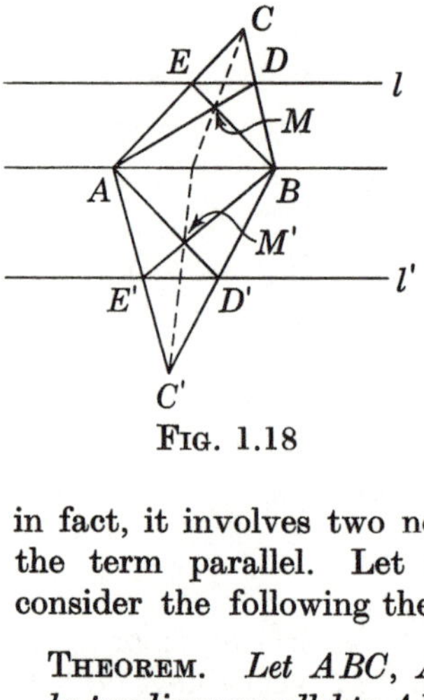

FIG. 1.18

We now have a theorem more general than the theorem on the medians, but all the same, it is not a projective theorem: in fact, it involves two nonprojective terms, the term mid-point and the term parallel. Let us try to get rid of these terms. First consider the following theorem.

THEOREM. *Let ABC, ABC' be two triangles (Fig. 1.18). Let l, l' be two lines parallel to AB, intersecting CA, CB, $C'A$, $C'B$, respectively, in E, D, E', D'; and let $M = AD \cdot BE$, $M' = AD' \cdot BE'$. Then CM and $C'M'$ meet on AB.*

In fact, CM and $C'M'$ must pass through the mid-point of AB. Note that this theorem is essentially the same as the previous theorem: it follows directly from it; but also vice versa, we see that however l is varied in Fig. 1.18, CM must cut AB in the same point, and to see that this is the mid-point, we take l so that D is the mid-point of CB, and then use the theorem on the medians. The present formulation has the advantage that one of the nonprojective terms, mid-point, has been eliminated.

Let us now project Fig. 1.18 so that the image of l is not parallel to AB. In the image figure, l, l', AB, or their images rather, meet in a point. We ought to write a bar, say, over the letter designating a point in order to indicate its image; for example, write $\bar{D}$ to indicate the image of D. But, for brevity, let us not do this. We then get Fig. 1.19.

In order to discuss this figure, we introduce for convenience several new terms.

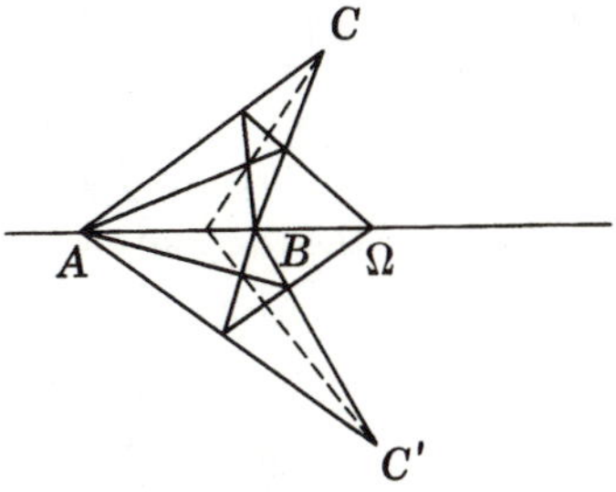

FIG. 1.19

DEFINITION. By a *triangle* one means a figure consisting of three noncollinear points and the three lines joining pairs of these points.

DEFINITION. By a *complete quadrangle* or *complete 4-point* one means a figure consisting of four points, no three of which are collinear, and of the six lines joining pairs of these points. If l is one of these lines, called a *side* (there are six sides), then it lies on two of the vertices, and the line joining the other two vertices is called the *opposite side* (to l). Thus the six sides can be arranged into three pairs of opposite sides. The intersection of two opposite sides is called a *diagonal point*: there are three diagonal points (Fig. 1.20).

DEFINITION. We will say that Ω' is the *fourth harmonic of* Ω *with respect to* A, B [and will write $H(A, B; \Omega, \Omega')$ to express this fact], if A, B are two vertices of a complete 4-point, Ω is the diagonal point on AB, and Ω' is the point where the line joining the other two diagonal points cuts AB.

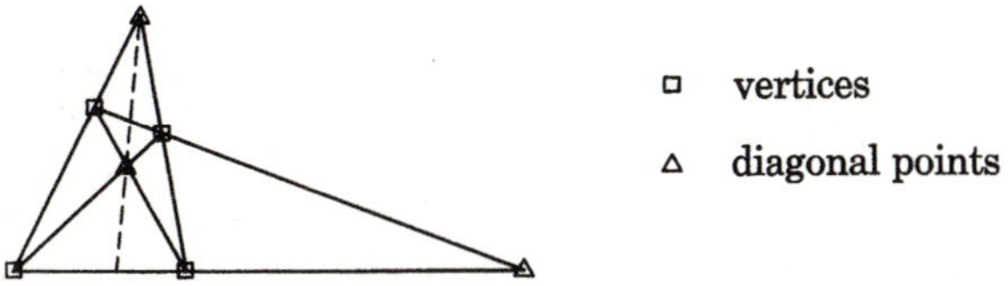

FIG. 1.20

Given A, B, Ω, it is clear that we can always construct a point Ω' such that $H(A, B; \Omega, \Omega')$; but the question immediately occurs: Do we always come to the same fourth point Ω', or do different constructions yield, possibly, different fourth harmonics? An examination of Fig. 1.19 suggests to us that the answer is that we always reach the same fourth point. To prove this, we have to project that figure back into a figure appropriate to the last theorem. Let us go through the argument. Given the collinear points A, B, Ω, we take a point C not on $AB\Omega$, draw CA, CB and a line through Ω cutting CA, CB in E, D, respectively; let $AD \cdot BE = M$ (Fig. 1.21). Then considering the complete 4-point $A\ B\ D\ E$, we see that CM will cut AB in a fourth harmonic of Ω with respect to A, B. We go through the construction again, with $\bar{C}$, $\bar{D}$, $\bar{E}$, $\bar{M}$. Now $\bar{C}\bar{M}$ will go through a fourth harmonic of Ω with respect to A, B. The question is: Do CM and $\bar{C}\bar{M}$ cut AB in the same place? If not, there is not much point in defining fourth harmonic. If they do, then any other construction for the fourth harmonic yields

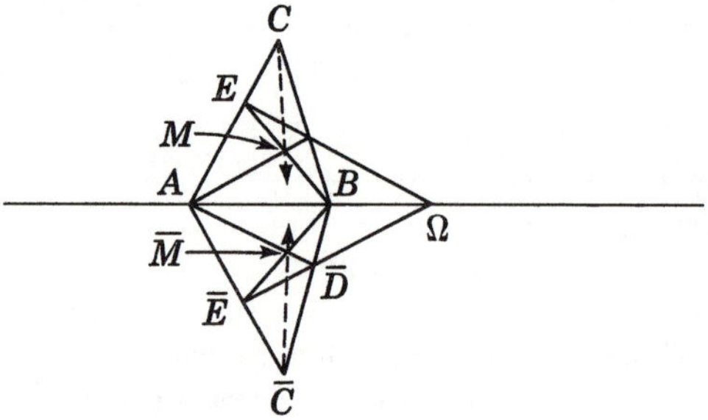

FIG. 1.21

the same point (because *any two* yield the same point). For the proof, project the figure so that Ω disappears. In the projected figure, ED and $\bar{E}\bar{D}$ are parallel to AB, so CM and $\bar{C}\bar{M}$ meet on AB (or rather C^*M^* and $\bar{C}^*\bar{M}^*$ meet on A^*B^*, where * indicates an image; but we will drop the stars for brevity). Now project back and see that in the initial figure, CM and $\bar{C}\bar{M}$ meet on AB. Q.E.D.

We state these facts as a theorem. This theorem is a *projective* theorem.

THEOREM. *Given three collinear points A, B, Ω. There is one and only one point Ω' such that $H(A, B; \Omega, \Omega')$.*

The next point is to prove that *fourth harmonic* is a projectively invariantive concept; that is what the following theorem says.

THEOREM. *If A, B, C, D are sent by a projectivity into A', B', C', D', and if $H(A, B; C, D)$, then $H(A', B'; C', D')$.*

PROOF. Construct a complete quadrangle exhibiting the fact that $H(A, B; C, D)$. Let this figure be in a plane π; and consider a projection from this plane π to another plane $\bar{\pi}$. The points A, B, C, D map into points, $\bar{A}$, $\bar{B}$, $\bar{C}$, $\bar{D}$, and the complete quadrangle exhibiting the fact that $H(A, B; C, D)$ maps into a complete quadrangle showing that $H(\bar{A}, \bar{B}; \bar{C}, \bar{D})$. We have proved that a *harmonic tetrad* maps under a projection into a *harmonic tetrad.* Any perspectivity can be regarded as a cross-section of a projection (see §10, first paragraph); therefore under a perspectivity, a harmonic tetrad goes into a harmonic tetrad. A similar conclusion for a projectivity, i.e., a product of perspectivities, is immediate.

COROLLARY. *There exist projectively nonequivalent collinear tetrads, i.e., there exist collinear points A, B, C, D and collinear points A', B', C', D'* such that A, B, C, D cannot be sent into A', B', C', D' by a *projectivity.*

PROOF. Take A, B, C, D such that $H(A, B; C, D)$. Now take collinear points A', B', C', X, D' such that $H(A', B'; C', X)$ and such that $D' \neq X$. By the uniqueness of the fourth harmonic we know, then, that $H(A', B'; C', D')$ is not the case. Then A, B, C, D cannot by a projectivity be sent into A', B', C', D'.

Exercise. Let $ABA'B'$ be a complete quadrangle; let $P = AA' \cdot BB'$, $Q = AB' \cdot A'B$, $R = AB \cdot A'B'$ be the diagonal points; and let $D' = PQ \cdot A'B'$, $D = PQ \cdot AB$. We have $H(A, B; R, D)$ and $ABRD \overset{P}{\underset{\wedge}{-}} A'B'RD'$. Hence by the preceding theorem we also have $H(A', B'; R, D')$. Draw this same conclusion directly from the figure, rather than by the above theorem. Also without using that theorem show that $H(P, Q; D', D)$.

Here is another proof that, under a projectivity, harmonic tetrads go into harmonic tetrads. This proof uses only the plane, not space. Later on, when we set up our axioms, these will be for plane geometry, partly to simplify the drawing of pictures. Many of the things we say in this part of the book will carry over to the second part. The following proof, for example, will carry over immediately; the first proof not at all or, at any rate, not immediately.

Now we come to the proof. We have to prove $(ABCD) \overset{}{\underset{\wedge}{=}} (A'B'C'D')$ and $H(A, B; C, D)$ imply $H(A', B'; C', D')$. Instead of a projectivity, it is sufficient to consider a single perspectivity. Suppose then that $(ABCD) \overset{P}{\underset{\wedge}{-}} (A'B'C'D')$; say this perspectivity is from l to l': $(l) \overset{P}{\underset{\wedge}{-}} (l')$. The lines l and l' may be in some very special position, for example l' might be passing through A. Let us introduce a third line $\bar{l}$, and consider $(l) \overset{P}{\underset{\wedge}{-}} (\bar{l}) \overset{P}{\underset{\wedge}{-}} (l')$, where $\bar{l}$ is in some general position. It is then sufficient to consider the perspectivities $l \overset{P}{\underset{\wedge}{-}} (\bar{l})$ and $(\bar{l}) \overset{P}{\underset{\wedge}{-}} (l')$; and, in fact, we only have to consider one of them, the considerations for the other being quite the same. So far the considerations have been largely preparatory: we need consider only a single perspectivity $(l) \overset{P}{\underset{\wedge}{-}} (l')$, and may further suppose that l and l' lie in no special relation to the points $A, \cdots, D'$ (Fig. 1.22). Consider the line CD', and let $A_1 = CD' \cdot AA'$, $B_1 = CD' \cdot BB'$. *By the exercise on page* 19, $H(A, B; C, D)$ implies $H(A_1, B_1; C, D')$, and this, in turn, by the same exercise implies $H(A', B'; C', D')$. Q.E.D.

12. Further theorems on harmonic tetrads.

THEOREM. *If $H(A, B; C, D)$ and $H(A', B'; C', D')$, then $(ABCD) \overline{\overline{\wedge}}$ $(A'B'C'D')$.*

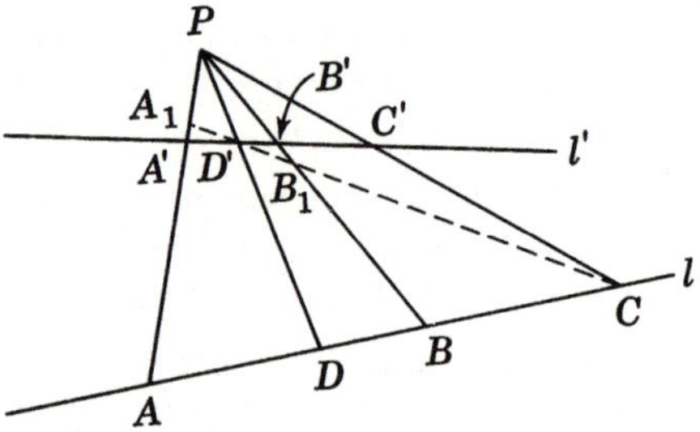

FIG. 1.22

PROOF. We know we can send A, B, C into A', B', C' by a projectivity. Let T be such a projectivity and let $(D)T = X$. By the theorem on page 18, we have $H(A', B'; C', X)$. We also have $H(A', B'; C', D')$, by hypothesis. By the uniqueness of the fourth harmonic we have $X = D'$, so T sends A, B, C, D into A', B', C', D'. This completes the proof.

The order in which the letters $A, B, C, D,$ occur in the symbol $H(A, B; C, D)$ is, of course, important. Nevertheless, given $H(A, B; C, D)$, various other arrangements of the letters also yield harmonic tetrads. For example, we also have $H(B, A; C, D)$.

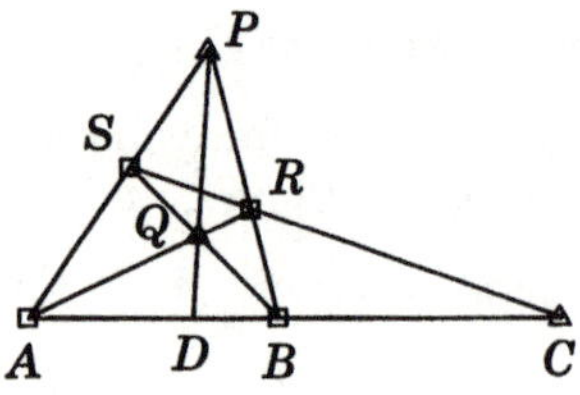

FIG. 1.23

THEOREM. *If $H(A, B; C, D)$, then $H(B, A; C, D)$.*

This theorem is almost trivial. Consider a quadrangle exhibiting the fact that $H(A, B; C, D)$ (Fig. 1.23). The points A

and B are vertices of this quadrilateral. But so are B and A and therefore the same quadrangle shows that $H(B, A; C, D)$.

Just a little less trivial is the following.

Theorem. *If* $H(A, B; C, D)$, *then* $H(A, B; D, C)$.

proof. Let the complete quadrangle $A\ B\ R\ S$, with diagonal points P, Q, C exhibit the fact that $H(A, B; C, D)$. Now consider the complete quadrangle $ABPQ$. One concludes that $H(A, B; D, C)$.

Exercises

1. From the above theorem and another preceding theorem, if $H(A, B; C, D)$, then $A\ B\ C\ D \,\overline{\wedge}\, A\ B\ D\ C$. Show this directly from the figure exhibiting $H(A, B; C, D)$.

2. Prove that if $H(A, B; C, D)$, then $H(C, D; A, B)$. [How does one proceed? First, clearly, *one needs Fig.* 1.23 *showing* $H(A, B; C, D)$. Now as a general principle one will not draw in any further lines unless one has to, or unless one sees a good reason to: all this principle amounts to is that we should not mess up the figure. Now can one avoid drawing auxiliary lines? *What are we looking for?* We are looking for a quadrangle having C, D as vertices, and having other properties besides. Now note that in Fig. 1.23, there are *three* lines through each *vertex*. If C, D are to be vertices of some complete quadrangle showing $H(C, D; A, B)$, then we shall need at least three lines through each of C and D. Hence, it is clear that one will have to draw at least two auxiliary lines. Trying to do this in the tidiest manner possible, we draw DS and PC (other possibilities are equally good). Let $DS \cdot PC = M$, $DP \cdot SC = N$. Hint: Consider point X such that $H(S, P; A, X)$.]

3. Let lines l, m meet in O, and let P be a point not on l or m. Through P draw a line q and let q intersect l in L, m in M. Let Q be such that $H(L, M; P, Q)$. Prove that the locus of Q is a straight line through O (i.e., as q varies about P, the point Q varies on a line through O).

13. The cross-ratio. As we have just seen, not every set of four collinear points can be sent by a projectivity into a given set of four collinear points. The next question then is: Given two sets of collinear points (A, B, C, D) and (A', B', C', D'), under what circumstances is it true that $(ABCD) \,\overline{\wedge}\, (A'B'C'D')$? That is, we are looking for a character of tetrads of collinear points such that $(ABCD) \,\overline{\wedge}\, (A'B'C'D')$ if and only if $(ABCD)$ and $(A'B'C'D')$ have the same character. The question could be explicated further, but will become fairly clear when we see the answer.

We have already noted that the ratio CA/CB in which a point C divides AB is not a projective invariant; and, in fact, there is no projective invariant for triads of collinear points (since any such triad is projectively equivalent to any other). For four collinear points $A,B,$ C,D, let us now consider the number

$$\frac{CA/CB}{DA/DB}.$$

This number is called the cross-ratio of C, D with respect to A, B and is written $R(A, B; C, D)$. We shall see that *two collinear tetrads* (A, B, C, D), (A', B', C', D') *are projectively equivalent, i.e.,* $(ABCD) \overline{\overline{\wedge}} (A'B'C'D')$, *if and only if* $R(A, B; C, D) = R(A', B'; C', D')$.

Let O be a point and let a, b be two rays through O (Fig. 1.24). These rays make various angles with each other, one of which (in absolute value) is not greater than 180°. Let (ab) designate this angle. We fix an orientation about O, i.e., we will regard (ab) as positive or negative in accordance with the way a must be rotated through $|(ab)|$ to fall on b; let us say (ab) is positive if this rotation is counterclockwise, negative if clockwise (the *opposite orientation* would make (ab) positive

FIG. 1.24

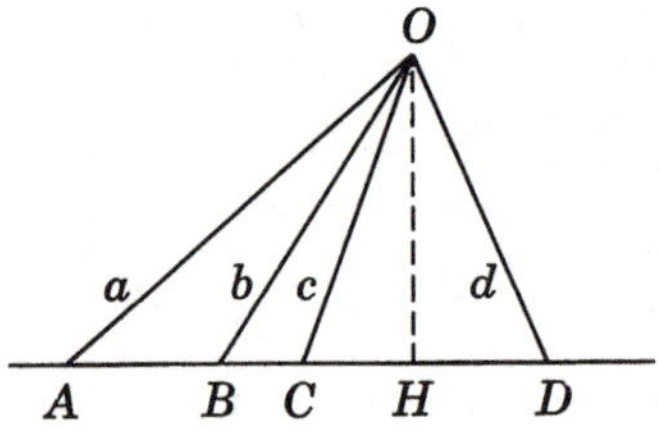

FIG. 1.25

if one had to rotate a *clockwise* to b). The above is ambiguous if $|(ab)| = 180°$. We shall be interested in sin (ab), so also in this case there will be no ambiguity.

Let now c be a third ray through O, and define the ratio in which c divides a, b to be $\sin (ca)/\sin (cb)$: here $c = b$ is not allowed, nor is c directly opposite to b allowed. Note that $\sin (ca)/\sin (cb)$ does not depend on the orientation selected. (The following remark will be useful in a moment. Let A, B, C be three (distinct) collinear points, O a point not on line ABC, and let $a = OA$, $b = OB$, $c = OC$. Then CA/CB and $\sin (ca)/\sin (cb)$ are both positive or both negative.)

Let now a, b, c, d be four rays through O on four lines and define

$$R(a, b; c, d) = \frac{\sin (ca)}{\sin (cb)} \Big/ \frac{\sin (da)}{\sin (db)}.$$

LEMMA. *Let A, B, C, D be four collinear points (Fig. 1.25) and let O be a point not on line $ABCD$; call $OA=a$, $OB=b$, $OC=c$, $OD=d$. Then $R(A, B; C, D)=R(a, b; c, d)$.*

PROOF. Drop the perpendicular OH on line $ABCD$. For a moment do not worry about signs in the following. We have

$$R(A, B; C, D) = \frac{\dfrac{CA}{CB}}{\dfrac{DA}{DB}} = \frac{\dfrac{\frac{1}{2}\,OH\cdot CA}{\frac{1}{2}\,OH\cdot CB}}{\dfrac{\frac{1}{2}\,OH\cdot DA}{\frac{1}{2}\,OH\cdot DB}}$$

$$= \frac{\dfrac{\text{area } COA}{\text{area } COB}}{\dfrac{\text{area } DOA}{\text{area } DOB}} = \frac{\dfrac{\frac{1}{2}\,OA\cdot OC \sin(ca)}{\frac{1}{2}\,OB\cdot OC \sin(cb)}}{\dfrac{\frac{1}{2}\,OA\cdot OD \sin(da)}{\frac{1}{2}\,OB\cdot OD \sin(db)}} = \frac{\dfrac{\sin(ca)}{\sin(cb)}}{\dfrac{\sin(da)}{\sin(db)}}$$

Now pay attention to signs. What we have really proved so far is that $R(A, B; C, D)=\pm R(a, b; c, d)$. By the remark just preceding the lemma, CA/CB and $\sin(ca)/\sin(cb)$ have like signs, and likewise DA/DB and $\sin(da)/\sin(db)$ have like signs. So $R(A, B; C, D)$ and $R(a, b; c, d)$ have like signs. This completes the proof.

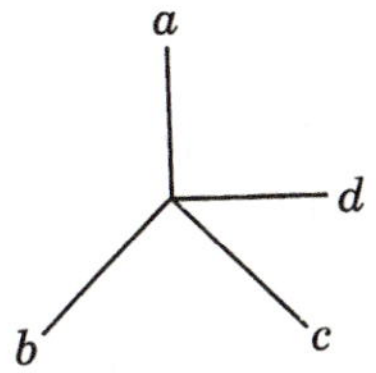

FIG. 1.26

Let us start with four rays a, b, c, d through a point O. Let $\bar{a}$, $\bar{b}$, $\bar{c}$, $\bar{d}$ be the rays that together with a, b, c, d make straight lines. Note that $\sin(ca)$ and $\sin(c\bar{a})$ have opposite signs: since a occurs twice in the expression of $R(a, b; c, d)$ by sines, we see that $R(a, b; c, d)=R(\bar{a}, b; c, d)$. Similarly, we may replace any of the letters in $R(a, b; c, d)$ by the corresponding barred letters. Thus we may speak of the cross-ratio of four lines, and not just of four rays. Now it may not be possible to cut the rays a, b, c, d by a straight line. For example, in Fig. 1.26 it is not possible to cut the rays a, b, c with a straight line. But as we have just seen, $R(a, b; c, d)$ depends not so much on the *rays* a, b, c, d as on the corresponding full *lines*. Hence, if these lines cut a secant in points A, B, C, D we shall have $R(A, B; C, D)=R(a, b; c, d)$.

THEOREM. *If $(ABCD) \;\overline{\wedge}\; (A'B'C'D')$, then $R(A, B; C, D)=R(A', B'; C', D')$.*

PROOF. A projectivity is a product of perspectivities. We have to show that the cross-ratio is unchanged under a product of perspectivities. Now if the cross-ratio is unchanged under any perspectivity, then obviously it will also be unchanged under a product of perspectivities. In short, it is sufficient to consider a perspectivity instead of a projectivity. Let then $(ABCD) \overset{O}{\underset{\wedge}{-}} (A'B'C'D')$ (Fig. 1.27). By the lemma, $R(A, B; C, D) = R(a, b; c, d) = R(A', B'; C', D')$. Q.E.D.

Conversely, *is it true that* $R(A, B; C, D) = R(A', B'; C', D')$ *implies* $(ABCD) \mathrel{\overline{\overline{\wedge}}} (A'B'C'D')$? At any rate, we can send A, B, C into A', B', C' by a projectivity. Let us do this, and suppose D falls at X. We want, if possible, to prove $X = D'$. By the theorem above, $R(A, B; C, D) = R(A', B'; C', X)$. And we are given $R(A, B; C, D) = R(A', B'; C', D')$. So $R(A', B'; C', X) = R(A', B'; C', D')$, or

$$\frac{C'A'}{C'B'} \bigg/ \frac{XA'}{XB'} = \frac{C'A'}{C'B'} \bigg/ \frac{D'A'}{D'B'},$$

whence

$$\frac{XA'}{XB'} = \frac{D'A'}{D'B'}$$

and $X = D'$ by a previous theorem. Q.E.D.

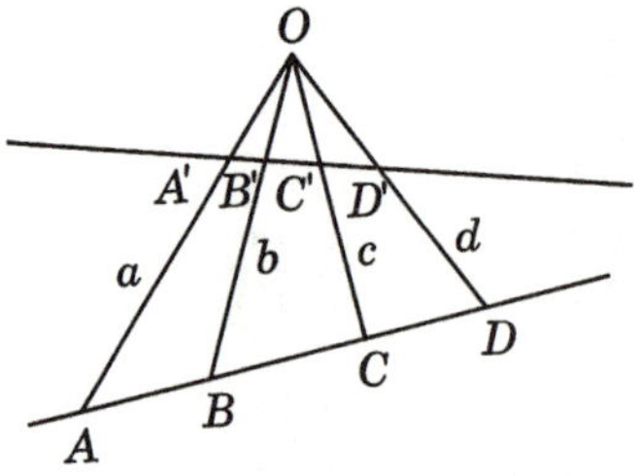

FIG. 1.27

The last point made can be stated as follows: Given three collinear points A, B, C. *There is at most one point X such that $R(A, B; C, X)$ has a given cross-ratio r.* Conversely, is it true that there exists at least one point X such that $R(A, B; C, X) = r$? Consider for a moment the cross-ratio $R(A, B; C, D)$ of four distinct points. Since $C \neq D$, we have $CA/CB \neq DA/CB$, so $R(A, B; C, D) \neq 1$; since $C \neq A$, we have $R(A, B; C, D) \neq 0$; since $DA/DB \neq 1$, we could not have $R(A, B; C, D) = CA/CB$. In short, a cross-ratio is never equal to 0 nor equal to 1 (if A, B, C, D are distinct), and $R(A, B; C, D) \neq CA/CB$. Avoiding these three values of r, however, we get an affirmative answer to our question.

THEOREM. *Given three distinct collinear points A, B, C. For any number r, $r \neq 0$, $r \neq 1$, $r \neq CA/CB$, there is one and only one point X such that $R(A, B; C, X) = r$.*

PROOF. We have

$$\frac{CA}{CB} \bigg/ \frac{XA}{XB} = r, \quad \text{or} \quad \frac{XB}{XA} = r \cdot \frac{CB}{CA}.$$

By a previous discussion, if

$$r \cdot \frac{CB}{CA} \neq 1,$$

there is one and only one such point X. Since $r \neq CA/CB$, also $r \cdot CB/CA \neq 1$, and our one and only X exists.

14. Fundamental theorem of projective geometry. Given three (distinct) collinear points A, B, C and three (distinct) collinear points A', B', C'. We know that there exists at least one projectivity sending A, B, C, respectively, into A', B', C'. *Can there be more than one?* Let T, U be projectivities sending A, B, C, respectively, into A', B', C'. We write $(A)T = A'$, $(B)T = B'$, $(C)T = C'$, and similarly $(A)U = A'$, $(B)U = B'$, $(C)U = C'$. Now let us be quite clear as to what is meant by $T \neq U$ or $T = U$. The projectivities T and U are products of perspectivities. In saying $T = U$ or $T \neq U$, we do not have regard for the way T and U are composed of perspectivities, but only have regard for the effects that T and U have upon the points of line ABC, i.e., $T = U$ if $(X)T = (X)U$ for every point on line ABC. And to say that $T \neq U$ is to say that for at least one point X on line ABC we have $(X)T \neq (X)U$.

FUNDAMENTAL THEOREM. *There is one and only one projectivity sending three given collinear points A, B, C, respectively, into three given collinear points A', B', C'.*

(The "one" we already have: it is the "only one" that is the important point.)

PROOF. Let T, U be projectivities sending A, B, C, respectively, into A', B', C'. To prove: $T = U$. Let X be any point on line $A\ B\ C$ ($X \neq A$, $X \neq B$, $X \neq C$). Then $R(A, B; C, X) = R(A', B'; C', (X)T)$; and $R(A, B; C, X) = R(A', B'; C', (X)U)$. Thus $R(A', B'; C', (X)T) = R(A', B'; C', (X)U)$, whence $(X)T = (X)U$ and $T = U$. Q.E.D.

Note the following fact about the Fundamental Theorem: it is stated in projective terms (i.e., later, in building up projective geometry, we are going to retain the concepts "points," "line," "perspectivity," and "projectivity"). It is true that the proof has made use of nonprojective terms (namely "distance"), but the theorem is itself in suitable terms for a strictly projective study.

Earlier we gave an exercise to prove Pappus' Theorem. The proof involved *distances*, and so was not projective in nature. We shall show how to prove Pappus' Theorem using the Fundamental Theorem; but first we prove the following consequence of the Fundamental Theorem.

THEOREM A. *Let T be a* projectivity *from a line l to a line $l' \neq l$ that leaves $O = l \cdot l'$ invariant, i.e., $(O)T = O$. Then T is a* perspectivity.

PROOF. Take two points A, B on l (Fig. 1.28). Let $A'=(A)T$, $B'=(B)T$. Let $E=AA'\cdot BB'$. Let U be the perspectivity $l \overset{E}{\underset{\wedge}{-}} l'$. Then $(O)U=O$, $(A)U=A'$, $(B)U=B'$; therefore T and U have the same effects on three distinct points. By the Fundamental Theorem, $T=U$, so T is a perspectivity.

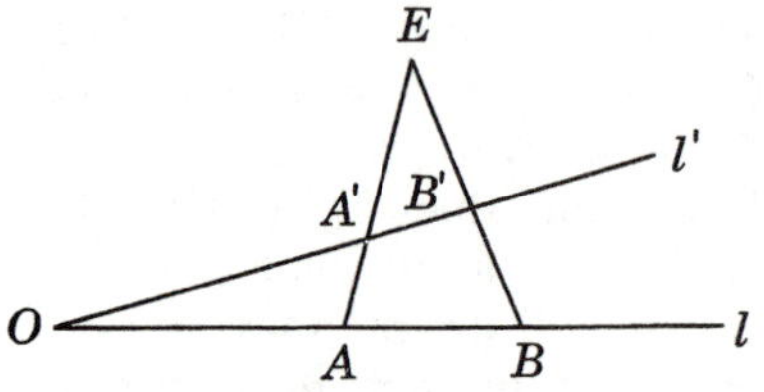

FIG. 1.28

This last theorem suggests a method that frequently can be applied to prove three points D, E, F collinear. Suppose one has a geometric situation in which one suspects that three of the points D, E, F are collinear. Consider two of them, say D and F. First see whether there is any projectivity, from a line l to a line l', sending D into F. Second, if one has found such a projectivity, check to see whether it leaves $l\cdot l'$ invariant. If it does, then it is a perspectivity. Supposing it is, compute its center, hoping that it will be E. If it is, then D, E, F are collinear, since the join of the corresponding points D, F must go through the center. However, all this is very general; let us see if it can be applied to prove Pappus' Theorem.

Draw a figure (Fig. 1.29). Consider F and D. These are to be on lines l and l'. What shall lines l and l' be? The line l, if we are going to introduce no auxiliary lines, must be AB' or BA'; and l' must be

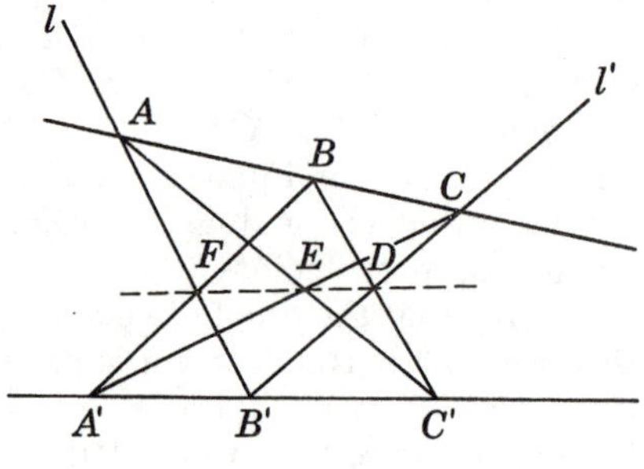

FIG. 1.29

CB' or $C'B$. Now $l\cdot l'$ comes into consideration, and again taking a simplest possibility, $l=AB'$, $l'=CB'$, so $l\cdot l'=B'$. We are going to consider a projectivity from l to l'. To each point X on AB' there is to correspond a point X' on CB': X varies, and to F shall correspond D. Think of F as X, and imagine F to vary on AB'. Keep A', B', C',

A, C fixed, and vary F? What happens? B varies on AC and D varies on $B'C$. The variation in F causes a variation in D. How? Via B. Now $F \to B$ is a *perspectivity* from l to m $(=AC)$, with center A'. And $B \to D$ is a perspectivity from m to l', with center C'. The product of these two perspectivities sends F into D. So, first, we see a projectivity sending F into D. Second, is the projectivity a perspectivity? We must examine B'. What happens to B' under the perspectivity $(l) \overset{A'}{\underset{\wedge}{-}} (m)$? Well, $B' \to W = AB \cdot A'B'$. And what happens to W under the perspectivity $(m) \overset{C'}{\underset{\wedge}{-}} (l')$? We have $W \to B'$. So our projectivity sends B' into B', hence is a perspectivity. Third, what is the center of this perspectivity? Compute it. The center must lie on lines joining corresponding points. Of course, F and D correspond, but let us save this information. Try another point on l. The point A looks promising. What happens to A? Well, $A \to A \to S = AC' \cdot B'C$. Therefore the center must be on AS, which is AC'. This fits in with the hope that E is the center. We need another pair of corresponding points, and the line joining them, we hope, passes through E. $A'C$ is a line through E; let it cut AB' in R. If to R corresponds C, then RC passes through the center. Does R go into C? Yes: $R \to C \to C$. So $RC \cdot AS = E$ is the center. $F \to B \to D$, so FD passes through the center; F, E, D are collinear. Q.E.D.

Exercise. Prove Desargues' Theorem using the Fundamental Theorem.

15. Further remarks on the cross-ratio. Let A, B, C, D be four (distinct) collinear points. The cross-ratio of four points depends upon the order in which they are taken; for example, $R(A, B; C, D)$ and $R(A, B; D, C)$ are not usually the same. We now ask: What are all the possible cross-ratios for four points, and what are the relations between these cross-ratios? First, we have to consider in how many ways we can take the four points, i.e., how many ordered ways. We can take the *first* point in 4 ways, then the *second* in 3 ways, leaving 2 ways for the *third* choice, and only 1 way for the *fourth* choice. This gives us $4 \cdot 3 \cdot 2 \cdot 1 = 24$ ways; and hence certainly at most 24 cross-ratios: but some of these may be equal. Incidentally, to write down the 24 cross-ratios systematically, think of the four letters as composing a *word*, and order "words" lexicographically. Thus we would start:

$A \quad B \quad C \quad D$
$A \quad B \quad D \quad C$
$A \quad C \quad B \quad D$
$A \quad C \quad D \quad B$
$A \quad D \quad B \quad C$, etc.

Let us now consider the relations between the various cross-ratios. We have $R(A, B; C, D) = (CA/CB)/(DA/DB) = (CA \cdot DB)/(CB \cdot DA)$. Interchange C and D. Then clearly the cross-ratio is inverted.

THEOREM. $R(A, B; D, C) = 1/R(A, B; C, D)$.

This theorem, incidentally, tells us what the cross-ratio of a harmonic tetrad must be. First, recall that if $H(A, B; C, D)$ and $H(A', B'; C' D')$, then $(ABCD) \overline{\wedge} (A'B'C'D')$, whence $R(A, B; C, D) = R(A', B'; C', D')$. So *all harmonic tetrads must have the same cross-ratio.* Let this be x. What is x? Well, if $H(A, B; C, D)$, then also $H(A, B; D, C)$, so $x = R(A, B; C, D) = R(A, B; D, C)$. But quite generally $R(A, B; D, C) = 1/R(A, B; C, D)$. So we have $x = 1/x$, whence $x^2 = 1$ and $x = \pm 1$. Now 1 is not a cross-ratio, so $x = -1$. Hence we have the following theorem.

THEOREM. *If $H(A, B; C, D)$, then $R(A, B; C, D) = -1$.*

THEOREM. *Conversely, if $R(A, B; C, D) = -1$, then $H(A, B; C, D)$.*

PROOF. Let A', B', C', D' be a harmonic tetrad. Then $R(A', B'; C', D') = -1$. Hence, by a previous theorem, $(A'B'C'D') \overline{\wedge} (ABCD)$. So, also by a previous theorem, $H(A, B; C, D)$. Q.E.D.

Resume the previous discussion: A, B, C, D a general tetrad of collinear points. In $R(A, B; C, D)$, interchange A and B; then one reciprocates both the numerator and denominator of $(CA/CB)/(DA/DB)$, and so inverts the whole cross-ratio. Hence:

THEOREM. $R(B, A; C, D) = 1/R(A, B; C, D)$.

Exercise. Using cross-ratios, prove that if $H(A, B; C, D)$, then $H(B, A; C, D)$. (Our previous way of doing this was better, as it was entirely projective. This way shows us how cross-ratios may point the way to truth.)

Thus, interchanging the first two letters inverts the cross-ratio, and interchanging the last two letters inverts the cross-ratio. Combining these two facts, we obtain the following theorem.

THEOREM. $R(A, B; C, D) = R(B, A; D, C)$.

This theorem shows that the 24 ordered tetrads will yield at most 12 cross-ratios.

So far we have interchanged the first two letters, and the last two, but have not interchanged letters from the two pairs. Let us then consider $R(A, C; B, D)$. We have $R(A, C; B, D) = (BA/BC)/(DA/DC) = (BA \cdot DC)/(BC \cdot DA)$. Except for sign, the denominator here is the same as above for $R(A, B; C, D)$. To study these cross-ratios further, set up a coordinate system: let a, b, c, d be the coordinates of A, B, C, D (Fig. 1.30). Then

$$CA \cdot DB = (a-c)\cdot(b-d) = ab-ad-cb+cd$$

and

$$BA \cdot DC = (a-b)\cdot(c-d) = ac-ad-bc+bd.$$

subtracting,

$$(CA \cdot DB)-(BA \cdot DC) = ab-ac+cd-bd = (a-d)(b-c) = DA \cdot CB.$$

Therefore

$$R(A, B; C, D)+R(A, C; B, D) = \frac{CA \cdot DB}{CB \cdot DA} - \frac{BA \cdot DC}{CB \cdot DA}$$

$$= \frac{CA \cdot DB - BA \cdot DC}{CB \cdot DA} = \frac{DA \cdot CB}{CB \cdot DA} = 1.$$

THEOREM. $R(A, C; B, D) = 1 - R(A, B; C, D)$. *In words: Interchanging the middle two letters gives the difference from* 1.

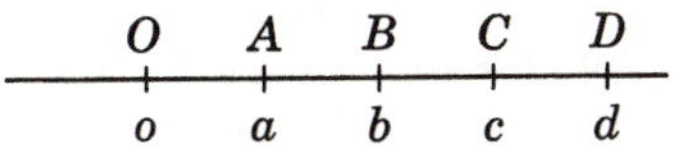

FIG. 1.30

COROLLARY. $R(D, B; C, A) = 1 - R(A, B; C, D)$, *i.e., interchanging the outer two letters also gives the difference from* 1.

PROOF. We first put the outer letters in the middle: $R(A, B; C, D) = R(B, A; D, C)$. Now we can interchange the middle letters: $R(B, D; A, C) = 1 - R(B, A; D, C)$, whence $R(D, B; C, A) = 1 - R(A, B; C, D)$.

COROLLARY. $R(A, B; C, D) = R(D, C; B, A)$, *i.e., interchanging the middle letters and also the other letters leaves the cross-ratio invariant.*

PROOF. $1-(1-x) = x$.

THEOREM. *The cross-ratio of A, B, C, D taken in any given order is equal to a cross-ratio of these points with A taken first.*

PROOF. Suppose A is in the second place, and say we are considering $R(C, A; B, D)$. Then $R(C, A; B, D) = R(A, C; D, B)$. Suppose A is in the third place, and say we are considering $R(C, B; A, D)$. Then

$R(C, B; A, D)=R(D, A; B, C)=R(A, D; C, B)$. A similar argument holds if A is in the fourth place.

This theorem shows that the 24 ordered tetrads will yield at most 6 cross-ratios. We need consider only $R(A, B; C, D)$, $R(A, B; D, C)$, $R(A, C; B, D)$, $R(A, C; D, B)$, $R(A, D; B, C)$, $R(A, D; C, B)$. Let us try to calculate these six in terms of one of them, say the first: call the first λ. We have

$$\begin{aligned}
R(A, B; C, D) &= \lambda.\\
R(A, B; D, C) &= 1/R(A, B; C, D) = 1/\lambda.\\
R(A, C; B, D) &= 1-R(A, B; C, D) = 1-\lambda.\\
R(A, C; D, B) &= 1/R(A, C; B, D) = 1/(1-\lambda).\\
R(A, D; B, C) &= 1-R(A, B; D, C) = 1-1/\lambda.\\
R(A, D; C, B) &= 1-R(A, C; D, B) = 1-1/(1-\lambda) = -\lambda/(1-\lambda).
\end{aligned}$$

The question now is: Is 6 really the final answer, or can some of these be equal? To see whether two of these can be equal, equate their expressions in λ. Since the first cross-ratio, $R(A, B; C, D)$ was an *arbitrary* one of the six, we need only inquire whether any of the last five can equal the first. For the second to equal the first, we need

$$\lambda = 1/\lambda \qquad \text{or} \qquad \lambda^2 = 1.$$

This can happen only for special values of λ, since by the Fundamental Theorem of Algebra, any polynomial equation of degree n has at most n roots. In the present case, $\lambda = \pm 1$, and even $\lambda = +1$ is excluded. The value $\lambda = -1$ corresponds to a harmonic tetrad.

Exercise. If $H(A, B; C, D)$, then how many cross-ratios are there?

In the same way, we ought to see what happens upon placing the third equal to the first. This gives $\lambda = 1-\lambda$, whence $\lambda = 1/2$. Then, however, $R(A, D; B, C)=1-1/\lambda=1-2=-1$, so $H(A, D; B, C)$: again we have, for an appropriate order, a harmonic tetrad.

Exercise. Study the three remaining cases.

Summing up we can say: In general there are exactly 6 cross-ratios, but for some special cases there may be less.

16. Construction of the projective plane. By now it is probably fairly clear what type of theorem we would like to study in projective geometry. The existence of parallel lines in the plane is the main cause of trouble. We have been, in the last few pages, simply neglecting the existence of parallel lines; except for this neglect, most of the considerations were quite rigorous. We will, therefore, try to construct an object that shall have the following properties: it should be very much like the ordinary, Euclidean plane, but the parallel postulate should not hold; rather, any two lines should meet in one and only one point. To achieve this end, we are going to start with an ordinary Euclidean plane and *modify* the meaning of the words point, line, etc., so that the parallel postulate will fail and so that any two lines meet in one and only one point; but the modification should not be so drastic that the resulting object, which we will call the projective plane, would bear no obvious relation to the plane we start with.

In the following we are making a *construction*, and the material is the ordinary plane; the procedure is *not* the *axiomatic* one we shall study shortly. Meanwhile the ordinary plane rests on an axiomatic foundation that we are not examining. Nonetheless, in order to be fairly exact, let us put down precisely what we are going to use in the construction. The main propositions used will be:

I. On any two points there is one and only one line.

II. Given a line and a point not on the line, there is one and only one line on the given point not meeting the given line.

The term "Euclidean" usually has the connotation that the concept of distance is involved; to avoid this connotation, one speaks of *the affine plane*. That is, in what immediately follows, we use propositions I and II, but in no manner use the concept of distance. One may say that one is starting from the affine plane.

Consider two parallel lines in the affine plane. They do not meet. We are going to so modify the meaning of the word *line* that the lines will meet. In order to distinguish the kind of lines we start with from the kind we finish with, which will be *like but not the same as* those we start with, we will call the second kind *Lines*, i.e., write the word with a capital letter. In speaking, we may say *projective line*, setting this phrase in opposition to *affine line*. We will also modify the concept of point, speaking then of *projective* points and *affine* points, and writing *Point* for a point of the new kind.

For each line we are going to introduce a new object, a Point, a projective point; and for parallel lines, the same object; for nonparallel lines, different objects. What are these new objects? Anything at all, just so long as for parallel lines we take the same object and for intersecting lines, different objects. For concreteness' sake, we could take the various pencils of parallel lines as the new objects, and would then say that on the given line l there now lies the parallel pencil determined by l; but such definite choice is not necessary and is even inconvenient. Let us call these new objects *Ideal Points*. (By a *pencil of lines* one means all the lines parallel to a given line or all the lines through a given point.)

DEFINITION. By a *Point*, or *projective point*, we will mean either an ordinary point or an Ideal point.

Suppose now we take as the definition of a Line, or projective line, the following:

A Line, or projective line, shall consist of an ordinary line and the Ideal Point that it determines.

(This will not be the final definition: we want to see what happens with the above definition.)

With these definitions, do any two Lines meet? *Yes.* Take two Lines. Each Line consists of an ordinary line l and an Ideal Point P; let us write this line, notationally, as (l, P). Now take two Lines: (l, P), (m, Q). Here l and m are affine lines that may or may not meet. (a) If l and m do not meet, then they are parallel, so $P = Q$, and (l, P) and (m, Q) meet in one and only one Point, namely, P. (b) If l and m do meet, say in a point R, then $P \neq Q$, and (l, P) and (m, Q) meet in one and only one Point, namely R. Therefore we would have, instead of II, the proposition:

II′. *On any two Lines there is one and only one Point.*

This is what we want: but we also want our old proposition I.

I. *On any two Points there is one and only one Line.*

Do we have it? *No,* because each Line, according to the above definition, has on it just one Ideal Point; therefore, if we take two Ideal Points, there could be no Line on both of them. We get around this difficulty by defining Line as follows.

DEFINITION. A *Line*, or *projective line*, shall consist of an ordinary line and the Ideal Point which it determines: also the totality of Ideal Points shall be said to constitute a Line.

This new Line, consisting of all the Ideal Points, we may call the Ideal Line: any other Line has on it just one Ideal Point.

Now do we have proposition I? We have to take two Points, P and Q, and see if there is one and only one Line on them. If P and Q are both Ideal Points, then the Ideal Line, and no other, is on both P and Q. Suppose P is Ideal, but Q is not. The Ideal Point P is determined by a pencil of parallel lines, and we are asking how many of them are on Q. Obviously one, and only one. Now, what if neither P nor Q is Ideal. Then P, Q determine an ordinary line l; l determines a Line (l, R); and (l, R) is the one and only Line on P and Q.

So we have I. But how about II′? In checking this before, we were not using our final definition of Line, so we have to go through the proof again, if we can.

Exercise. Can we?

It turns out that we can, and so the object we have just constructed has the basic properties I and II′ and is called a projective plane.

17. Previous results in the constructed plane. Starting with the ordinary, Euclidean plane, we introduce the projective plane as before. All of our previous considerations could now be carried over, in a rigorous manner, to the projective plane. Let us see how to get the Fundamental Theorem—once we had that, Desargues' Theorem and Pappus' Theorem (as well as the theorems on harmonic tetrads) would follow as before. The Fundamental Theorem was founded on the cross-ratio. Now, although cross-ratio was founded on the concept of distance, it is nonetheless projective, and we can introduce the cross-ratio into the projective plane. For four collinear Points A, B, C, D, none of which is Ideal, we define the cross-ratio as before. Suppose

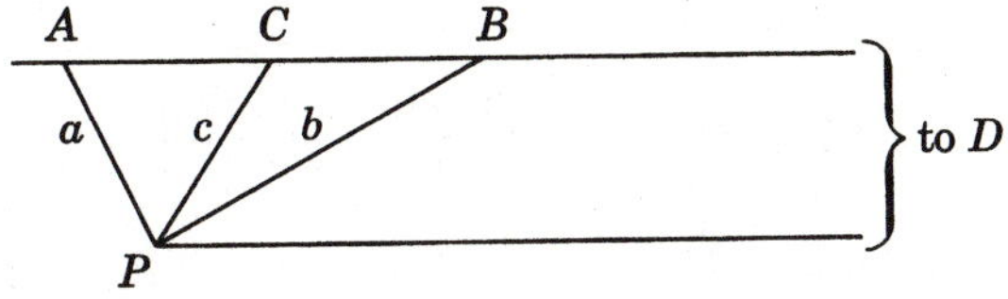

FIG. 1.31

now that A, B, C, D are all ideal. Take a Point P that is not Ideal. Let a, b, c, d be lines through P and the Points A, B, C, D; each of these Lines consists of an affine line and an Ideal Point—for the sake of simplicity in notation, let us designate these affine lines by a, b, c, d also. Then let us define $R(A, B; C, D)$ to be equal to $R(a, b; c, d)$ in the previously defined sense, i.e.,

$$R(A, B; C, D) = \frac{\sin (ca)}{\sin (cb)} \Big/ \frac{\sin (da)}{\sin (db)}.$$

For this definition to be acceptable, we have to see that the cross-ratio does not depend on the choice of P. If P' is another choice for P, however, and $a' = P'A$, etc., then $\sin (c'a') = \sin (ca)$, for example, and one sees that one comes out with the same value for $R(A, B; C, D)$. Let now one of A, B, C, D be Ideal, say D. Again take a point P, not Ideal, and consider the lines $a = PA$, $b = PB$, etc., and define $R(A, B; C, D)$ as $R(a, b; c, d)$ (Fig. 1.31); and again one must check that this value is independent of P. First, not worrying about signs, and recalling that the area of a triangle is given by one-half the product of any two sides and the sine of the included angle, we have:

$$R(AB, CD) = \frac{\sin (ca)}{\sin (cb)} \Big/ \frac{\sin (da)}{\sin (db)}$$

$$= \frac{2(\text{area } CPA)/CP \cdot AP}{2(\text{area } CPB)/CP \cdot BP} \Big/ \frac{2(\text{area } CPA)/(CA \cdot AP)}{2(\text{area } CPB)/(CB \cdot BP)} = \frac{CA}{CB}.$$

Therefore, except possibly for sign, $R(A, B; C, D) = CA/CB$. As for the sign, $\sin(da)/\sin(db)$ is certainly positive and $\sin(ca)/\sin(cb)$ clearly has the same sign as CA/CB. So the sign is right, too.

Thus we can at least define cross-ratio, but we still have to see that it is invariant under a projectivity. It is sufficient to see that $(ABCD) \overset{P}{\underset{\wedge}{-}} (A'B'C'D')$ implies $R(A, B; C, D) = R(A', B'; C', D')$. If P is not an ideal point and if a, b, c, d are the lines PA, PB, PC, PD, then

$$R(A, B; C, D) = \frac{\sin (ca)}{\sin (cb)} \Big/ \frac{\sin (da)}{\sin (db)}$$

either by definition if at least one of the points A, B, C, D is ideal, or by previous results if none of the points A, B, C, D are ideal. That $R(A, B; C, D) = R(A', B'; C', D')$ follows immediately. There is still the case to consider that P is ideal. We will not write out the details of the proof, which is entirely straightforward.

With the cross-ratio in hand, we could easily make rigorous all, or almost all, of our previous considerations. Even so, we would, in this way, always have with us, in the foundation at any rate, concepts foreign to the center of our interest, namely, the projectively invariantive type of theorem. The most satisfying way to proceed now is to start all over again and lay the foundations of projective geometry axiomatically —and what that means, we shall see.

18. Analytic construction of the projective plane. Before doing that, however, there is one topic that may most conveniently be dealt with here. Starting with the ordinary, Euclidean plane, we introduce coordinates in the familiar manner. Each point then has an ordered pair of numbers (x, y) as coordinates, different points have different coordinates, and a line consists of all the points whose coordinates (x, y) satisfy a given linear equation $ax + by + c = 0$, where $a \neq 0$ or $b \neq 0$. To free ourselves from any doubt as to what Euclid's axioms were, we could proceed as follows: we *define* a point to be an ordered pair of numbers (x, y), and a line to be the set of solutions (x, y) to a linear equation $ax+by+c=0$, where $a\neq 0$ or $b\neq 0$. It is a simple algebraic exercise to prove that with point and line thus defined, the following propositions hold.

I. On any two points, there is one and only one line.
II. Given any line and a point not on the line, there is one and only one line through the point not meeting the given line.

Exercise. Translate these propositions into directly algebraic statements.

Solution. To say that there is a line $ax+by+c=0$ through points (x_1, y_1), (x_2, y_2) is to say that the system of simultaneous equations

$$\begin{aligned} ax_1+by_1+c &= 0 \\ ax_2+by_2+c &= 0, \end{aligned}$$

in the unknowns a, b, c, has a solution with $a\neq 0$ or $b\neq 0$. Assume that the points (x_1, y_1), (x_2, y_2) are distinct, i.e., that $x_1\neq x_2$ or $y_1\neq y_2$. The statement that there is only one line through the points comes to saying that if (a, b, c) and (a', b', c') are solutions of the above equations, with $a\neq 0$ or $b\neq 0$ and $a'\neq 0$ or $b'\neq 0$, then any solution (x, y) of either of the equations $ax+by+c=0$, $a'x+b'y+c'=0$ is a solution of the other. (Under these circumstances, one can prove that (a', b', c') is a multiple of (a, b, c), i.e., that there is a number λ such that $a'=\lambda a$, $b'=\lambda b$, $c'=\lambda c$. Thus all equations of a given line are related simply to one another: each is a multiple of the others. Because of this, one speaks of *the* equation of a line, although this is not an exact way of speaking, since a line has many equations.)

As to proposition II, this can be stated as follows. Let a, b, c, x_1, y_1 be numbers with $a\neq 0$ or $b\neq 0$ and $ax_1+by_1+c\neq 0$. Then there are numbers a', b', c' with $a'\neq 0$ or $b'\neq 0$ such that $a'x_1+b'y_1+c'=0$ and such that the equations $ax+by+c=0$ and $a'x+b'y+c'=0$ have no common solution; and if a'', b'', c'' have the properties just stated for a', b', c', then any solution of either of the equations $a'x+b'y+c'=0$ and $a''x+b''y+c''=0$ is a solution of the other.

In other words, points and lines, thus defined, together constitute an affine plane. To this plane can be adjoined new points, the so-called Ideal Points, so that the resulting object is a projective plane. We did this, before, synthetically, and want to show now how to do it analytically; the only difference is that now we want to define the new point to be added to a line in terms of the equation of the line.

Let $ax+by+c=0$ be a line (or equation of a line). Here we think of a, b, c fixed and x and y as variables, or letters. One also has occasion, however, to consider (x, y) as fixed, and a, b, c as variables. For example, suppose one asks about the lines through the *given* point (u, v); any such line would be of the form $ax+by+c=0$ and we would be concerned to know those triples (a, b, c) for which $au+bv+c=0$.

Consider equations of the form $ax+by+c=0$ solely from the algebraic point of view for a moment. If a, b, c are understood to be fixed numbers, and x and y letters, this equation is of the first degree in x and y, because each term involves x and y multiplied together only to the first power at most (dx^py^q is said to be of degree $p+q$ in x and y, if d is fixed and not equal to zero—so ax and by are of degree 1, and c is of degree 0, provided they are different from zero: *zero* is not given a degree). If we regard x, y as fixed, and a, b, c as variables, then $ax+by+c=0$ is also of the first degree in a, b, c. If all the terms of an equation are of equal degree, we say it is homogeneous—otherwise, nonhomogeneous. Thus $ax+by+c=0$ is nonhomogeneous in x and y (provided $c \neq 0$), but is homogeneous in a, b, c.

In studying the equation $ax+by+c=0$, it is sometimes convenient to bring it into association with the equation $aX+bY+cZ=0$, which has the advantage of being homogeneous. Any information we have on $aX+bY+cZ=0$ could be translated into information on $ax+by+c=0$, and vice versa. Suppose, for example, that (X, Y, Z) is a triple of numbers satisfying the equation $aX+bY+cZ=0$; and suppose $Z \neq 0$. Then X/Z, Y/Z will be a pair of numbers satisfying $ax+by+c=0$, i.e., $a(X/Z)+b(Y/Z)+c=0$. Conversely, if (x, y) is a pair of numbers satisfying $ax+by+c=0$, then $(x, y, 1)$ will be a triple satisfying $aX+bY+cZ=0$, as will $(\lambda x, \lambda y, \lambda)$, and in fact any triple (X, Y, Z) will be a solution if $X/Z=x$, $Y/Z=y$. There are further solutions to $aX+bY+cZ=0$, which do not correspond to solutions of $ax+by+c=0$, namely, those for which $Z=0$. The solution $(0, 0, 0)$ will, of course, give us no information on $ax+by+c=0$, because *every* equation of the form $aX+bY+cZ=0$ has $(0, 0, 0)$ as a solution, and this solution is therefore called the *trivial* solution: any other solution $(X, Y, 0)$, with $X \neq 0$ or $Y \neq 0$, would yield information on the equation $ax+by+c=0$; it would tell us, in fact, that a and b are such that $aX+bY=0$.

Let us now consider the situation also from the geometric point of view. We are interested in (ordered) pairs of numbers (x, y) and also (ordered) triples (X, Y, Z): let us associate the pair (x, y) with a triple (X, Y, Z) if $Z \neq 0$, $X/Z=x$, and $Y/Z=y$. The pair (x, y) represents, or *is*, a point. If $X/Z=x$, $Y/Z=y$, then the triple (X, Y, Z) may also *represent* the point, because given these three numbers, and knowing that they are related to x and y in the above way, we can say with what point the triple is associated. That being so, we may speak of (X, Y, Z) as being *homogeneous coordinates* of the point (x, y). Note that if (X, Y, Z) are homogeneous coordinates of a point, then $(\lambda X, \lambda Y, \lambda Z)$, if $\lambda \neq 0$, are also homogeneous coordinates of the *same* point. If we want to speak of the point with coordinates (X, Y, Z) we *may* sometimes say "the point (X, Y, Z)," but this is not an exact expression. We will, however, refer to the point with coordinates X, Y, Z as "the point

$P\begin{pmatrix}X\\Y\\Z\end{pmatrix}$": this is exact—$P\begin{pmatrix}2\\3\\5\end{pmatrix}$ and $P\begin{pmatrix}4\\6\\10\end{pmatrix}$, for example, are the *same* point.

Let us separate all ordered triples of numbers (not all of which are zero) into classes, putting two triples (X, Y, Z), (X', Y', Z') into the same class if there is a $\lambda \neq 0$ such that $X' = \lambda X$, $Y' = \lambda Y$, $Z' = \lambda Z$. In this way, all the triples corresponding to a point form a single class. Note that if (X, Y, Z) is a solution to the equation $aX + bY + cZ = 0$, then so is $(\lambda X, \lambda Y, \lambda Z)$: together with a nontrivial solution (X, Y, Z) we get a whole *class* of solutions to $aX + bY + cZ = 0$. It is convenient to speak of these *classes* as solutions, but note that the word "solution" now has a slightly different meaning—in particular, in this new sense, we do not refer to $(0, 0, 0)$ as being a solution at all.

Let now a line be given by the equation $ax + by + c = 0$, and consider along with it the corresponding homogeneous equation $aX + bY + cZ = 0$. By a solution of this homogeneous equation we mean a class of triples, as explained above. Then every solution (x, y) of $ax + by + c = 0$ yields a solution $\{(\lambda X, \lambda Y, \lambda Z)\}$ of $aX + bY + cZ = 0$ for which $Z \neq 0$; and conversely, every solution $\{(\lambda X, \lambda Y, \lambda Z)\}$ of $aX + bY + cZ = 0$ for which $Z \neq 0$ yields a solution (x, y) of $ax + by + c = 0$. (The symbol $\{\cdots\}$ is used to indicate a whole set of objects, and between the braces is placed either the whole set or a typical element of the set.) Thus there is a one-to-one correspondence between the solutions of $ax + by + c = 0$ and those of $aX + bY + cZ = 0$ in which the third coordinate does not equal zero. How many solutions of $aX + bY + cZ = 0$ are there for which $Z = 0$? Recall, $(0, 0, 0)$ does not count as a solution. Recall also that, since $ax + by + c = 0$ represents a line, we have $a \neq 0$, or $b \neq 0$. Therefore $(-b, a, 0)$ is a solution, or rather, we should say, represents a solution of $aX + bY + cZ = 0$. Either $a \neq 0$ or $b \neq 0$; let us proceed under the assumption that $a \neq 0$—were $a = 0$, then $b \neq 0$, and the considerations would be quite the same except for notation. Now we are looking for triples $(X, Y, 0)$ satisfying $aX + bY + cZ = 0$. Were $Y = 0$, then $aX + b \cdot 0 + c \cdot 0 = 0$, so also $X = 0$, but $(0, 0, 0)$ is out. So $Y \neq 0$ for any solution $(X, Y, 0)$. Fixing $Y \neq 0$, we solve for X: $X = -bY/a$. Therefore, every solution with $Z = 0$ must be of the form $((-b/a)\,Y, Y, 0)$, and these, we observe, constitute precisely one class, i.e., there is just one solution more to the equation $aX + bY + cZ = 0$ than was already considered previously. (Note that the word "solution" is used in two senses—and that with each use of the word one must understand from the context in what sense it is being used.)

Let us now call the classes that are the solutions of $aX + bY + cZ = 0$ *Points*. There is then a close relation between the points of $ax + by + c$

$=0$ and the Points of $aX+bY+cZ=0$: they are, in fact, in one-to-one correspondence except for just one Point. The equation $aX+bY+cZ=0$ has, so to speak, room for just one more solution than does $ax+by+c=0$. Let us take this extra solution, this Point with $Z=0$, as the object to be specified as the Ideal Point on the line $ax+by+c=0$. We must here observe that a line does not have only *one* equation, but if $ax+by+c=0$ is *an* equation, then $\lambda ax+\lambda by+\lambda c=0$, $\lambda \neq 0$, is also an equation, and conversely, if $ax+by+c=0$ and $a'x+b'y+c'=0$ are equations of the same line (i.e., have the same set of solutions), then there is a λ, $\lambda \neq 0$, such that $a'=\lambda a$, $b'=\lambda b$, $c'=\lambda c$. Thus the solution of $aX+bY+cZ=0$ for which $Z=0$ is the same as that of $\lambda aX+\lambda bY+\lambda cZ=0$ for which $Z=0$. That is, we have specified just one Ideal Point to each line. Also, distinct lines $ax+by+c=0$ and $a'x+b'y+c'=0$ do not have a solution in common, i.e., are parallel, if and only if there is a $\lambda \neq 0$ such that $a'=\lambda a$, $b'=\lambda b$. Thus the same Ideal Point is assigned to parallel lines, different Ideal Points to intersecting lines.

We now come to the following curious consideration. The lines $ax+by+c=0$ give rise to almost all the equations of the form $aX+bY+cZ=0$—the only one, or ones, that they do not give rise to is the equation $cZ=0$: since this equation and the equation $Z=0$ have precisely the same solutions, we may say that there is just one equation $aX+bY+cZ=0$ not arising from a line, namely, the equation $Z=0$. Now this equation is precisely the condition (or equation) on the coordinates of an Ideal Point. Hence, if by a *Line* we mean the set of Points satisfying an equation $aX+bY+cZ=0$, with not all $a, b, c=0$, then we have coordinates for *all* the Points of the projective plane and linear equations for *all* the Lines.

19. Elements of linear equations. In the above analytical construction we omitted a number of details that we proceed to fill in:

THEOREM. *If $x_1 \neq x_2$ or $y_1 \neq y_2$, then the simultaneous system of equations:*

$$(1) \qquad ax_1+by_1+c = 0$$
$$(2) \qquad ax_2+by_2+c = 0$$

in the unknowns a, b, c have a solution with $a \neq 0$ or $b \neq 0$. Moreover, if (a, b, c) and (a', b', c') are solutions with $a \neq 0$ or $b \neq 0$ and $a' \neq 0$ or $b' \neq 0$, then there is a number λ such that $a'=\lambda a$, $b'=\lambda b$, $c'=\lambda c$.

PROOF. Observe first that if (a, b, c) is a solution, then also $(\lambda a, \lambda b, \lambda c)$ is a solution for every $\lambda \neq 0$. This leads us, just as in the previous section, to put two triples (a, b, c), (a', b', c') into the same class if one is a multiple (by a $\lambda \neq 0$) of the other: we then speak of the class as a solution, and the theorem can then be restated as saying that (1) and (2) have one and only one nontrivial solution.

We proceed to solve (1) and (2) simultaneously in the familiar way. If (a, b, c) is a solution of (1) and (2), then, subtracting (1) from (2), we obtain:

$$(3) \qquad a(x_2-x_1)+b(y_2-y_1) = 0.$$

We have $x_2 \neq x_1$ or $y_2 \neq y_1$, say $x_2 \neq x_1$. Then we get

$$(4) \qquad a = -b\,\frac{y_2-y_1}{x_2-x_1}$$

and from (1),

$$(5) \qquad c = -by_1-ax_1 = -b\left[y_1-x_1\,\frac{y_2-y_1}{x_2-x_1}\right]$$

$$= \frac{-b[y_1(x_2-x_1)-x_1(y_2-y_1)]}{x_2-x_1}.$$

Thus a, b, c are $\dfrac{-b}{x_2-x_1}$ times, respectively, $y_2-y_1, -(x_2-x_1)$, $y_1(x_2-x_1)-x_1(y_2-y_1)$. We reach a similar conclusion if after (3) we proceed under the assumption $y_2 \neq y_1$, rather than $x_2 \neq x_1$. So far, then, we have established that there is at most one nontrivial solution of (1) and (2), namely, the solution, if any, must be the class of $(y_2-y_1, -(x_2-x_1), y_1(x_2-x_1)-x_1(y_2-y_1))$.

To see that there is a solution we check that $(y_2-y_1, -(x_2-x_1), y_1(x_2-x_1)-x_1(y_2-y_1))$ satisfies (1) and (2). It is not the trivial solution since $y_2-y_1 \neq 0$ or $-(x_2-x_1) \neq 0$.

COROLLARY. *Through two points* (x_1, y_1), (x_2, y_2) *there passes one and only one line.*

THEOREM. *Let* a, b, c; a', b', c' *be numbers with* $a \neq 0$ *or* $b \neq 0$ *and* $a' \neq 0$ *or* $b' \neq 0$. *If* $ab'-a'b \neq 0$, *then the simultaneous system*

$$(1) \qquad ax+by+c = 0$$

$$(2) \qquad a'x+b'y+c' = 0$$

in the unknowns x, y *has exactly one solution. If* $ab'-a'b=0$, *then either the system has no solution; or if it has a solution, then it has at least two solutions (whence by the previous theorem* (a', b', c') *is a multiple of* (a, b, c) *and every solution of either* (1) *or* (2) *is a solution of the other).*

PROOF. Multiplying (1) by b' and (2) by b and subtracting, we get

$$(3) \qquad (ab'-a'b)x+b'c-bc' = 0.$$

Multiplying (1) by a' and (2) by a and subtracting, we get

$$(4) \qquad (a'b-ab')y+a'c-ac' = 0.$$

If $ab'-a'b \neq 0$, then from (3) and (4) we see that if (x, y) satisfies (1) and (2) then $x=(-b'c+bc')/(ab'-a'b)$ and $y=(ac'-a'c)/(a'b-ab')$. Thus there is at most one solution to (1) and (2): conversely, one checks directly that these values of (x, y) satisfy (1) and (2).

If $ab'-a'b=0$, then there is a number $\lambda \neq 0$ such that $a'=\lambda a$, $b'=\lambda b$: in fact, $a \neq 0$ or $b \neq 0$; if $a \neq 0$ we can write $b'=\dfrac{a'}{a}b$ and $a'=\dfrac{a'}{a}a$, so $\lambda=a'/a$; a similar result holds if $b \neq 0$. Dividing (2) by λ (which does not equal zero since $a' \neq 0$ or $b' \neq 0$), we get an equation (5) having the same solutions as (2):

$$(5) \qquad ax+by+d = 0, \quad \text{where} \quad d = c'/\lambda.$$

If $d \neq c$, then clearly (1) and (5), hence also (1) and (2), have no common solution. If $d=c$, then every solution of (1) is a solution of (2); in this case it remains to see that (1) has at least two solutions. If $a \neq 0$, then placing $y=0$, 1 successively in (1), we solve for x to get two solutions. Similarly if $b \neq 0$.

From the proof, any line not meeting $ax+by+c=0$ has an equation of the form $ax+by+d=0$. Hence if (x_1, y_1) is not on $ax+by+c=0$, there is one and only one line on (x_1, y_1) that does not meet $ax+by+c=0$, namely, $ax+by-ax_1-by_1=0$.

In sum, we have proved in detail propositions I and II of the previous section.

Exercises

1. The following properties of points, lines, and planes are familiar from high school solid geometry.

1. On any two points there is one and only one line.
2. On any three noncollinear points there is one and only one plane.
3. If two points of a line are on a plane, then the line is on the plane.
4. Two planes cannot meet in a single point.
5. Not all points are on the same plane.

6A. Every plane is an affine plane. (Since each plane is affine, the Parallel Postulate holds in each of them. Parallel lines, then, are coplanar lines that do not meet; in addition, a line is said to be parallel to itself.)

7A. Lines parallel to the same line are parallel to each other.

These properties characterize a so-called affine 3-space. A projective 3-space is characterized by conditions 1 through 5, but this time we require:

6P. Every plane is a projective plane.

The exercise is as follows. Given an affine 3-space, show how to introduce ideal points to obtain a projective 3-space.

2. Introduce the ideal points for projective 3-space using coordinates.

CHAPTER II

THE AXIOMATIC FOUNDATION

1. Unproved propositions and undefined terms. One way to establish a theorem is to prove it, and that means to show how it follows from previous theorems, i.e., theorems we already regard as established. If now we demand that these previous theorems be proved, we have to go back to still earlier theorems, and so on. It becomes clear that if we are going to prove anything, there must also be propositions that we regard as true but for which we demand no proof. In order to go forward, we must stop going backward. When certain propositions are laid down as the starting point of a deductive theory, and no proofs are required for these propositions, then these propositions are called *axioms*.

Axioms are sometimes said to be obvious truths. Sometimes people say that a thing is obvious when they are in no position to offer a proof. We, too, will use this convenient word, but only to indicate that certain details of a proof are being left out, and that anyone who has otherwise followed the argument can easily supply the missing details for himself. In this sense, it is incorrect to speak of axioms as being obvious, because there is no question of proof involved.

Just as it is with propositions, so is it with definitions. To define an object or term is to give its meaning in terms of other objects and terms, and to define these would mean to relate them to still other objects and terms, and so on. Again it is clear that if anything is to be defined, there must also be undefined terms.

2. Requirements on the axioms and undefined terms. The above seems to leave the axioms in such a nebulous state that one can ask what it is that makes us choose certain statements as axioms and not others. Part of these doubts can be easily laid to rest: the axioms we choose will depend upon the kind of things we want to study. If we want to study algebra, our axioms will have an algebraic sound to them: and if we want to study geometry, then they will have a geometric sound. The axioms are thus *roughly* determined by the subject matter we want to consider and by the theorems that we would like to establish. It is for this reason that a preliminary and not wholly rigorous consideration of various geometric theorems was entirely in order.

Actually there are absolutely no *requirements* on an axiom system, but only *desiderata.* What these are we shall see in a moment. Meanwhile, to make the argument definite, let us write down the axioms that will mainly interest us in the near future.

3. Undefined terms and axioms for a projective plane. The kind of thing we roughly have in mind for study is the projective plane. As *undefined terms* we take *point*, *line*, and a relation *on.* There is no use in asking what we mean by a point, because we are taking points as undefined; and the same is true for lines. We have a certain set of objects, absolutely unrestricted in character, and these are to be called points; another set of objects, also unrestricted, are to be called lines. We also have an *undefined relation* between points and lines: a point P and a line l may be in such a relation that we say *P is on l*; again, we do not ask what this means, taking it as undefined. If P is on l we will also say *l is on P*: this is already a definition, but it is of such a trifling nature that we will not label it. We will also say that *l passes through P* if P is on l—we do not need this extra phrase, but there is nothing illogical in its use. If l and m are lines, and P is on l and P is on m, we will say that l and m meet at P, or cross at P, or intersect in P; if points P and Q are on l, we will say that l joins P and Q. If it is not the case that P is on l we will say *P is not on l.*

We take the following as axioms:

0. *There exist at least one point and at least one line.*
1. *If P and Q are distinct points, then there is at least one line on both of them.*
2. *If P and Q are distinct points, then there is at most one line on both of them.*
3. *If m and n are distinct lines, then there is at least one point on both of them.*
4. *There are at least three points on any line.*
5. *Not all points are on the same line.*

4. Initial development of the system; the Principle of Duality. Before discussing the suitability of the above system, let us develop it somewhat. In the course of doing so, we shall draw figures. These figures are only intended to help us keep in mind the objects being spoken about. They may sometimes suggest the truth to us, but they may also suggest false things to us. The figure cannot be used as substitute for proof; nor are we arguing about the figure, but only about things which the figure may call quickly to mind.

THEOREM 1. *If m and n are distinct lines, then there is at most one point on both of them.*

PROOF. By *reductio ad absurdum.* Suppose there were two (distinct) points P and Q on both m and n. Then on these two (distinct) points there would be more than one line, contradicting Axiom 2. The proof is complete.

In this proof we have used *only* Axiom 2.

Note the relationship between Theorem 1 and Axiom 2. The statement of Theorem 1 can be obtained from the statement of Axiom 2 by the interchange of the words *point* and *line*. Given a statement couched entirely in terms of the words *point, line,* and *on* (and such general concepts as *existence, identity,* and others), we can obtain another statement, called its *dual*, by interchanging the words *point* and *line*. The dual of the dual statement is the original statement. Even if a given statement is true, we are not asserting, for the moment at any rate, that the dual statement is also true. But at least we can consider the dual statement, and perhaps prove it. We dualized Axiom 2 and proved the dual.

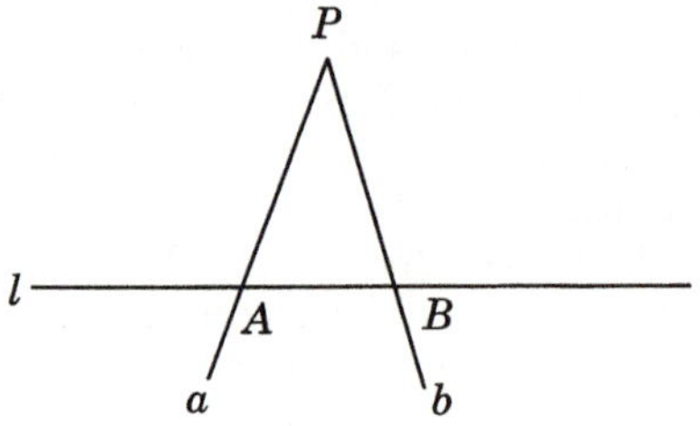

FIG. 2.1

Let us try dualizing the other axioms. Dualizing Axiom 0, we just get Axiom 0 back again: Axiom 0 is *self-dual*. The dual of Axiom 1 is Axiom 3, and this we have. The dual of Axiom 2 is Theorem 1. The dual of Axiom 3 is Axiom 1. The dual of Axiom 4 is the following, and is new.

THEOREM 2. *There are at least three lines on any point.*

There is no guarantee beforehand that this theorem is true, but it *is* true. Before giving the proof, we establish a lemma.

LEMMA. *Let the point P be not on the line l, and let A, B be two (distinct) points on l. If a is a line on P and A and b is a line on P and B, then a and b are distinct.*

PROOF. Since P is on a but not on l, we have $a \neq l$ (no axiom is used) (Fig. 2.1). By Theorem 1, no point other than A is on both a and l; B is on l, hence not on a. Then $b \neq a$, since B is on b but not on a.

In the proof we used only Theorem 1; in terms of the axioms, we have used only Axiom 2.

PROOF OF THEOREM 2. Let P be a point: to prove there exist three distinct lines through P. If we could find a line l *not* on P, we could easily complete the proof. In fact, suppose we had such a line l. On it take three points A, B, C, by Axiom 4. Let a, b, c be lines on P, and A, P and B, P and C, respectively: here Axiom 1 is used ($P \neq A$, since A is on l but P is not; and similarly $P \neq B$, $P \neq C$). By the lemma $a \neq b$, $a \neq c$, $b \neq c$. It thus remains to prove the following theorem—note that it is the dual of Axiom 5.

THEOREM 3. *Not all lines are on the same point.*

PROOF. Let P be a point (no axiom is used, because we agree to say that a theorem is true if the objects spoken about in the hypothesis do not exist—the theorem, reformulated, states that *if P is a point, then not all lines are on P*) (Fig. 2.2). Now let l be a line: here Axiom 0 *is* used, because if there are no lines, we cannot take one—but Axiom 0 furnishes us with lines to take. If l is not on P, which may be the case, then l is a line of the type sought. If l is on P, let Q be a point on l, $Q \neq P$: here Axiom 4 is used. And let R be a point not on l, by Axiom 5. Now $R \neq Q$, since Q is on l, but R is not. By Axiom 1, there is a line r on R and Q. Since R is on r, but not on l, we have $r \neq l$. By Axiom 2, then, P is not on r; and r is a line of the type sought.

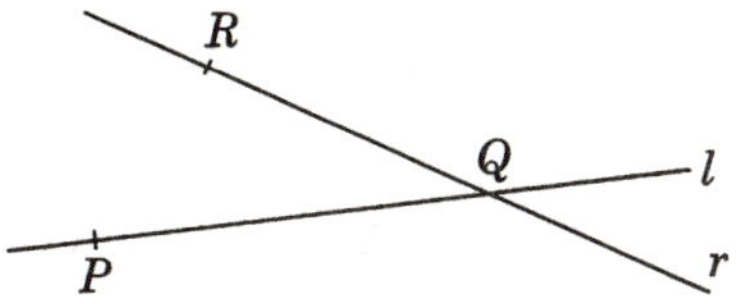

FIG. 2.2

All the axioms except Axiom 3 entered into the proof of Theorem 3. Axiom 3 entered into the proof of Theorem 2, and as Theorem 3 did, too, we have used all the axioms to prove Theorem 2. Later on, we shall not keep such careful track over which axioms are used, but it is a good idea to do so at first.

Note that now we have the dual of each axiom. The axioms and their duals are listed here in parallel columns.

Axiom 0	Axiom 0
Axiom 1	Axiom 3
Axiom 2	Theorem 1
Axiom 3	Axiom 1
Axiom 4	Theorem 2
Axiom 5	Theorem 3

We can now state the so-called principle of duality.

THE PRINCIPLE OF DUALITY. *If a theorem is deducible from the axioms, then its dual is also deducible from the axioms.*

In fact, to write out the proof of the dual statement we could just write down, statement by statement, the dual of the statements that prove the original statement.

Perspectivity, Projectivity. We define *perspectivity* and *projectivity* exactly as we did in the first part of the book. We can also prove, just as in that part, the following theorem.

THEOREM. *Given any three (distinct) collinear points A, B, C and any other three (distinct) collinear points A', B', C'; then there exists a projectivity sending A, B, C, respectively, into A', B', C'.*

The proof is quite as before, but since we are insisting on rigor, let us go through the proof again. As a first, special case consider the possibility that $A = A'$, but ABC and $A'B'C'$ are distinct lines (Fig. 2.3).

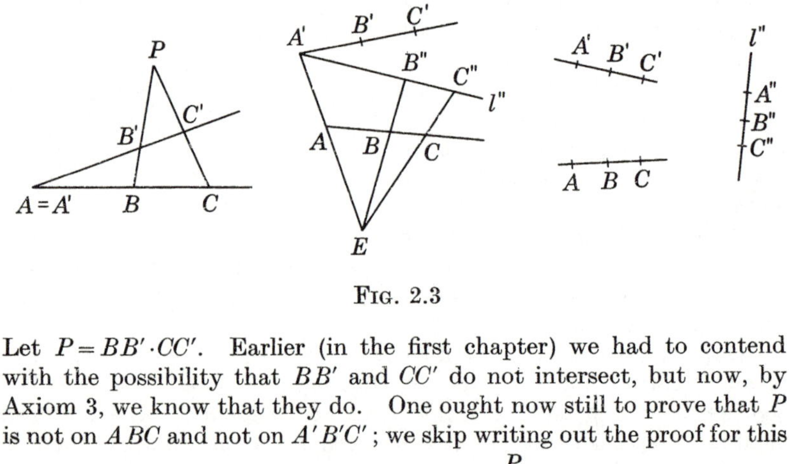

FIG. 2.3

Let $P = BB' \cdot CC'$. Earlier (in the first chapter) we had to contend with the possibility that BB' and CC' do not intersect, but now, by Axiom 3, we know that they do. One ought now still to prove that P is not on ABC and not on $A'B'C'$; we skip writing out the proof for this on the grounds that it is easy. Then $(ABC) \overset{P}{\underset{\wedge}{-}} (A'B'C')$. As a second case, let $A' \neq A$ and let A' be not on ABC. Let E be a point on AA', $E \neq A$, $E \neq A'$; here we use Axiom 5. Let l'' be a line on A', but $l'' \neq A'B'C'$, $l'' \neq A'AE$; here we use the dual of Axiom 5, or Theorem 3. Now we project A, B, C, into A', B'', C'' on l'' from center E, and then A', B'', C'' into A', B', C' as in the first case. Finally, let ABC, $A'B'C'$ be as given in the theorem. Let A'' be a point not on ABC and not on $A'B'C'$: one has to *prove* that such a point exists—we will not write out the proof, but one should think the proof through and see what axioms are involved. Let l'' be a line on A'' but not on A or A'; Theorem 2 is here used. Let B'', C'' be two further points on l''. By the second case, one can by a projectivity send A, B, C into A'', B'', C'', respectively; and by the same case, can send A'', B'', C'' in A', B', C', respectively. The product of the two projectivities sends A, B, C into A', B', C', respectively. Q.E.D.

How about the *dual* of the above theorem? What is its dual? To dualize a theorem, the theorem ought to be stated in terms of our primitive notions *point*, *line*, and *on*—then we would merely interchange the words *point* and *line*. But the above statement contains the word *projectivity*. This has been defined in terms of our primitive notions, but if now we are obliged to write its meaning out in full, what is the good of the term? A way out is to *dualize the definitions*. That is, we have introduced the term *perspectivity*: a perspectivity, as defined, relates points on one line with points on another. Let us write out the definition.

DEFINITION. Let l, l' be two lines, E a point not on l and not on l'. By a *perspectivity from l to l' with center E* we mean a transformation that associates to each point A on l the point $A' = EA \cdot l'$ on l'.

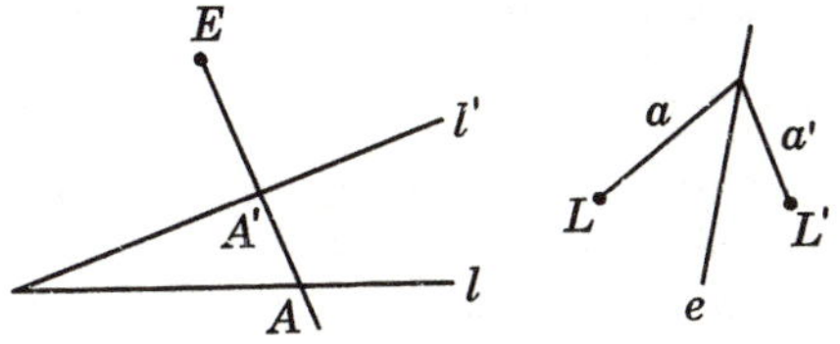

FIG. 2.4

What is the dual of the concept of a perspectivity? Instead of dealing with two lines l, l', we shall deal with two points L, L', and we make lines on L' correspond in a certain way with lines on L (Fig. 2.4). This correspondence, when we have written it down explicitly, we shall also call a perspectivity, but it will be between two pencils, rather than two lines (or *ranges*, to have a term dual to the term pencil). One will always understand from the context which type is meant. Now we *dualize* the above definition, i.e., we interchange in it the words *point* and *line*.

DEFINITION. Let L, L' be two points, e a line not on L and not on L'. By a *perspectivity from L to L' with axis e* we mean a transformation that associates to each line a on L the line $a' = ea \cdot L'$ on L'.

(We have also changed the word *center* to *axis*—this was not strictly necessary, but is convenient.)

These two definitions are duals of each other, and in dualizing we simply replace the one definition by the other. In verbal statements, we have to change *perspectivity* to *perspectivity*, i.e., there is a real change, but verbally there is none. The dual of our previous theorem now reads as follows.

THEOREM. *Given any three (distinct) concurrent lines a, b, c and any other three (distinct) concurrent lines a', b', c'; then there exists a projectivity sending a, b, c, respectively, into a', b', c'.*

Exercises

1. The above theorem follows from the Principle of Duality. Prove this theorem by dualizing, step by step, the proof of the dual theorem.

2. The complete 4-line or quadrilateral is dual to the complete 4-point or quadrangle. Write out the definition of a complete 4-line. Draw a figure. What is the diagonal triangle?

THEOREM. *If there are at least n points on some line l, then there are at least n points on any line l'.*

PROOF. Let E be a point not on l or l' (Theorem 2 is used), and consider the perspectivity $l \overset{E}{\underset{\wedge}{-}} l'$. By a lemma above, if A, B are distinct points on l, then EA and EB are distinct lines. Let these lines cut l' in A', B', respectively. Since E is not on l', we have $E \neq A'$, $E \neq B'$. Were $A' = B'$, the lines EA, EB would have two points in common, and that is not so; so $A' \neq B'$. In short, to distinct points on l there correspond by the perspectivity distinct points on l'. Therefore on l' there are at least n distinct points.

As we shall see shortly, it does not follow from the axioms that there are necessarily infinitely many points on a line. The following theorem is quite sensible.

THEOREM. *If there are at most n points on some line l, then there are at most n points on any line l'.*

This theorem follows easily from the preceding one.

Combining the two theorems we get the following.

THEOREM (P). *If there are exactly n points on one line, then there are exactly n points on every line.*

The dual of this theorem is the following.

THEOREM (Q). *If there are exactly n lines on one point, then there are exactly n lines on every point.*

In order not to confuse the n of Theorem (P) with the n of Theorem (Q) let us write:

THEOREM (Q'). *If there are exactly m lines on one point, then there are exactly m lines on every point.*

Theorem (Q′) does follow by duality from Theorem (P), but the equality $m = n$ does not follow by appeal to duality. The reason is that Theorem (P) involves, in addition to our axioms, an extra assumption, namely, that there are exactly n points on some one line. The conclusion drawn in Theorem (P) is not drawn from the axioms alone.

Although the equality $m = n$ does not follow from duality, nonetheless it is true that $m = n$. That is, the following is true.

THEOREM. *If there are exactly n points on some (hence any) line, then there are exactly n lines on any point.*

PROOF. Let l have on it exactly n points. Let P be a point not on l. Distinct lines on P cut l in distinct points, and, conversely, if A, B are distinct points on l, then PA, PB are distinct lines on P. Therefore there are exactly as many lines on P as points on l, namely, n. By a previous theorem, then, there are exactly n lines on any point.

THEOREM. *If there are exactly n points on a line, then altogether there are* $n^2 - n + 1$ *points.*

PROOF. Let P be a point; every point Q is on a line through P. Consider the lines through P. On each such line there are, not counting P, $n-1$ points. There are n such lines, so, not counting P, there are $n(n-1)$ points. Counting P, we have $n^2 - n + 1$ points.

Exercise. By the duality principle there are $n^2 - n + 1$ lines (given that some line has on it exactly n points). Draw the same conclusion without appeal to duality.

5. Consistency of the axioms. As we said before, there are absolutely no conditions on an axiom system, not even the condition that the axioms be consistent. Of course, the moment we found an inconsistency, the system would lose all interest for us. But if a system were actually inconsistent without our knowing it beforehand, the only way we could detect the inconsistency would be to study the system, and that means that we allow ourselves to study inconsistent systems.

Since, however, inconsistent systems are not interesting, one usually tries to pay some attention to this point at the very beginning. The question here arises, *How can one prove that a system of axioms is consistent?* One way is to *build a model* of the system; what this means we now explain.

To build a model of our axiom system (Axioms 0 to 5) means to *specify* certain objects as *points*, certain objects as *lines*, to specify a meaning for *on*, and to do this in such manner that the axioms become verifiable theorems. This will become clearer if we actually construct a model.

Take seven objects, any objects whatsoever—a tree, a book, a dream, etc., any seven different things—and designate them A, B, C, D, E, F, G. These objects we will call *points*: note that now, in building the model, *point* is *not* undefined—a point is one of the objects $A, \cdots, G$. By *lines* we will mean certain triplets of points, namely, the triplets of points in the columns of the following array.

$$\begin{array}{ccccccc} A & A & A & B & B & C & C \\ B & D & F & D & E & D & E \\ C & E & G & F & G & G & F. \end{array}$$

Thus the set $\{A, B, C\}$ is specified to be a line, likewise $\{C, D, G\}$, and there are five others. By *on* we will mean *is a member of*: thus C is a member of $\{C, E, F\}$, and we shall say that C is on the line $\{C, E, F\}$.

We now try to *prove* the statements 0 to 5 which we previously called (and in the future shall still call) axioms. In trying to build a model, we want the axioms to become true and proved theorems.

PROOF OF 0. A is a point, and $\{A, B, C\}$ is a line, so 0 is checked.

PROOF OF 1. We list all pairs of points. Before doing so, we ask how many pairs there are; in this way, we get a check on the list. We can pick a point in 7 ways, and a second point then in 6 ways, so altogether, there are $7 \cdot 6 = 42$ pairs if we insist on order, i.e., regard AB as different from BA. If we do not care about the order, and for present purposes that is the case, there are $7 \cdot 6/2 = 21$ pairs. Now list them. They are:

$$\begin{array}{cccccc} AB, & AC, & AD, & AE, & AF, & AG, \\ & BC, & BD, & BE, & BF, & BG, \\ & & CD, & CE, & CF, & CG, \\ & & & DE, & DF, & DG, \\ & & & & EF, & EG, \\ & & & & & FG. \end{array}$$

For each of these pairs we examine the preceding array to see whether some column contains the pair. Each of the 21 trials turns out to be successful. This proves 1.

PROOF OF 2. Here we could check that each of the above pairs occurs in just one column. We have to do this in one way or another and could do it by a direct enumeration of cases. But that would be inelegant. Here is a typical kind of argument which occurs in such problems. We have to see that if X, Y are two letters, then XY does not occur in more than one column; by 1, it occurs in at least one column. Suppose we were to write down all the pairs occurring in a given column. Since there are 3 letters in a column, we would be writing down 3 pairs $(3 = 3 \cdot 2/2)$. There are 7 columns, so we would be writing down $7 \cdot 3 = 21$ pairs. The question is: would we have written down any pair twice or more? By 1 we know that we would be writing down at least 21 *distinct* pairs, and since we would have written down only 21 pairs in any event, there could be no duplication.

PROOF OF 3. We have to check that every two columns contain a letter in common. There are 21 details to check.

PROOF OF 4. Check that no letter occurs twice in the same column. (This, incidentally, we know already from the argument given for 2.)

PROOF OF 5. There are only 3 points on a line, and there are 7 points, so they could not all be on one line.

The existence of the above model shows that our axiom system is consistent: because if we could deduce a contradiction from the axioms, then we could also deduce a contradiction from the existence of 7 distinct objects, namely, by forming them into a model as above and drawing the contradiction in the model. (If the existence of 7 objects involves a contradiction, then there do not exist 7 objects. We will not attempt to prove that there do exist 7 objects, and, to this extent, the consistency proof is incomplete.)

Exercise. Show that the diagonal points of any complete quadrangle in the 7-point plane are collinear.

6. Other models. How was the above model come upon? If Axiom 5 was to be satisfied, then at least three points would be on any line. Using the principle that we do not introduce into consideration more than we have to, we try to find a model with just three points on each line. By a theorem, there would have to be $3^2 - 3 + 1 = 7$ points altogether. So if a model with just 3 points on each line could be constructed at all, there would have to be 7 points. On any point, say A, by a theorem, there would have to be 3 lines, and on each of these, 2 points besides A. So 3 lines must be (except for notation)

$$\begin{matrix} A & A & A \\ B & D & F \\ C & E & G \end{matrix}$$

Now on B there must be 2 other lines: BDF is a possibility, as is BDG, but these two possibilities differ only in notation, nothing else. Take, say, BDF. This leaves BEG as the other possibility for a line on B. On C and D there must be a line, and one sees that the third point could only be G, i.e., CDG must be a line. Similarly CEF must be a line. By this time there are 7 lines. If a model (with 3 points on each line) could be built at all, there would be exactly 7 lines (by a theorem). So the lines must be, except for notation, ABC, ADE, etc. That is, nothing else would do, but it still remained to show that the above does yield a model.

Exercises

1. Try to find a model with just 4 points on each line. (The word "try" is used, because nothing guarantees beforehand that any such model exists. But it does.)

2. Using familiar facts from solid geometry, show that the following yields a model of our system: first fix a point O in space; by a "point" shall be meant a line (ordinary space line) through the point O; by a "line" shall be meant a plane through O; a "point" P shall be said to be on a "line" l if the line P is on the plane l.

3. Show that no 7 points of a 13-point plane (4 points per line) can form a 7-point plane. (Of course, three points A, B, C are to be called collinear in the proposed 7-point plane if and only if they are collinear in the 13-point plane.) Hence conclude that the diagonal points of any complete 4-point in the 13-point plane are not collinear.

Starting from the Euclidean plane, we can construct a model of the projective plane (i.e., of the above axiom system) by the adjunction of ideal points, as in the first part of the book. Although this really yields a model, this model is not as satisfactory for the purpose of showing consistency as the 7-point geometry, because in the present case the consistency would be made to depend upon that of Euclidean geometry, whereas above the consistency was made to depend only on the existence of 7 objects. We might, of course, lay down axioms for the affine plane and see that we would get a projective plane by the adjunction of ideal points. Let us do this.

AXIOMS FOR AN AFFINE PLANE.

A0. *There exist at least one point and at least one line.*

A1. *On any two points there is at least one line.*

A2. *On any two points there is at most one line.*

A3. *If l is a line and P a point not on l, then there is one and only one line l' on P that does not intersect l.* (*We may call this the* Parallel Postulate.)

A4. *There are at least two points on every line.*

A5. *Not all points are on the same line.*

As just remarked, any affine plane can be converted into a projective plane by the adjunction of ideal points. The consistency of our Axioms 0–5 would be established if we had the consistency of A0–A5. These are "known" theorems of ordinary Euclidean geometry, but it would be better, for a consistency proof, to build a model directly. Let then A, B, C, D be four objects, call them points, and let each pair be called a line (Fig. 2.5). One verifies Axioms A0–A5 easily. Calling two lines parallel if they do not meet, note that AD and BC are parallel (as are AB and CD, and AC and BD). How does the introduction of ideal points go in this case? For the pair of parallel lines AB and CD we have to introduce a new object E, an "ideal" point, that will be on the projective lines on AB and CD. For the pair AC and BD, introduce an object F as ideal point; and for AD and BC introduce G. Recall our

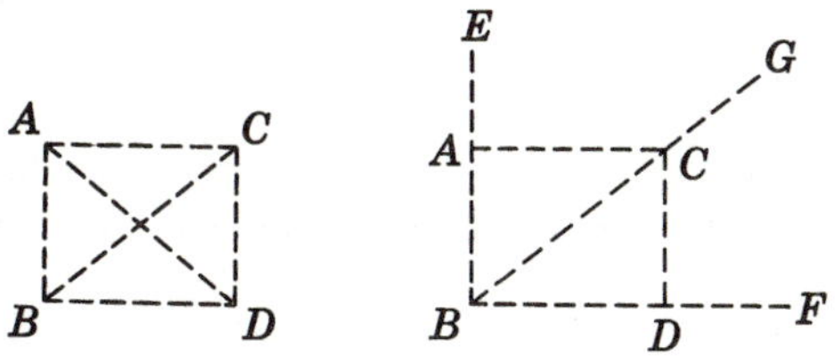

FIG. 2.5

previous considerations: E, F, G are collinear, in fact, form the ideal line. Thus we get a model of the projective plane with 7 points A, B, C, D, E, F, G, and with (A, B, E), (C, D, E), (B, D, F), (A, C, F), (B, C, G), (A, D, G), (E, F, G) as lines. This, except for notation, is our previous 7-point geometry.

Exercise. Construct an affine plane with 3 points on a line, and introduce ideal points to get a projective plane with 4 points on a line.

7. Independence of the axioms. As we have already said, there is no absolute requirement on an axiom system. Consistency is somewhat urgent, because the system would collapse upon being proved inconsistent. There is another property of the axioms, which is not very urgent, but is desirable: this is the property of being *independent*, and this means that none of the axioms should be deducible from the rest. As far as consistency is concerned, Euclid, for example, might well have included the Theorem of Pythagoras among his axioms. But one of the most fascinating results of geometry would thereby have been lost. For a similar reason, we do not want to include in our axioms anything that can be proved. Sometimes, if the point is trifling, but might otherwise add to clarity, we would not insist on independence.

How can one prove independence? How can one prove, for example, that Axiom 5 does not follow from the others? One way would be to build a model satisfying Axioms 0–4 and also satisfying the *negation* of Axiom 5. More generally, to prove the independence of a statement A from statements $B, C, D, \cdots$, it would be sufficient to construct a model in which $B, C, D, \cdots$ obtain but A does not obtain. Let us prove our Axioms 0–5 independent, going through them one at a time.

Independence of Axiom 5. Take three objects A, B, C. Call them points. Call the set of them a line (Fig. 2.6). Let *on* mean *is a member of*. One verifies at once that Axioms 0–4 obtain, but 5 fails. Note that Axiom 3 is true because its hypothesis cannot be fulfilled: we say that Axiom 3 is *vacuously* true.

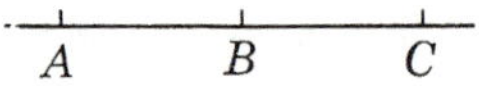

FIG. 2.6

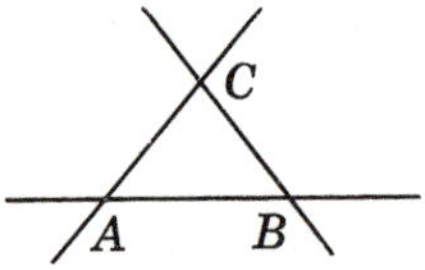

FIG. 2.7

Independence of Axiom 4. Take three objects A, B, C. Call them points, and pairs of them lines (Fig. 2.7). Let *on* mean *is a member of*. Axiom 4 fails, but the other axioms are immediately verified.

Independence of Axiom 3. The ordinary Euclidean plane satisfies all the axioms but Axiom 3. Instead of the Euclidean plane, the affine plane of 9 points constructed above (in an exercise) could be used here.

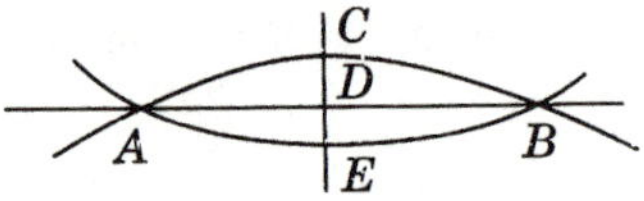

FIG. 2.8

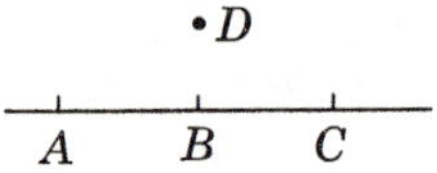

FIG. 2.9

Independence of Axiom 2. Take five objects A, B, C, D, E. Call them points, and call the following sets of lines: $\{A, C, B\}$, $\{A, D, B\}$, $\{A, E, B\}$, $\{C, D, E\}$ (Fig. 2.8). Let *on* mean *is a member of*. All the axioms but 2 check: 2 does not.

Independence of Axiom 1. Take four objects A, B, C, D. Call them points. Let the set $\{A, B, C\}$ be called a line (Fig. 2.9). Let *on* mean *is a member of*. All the axioms except 1 check: 1 does not. Note that Axiom 3 is vacuously true again.

Independence of Axiom 0. Take the empty set, call its members points; call them lines also. Let *on* mean *is a member of*. All the axioms but 0 are vacuously true. Axiom 0 fails.

If the axiom system is sufficiently simple, checking the independence of each axiom from the others is desirable. In more complicated situations one will merely guard against adding as an axiom something that can be proved on the basis of the axioms already laid down.

Exercises

1. Split Axiom 0 into two axioms:

AXIOM 0′. *There is at least one point.*

AXIOM 0″. *There is at least one line.*

Prove that Axioms 0′, 0″, 1, 2, 3, 4, 5 are *not* independent.

2. Find *all* the models satisfying Axioms 0, 1, 2, 3, 5, but for which Axiom 4 fails.

3. Show that the following Axiom 4* can replace Axioms 0, 4, and 5.

AXIOM 4*: *There exist four points, no three of which are collinear.*

8. Isomorphism. An axiom system consists of several sets $S_1, S_2, \cdots$ of undefined objects together with undefined relations $R_1, R_2, \cdots$ between them (plus axioms). For example, the axiom system we have been studying consists of two sets of objects S_1, S_2, one set called *points*, the other *lines*, together with the relation *incidence*—for later purposes we refer to such a system as a *system of type* Σ. Let $\mathscr{S}$ be a system made up from sets $S_1, S_2, \cdots$ through relations $R_1, R_2, \cdots$; let $\mathscr{S}'$ be another system made up from sets $S_1', S_2', \cdots$ through relations $R_1', R_2', \cdots$: we say $\mathscr{S}$ and $\mathscr{S}'$ are isomorphic if there is a one-to-one correspondence of S_1 with S_1', of S_2 with S_2', etc., in such a way that a relation R_i holds between certain elements of $S_1, S_2, \cdots$ if and only if R_i' holds between the corresponding elements of $S_1', S_2', \cdots$. For example, two systems of type Σ are isomorphic if the points and lines of one system are put in one-to-one correspondence, respectively, with the points and lines of the other in such way that a point and line of the first system are incident if and only if the corresponding point and line of the second system are incident.

Roughly speaking, two systems are isomorphic if they look alike, although made up of different materials.

THEOREM. *Any two 7-point geometries are isomorphic.*

PROOF. This follows from our previous considerations as a result of the fact that there is at any stage essentially only one way to proceed in converting seven objects into a geometry. In writing down a detailed proof, though, we will follow a different idea. Let then A, B, C be any three noncollinear points in the first plane; let A', B', C' be any three noncollinear points in the second plane. On BC there is a third point—call it D; on $B'C'$ there is a third point—call it D'. Similarly, let E be on CA; E' on $C'A'$; F on AB; F' on $A'B'$. Finally there is a seventh point G in the first plane and a seventh point G' in the second. The lines of the first plane are $\{A, B, F\}$, $\{A, G, D\}$, etc.; the lines of the second are $\{A', B', F'\}$, $\{A', G', D'\}$, etc. Then associating A with A', B with B', etc; and $\{A, B, F\}$ with $\{A', B', F'\}$, $\{A, G, D\}$ with $\{A', G', D'\}$, etc. we have an isomorphism.

REMARK. The above theorem is frequently expressed by saying that there is only one 7-point geometry, but what is meant is that there is only one up to form, and this last point is tacit.

Exercises

1. If π_1 and π_2 are two isomorphic systems of type Σ and one is a plane (i.e., satisfies our axioms), then so is the other.

2. Let $\mathscr{S}$ be a system of type Σ. Define another system $\mathscr{S}'$ of type Σ as follows: the points of $\mathscr{S}'$ are just the points of $\mathscr{S}$; a line l' of $\mathscr{S}'$ consists of the set of points on a given line l of $\mathscr{S}$; incidence in $\mathscr{S}'$ is defined through membership. Show that $\mathscr{S}$ and $\mathscr{S}'$ are isomorphic.

9. Further axioms. The above axioms (Axioms 0–5, p. 43) are called the axioms of alignment. For sake of explicitness it may be well to define a *projective plane.*

DEFINITION. A *projective plane* is the composite notion of two non-empty sets, the members of one of which are called *points*, those of the other *lines*, and a relation *on* between points and lines for which the axioms of alignment are verified.

But we are going to add further axioms to our system. If the theorem of Desargues holds in a projective plane, whether deduced or by assumption, we will call the plane *Desarguesian.* Similarly if Pappus' Theorem holds, we will call the plane *Pappian.* If the Fundamental Theorem holds, we will speak of the *classical* projective plane. Note that all three of these theorems can be *stated* in any projective plane: whether they are *true* is another matter.

It is difficult to do much on the basis of the axioms of alignment alone. There are some combinatorial theorems of the type already mentioned, but there is relatively little scope for further geometrical development. Here is a theorem that follows from the axioms of alignment alone.

THEOREM. *If A, B, C, D are four collinear points, then there exists a projectivity sending A into B, B into A, C into D, and D into C.*

PROOF. Let l be a line on A, $l \neq AB$ (Fig. 2.10). Let D'' be not on l and not on AB. Let $ABCD \overset{D''}{\underset{\wedge}{-}} AB'C'D'$, where B', C', D' are on l.

Let $C'' = DC' \cdot D''B$. Then $ABCD \overset{D''}{\underset{\wedge}{-}} AB'C'D' \overset{D}{\underset{\wedge}{-}} BB'C''D'' \overset{C'}{\underset{\wedge}{-}} BADC$ whence $ABCD \overline{\overline{\wedge}} BADC$. Q.E.D.

Exercise. Prove this theorem using the concepts of cross-ratio, etc., of the first part of the book.

It is not hard to see why one cannot go far on the basis of Axioms 0–5 alone. Almost the simplest type of geometric theorem one can think of concludes that under certain circumstances three lines are concurrent or three points are collinear. The proof of such a theorem presumably

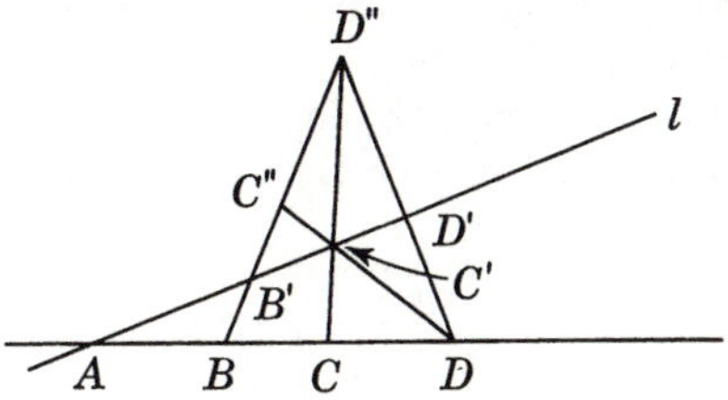

FIG. 2.10

would rest on another theorem of a similar type, but simpler. Now we do have axioms that enable us to conclude that sometimes three points are collinear. For example, if A, B, C are distinct and collinear and B, C, D are collinear, then A, C, D are collinear. All the same, our axioms *seem* too weak to prove something like Desargues' Theorem. The above is intended to make this conclusion plausible, but one can prove it. We build a model satisfying axioms of alignment but in which Desargues' Theorem fails.

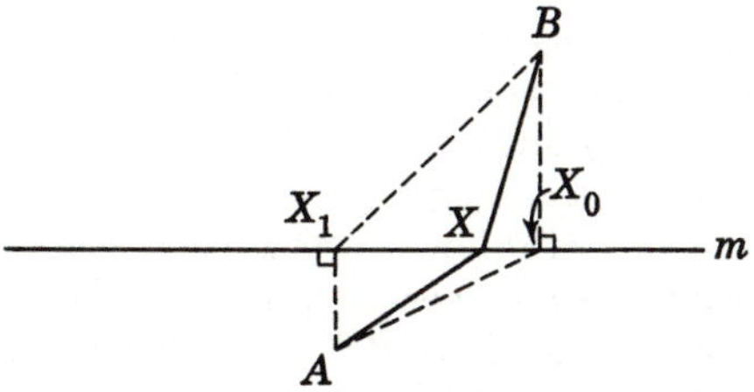

FIG. 2.11

Let m be a line in the ordinary, Euclidean plane: let m be the x-axis in a cartesian coordinate system, say (Fig. 2.11). We are going to define a certain *affine* plane in terms of this Euclidean plane. The *points* of this plane will be the same as the points of the Euclidean plane, but some *lines* will consist of two half-lines meeting on m: of the two half-lines, one lies in the *upper* part, the other in the *lower* part of the plane, and we require that the slope of the upper half be *positive* and twice the slope of the lower half. In addition, vertical and horizontal lines of the Euclidean plane and lines of negative slope shall be *lines* in the affine plane being constructed. With this definition of *line* one proves without much difficulty that Axioms A0–A5 are verified. Axiom A1 may call for some comment: suppose A, B are on opposite sides of m, as in Fig. 2.11. The question is: Is there a point X on m such that slope $BX = 2 \cdot$ slope AX? Let X vary on m: as X approaches X_0, slope BX/slope AX grows continually larger, getting bigger than 3, say. As X approaches X_1, slope BX/slope AX approaches zero. From the continuity of slope BX/slope AX one then concludes that there must be a point X on m such that slope BX/slope $AX = 2$. (We

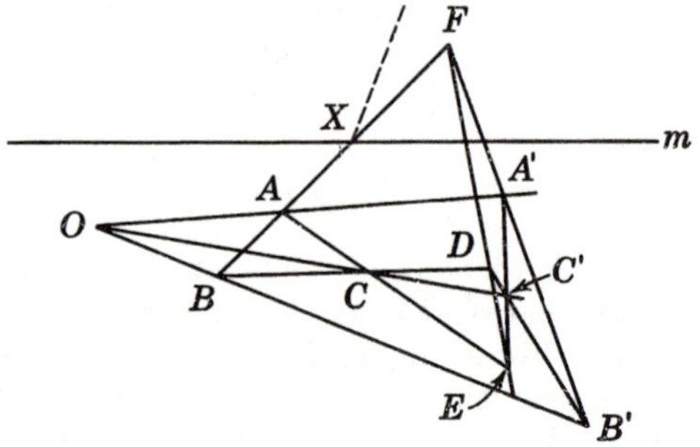

FIG. 2.12

omit further discussion of the axioms in order to save space.) Now convert this affine plane into a projective plane by adding ideal points. In this projective plane Desargues' Theorem fails. One sees this easily by arranging the Desargues' configuration as in Fig. 2.12, with F lying above m, the rest of the configuration below m, and moreover in such way that DE and $A'B'$ have negative slope, and AB positive slope. The figure described is drawn in the ordinary, Euclidean plane, so D, E, F are collinear. Now consider this configuration in the constructed plane. Triangles ABC, $A'B'C'$ are still centrally perspective. The lines $A'B'$ and DE in the constructed plane are the same as in the original plane (because of the negative slope). But BA, in the constructed plane, will not pass through $DE \cdot A'B' = F$ because of the break at $X = m \cdot AB$.

Since Desargues' Theorem cannot be proved from the axioms of alignment alone, it is entirely in order to add it on as an axiom. A little later we shall also add Pappus' Theorem as an axiom; meanwhile, let us just add Desargues'. Now whenever we add an axiom, all the questions previously considered come up again—for example, *consistency*. Are Axioms 0–5 and Desargues' Theorem consistent?

A nontrivial Desargues' configuration involves at least 10 distinct points. Hence Desargues' Theorem holds trivially in the 7-point geometry. This proves consistency; all the same one is not very content with the proof, because one feels that the Desargues' Theorem has not been given a chance to manifest itself. We can formulate our discontent in a strictly mathematical way: Is Desargues' Theorem consistent with Axioms 0, 1, 2, 3, 5 and the requirement that there be at least 4 points on a line? In the 13-point geometry it is easy to find nontrivially located centrally perspective triangles. Are such necessarily axially perspective?

THEOREM. *If there are exactly four points on each line, then Desargues' Theorem holds.*

PROOF. Recall that in the first chapter we proved that the Fundamental Theorem implies Desargues' Theorem. Rigor was not demanded at the time, but one sees now that the argument could be made entirely rigorous. Now in the 13-point geometry, the Fundamental Theorem does hold. For let l, l' be two lines; A, B, C, D the points on l; A', B', C', D' the points on l'. The question is: How many projectivities are there sending A, B, C, respectively, into A', B', C'? Can there be two? Let T, U be projectivities sending A, B, C, respectively, into A', B', C'. What is $(D)T$? It can only be D'. What is $(D)U$? It can only be D'. So $(D)T=(D)U$. Hence $T=U$ and the Fundamental Theorem is verified for the 13-point geometry. Desargues' Theorem for the 13-point geometry now follows.

Consistency is proved. How about *duality*? Remember, the principle of duality depends on our having as theorems the dual of each axiom. The question then becomes: Can one prove the dual of Desargues' Theorem on the basis of Desargues' Theorem and Axioms 0–5?

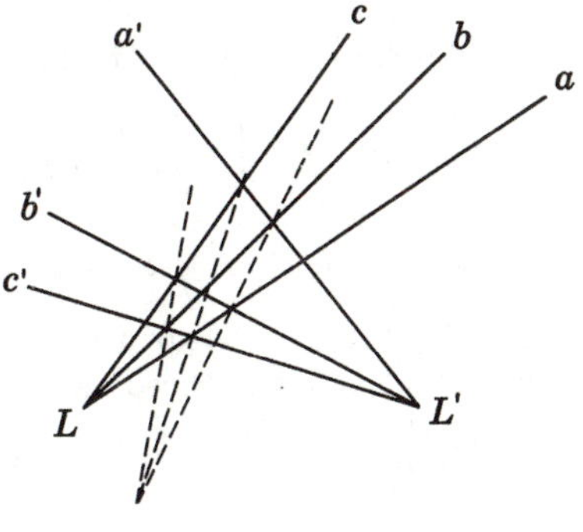

FIG. 2.13

Exercises

1. Prove that the answer to the last question is *yes*.

2. Using Pappus' Theorem, prove the dual of Pappus' Theorem (Fig. 2.13).

10. Consequences of Desargues' Theorem. We are now proceeding on the basis of Axioms 0–5 and Desargues' Theorem. On this meager basis it is possible to go quite far; for example, we can introduce coordinates into the plane. What that means requires some explanation; meanwhile, let us develop our geometry somewhat.

Quadrangular sets of points and of lines. Let P, Q, R, S be a complete quadrangle and let l be a line through none of the vertices. Then l cuts the sides in 6 points, which, however, need not be distinct. Let us select one of the vertices, say P, and divide the six sides into two sets: the set PQ, PR, PS and the set RS, SQ, QR (Fig. 2.14). Let these cut l in A, B, C and in A', B', C', respectively. (The notation is such that

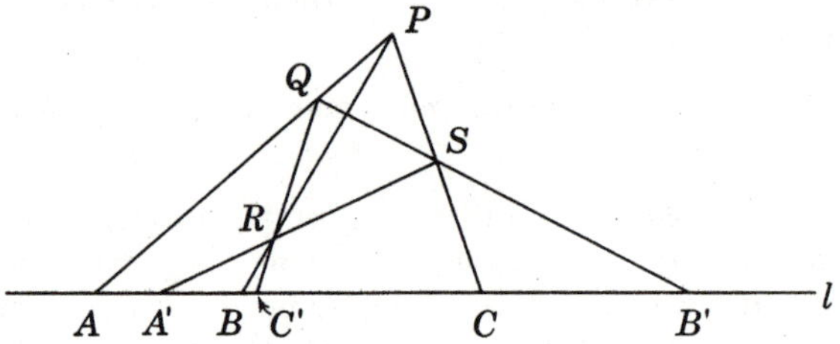

FIG. 2.14

A, A' lie on opposite sides of the complete quadrangle $PQRS$; similarly for B, B' and C, C'.) Under these circumstances we call ABC; $A'B'C'$ a *quadrangular hexad* and write $Q\begin{pmatrix} A & B & C \\ A' & B' & C' \end{pmatrix}$. Thus a quadrangular hexad is *by definition* a set of collinear points ABC; $A'B'C'$ having the abovementioned relation to some complete 4-point. The set A, B, C will be called a *point triple*; A', B', C' a *triangle triple*.

The order of the letters in the symbol $Q\begin{pmatrix} A & B & C \\ A' & B' & C' \end{pmatrix}$ is, of course, important, but we have the following theorem.

THEOREM. $Q\begin{pmatrix} A & B & C \\ A' & B' & C' \end{pmatrix}$ *implies* $Q\begin{pmatrix} A & B' & C' \\ A' & B & C \end{pmatrix}$.

PROOF. Let P, Q, R, S be a complete 4-point showing $Q\begin{pmatrix} A & B & C \\ A' & B' & C' \end{pmatrix}$ (see Fig. 2.14). Divide the six sides into the triple of lines passing through Q and the three sides of triangle PSR. Then one also sees that $Q\begin{pmatrix} A & B' & C' \\ A' & B & C \end{pmatrix}$.

Given $Q\begin{pmatrix} A & B & C \\ A' & B' & C' \end{pmatrix}$, we say A, A' correspond; also B, B' and C, C'.

THEOREM. *Given* $Q\begin{pmatrix} A & B & C \\ A' & B' & C' \end{pmatrix}$. *Then* A, B, C *are distinct*; A', B', C' *are distinct*; *if a point of the point triple is the same as a point of the triangle triple, then they correspond* (*i.e., for example,* $A = B'$ *is impossible, as is* $A = C'$).

PROOF. From the definition of a complete 4-point, no 3 vertices are collinear. So $PQ \neq PR$; for the same reason PQ, PR, PS are distinct. Since l does not pass through P, the points A, B, C are distinct. Similarly, A', B', C' are distinct. Since $Q\begin{pmatrix} A & B' & C' \\ A' & B & C \end{pmatrix}$, we see that $A = B'$ is impossible, as is $A = C'$.

This theorem tells us that in a quadrangular set of points certain of them are unequal. We will refer to the inequalities given by the last theorem as the *necessary inequalities on quadrangular hexads*.

THEOREM. *Given five collinear points* A, B, C, A', B'. *Then there exists a point* C' *such that* $Q\begin{pmatrix} A & B & C \\ A' & B' & C' \end{pmatrix}$, *provided* A, B, C, A', B' *satisfy the necessary inequalities* (*namely,* $A \neq B$, $A \neq C$, $B \neq C$, $A' \neq B$, $A' \neq C$, $B' \neq A$, $B' \neq C$).

PROOF. Take any point P not on l. Through A' take a line ($\neq A'P$, $\neq l$) cutting PB, PC in R, S, respectively. Let $Q = PA \cdot B'S$. Then $C' = QR \cdot l$ is a desired point C'.

REMARK. If $Q\begin{pmatrix} A & B & C \\ A' & B' & C' \end{pmatrix}$, then $BB'CA' \,\overline{\overline{\wedge}}\, BB'AC'$.

PROOF. Consider $l \overset{S}{\underset{\wedge}{-}} m \overset{Q}{\underset{\wedge}{-}} l$, where $l = AB$, $m = PB$ (see Fig. 2.14).

THEOREM. *With hypotheses of last theorem, there is only one point* C' *such that* $Q\begin{pmatrix} A & B & C \\ A' & B' & C' \end{pmatrix}$.

PROOF. One has to go through the construction given in the proof of the last theorem twice and see that one comes both times to the same point C'. Take any point $\bar{P}$ not on l. Through A' take a line ($\neq A'\bar{P}$, $\neq l$) cutting $\bar{P}B$, $\bar{P}C$ in $\bar{R}$, $\bar{S}$, respectively. Let $\bar{Q} = \bar{P}A \cdot B'\bar{S}$, and $\bar{C}' = \bar{Q}\bar{R} \cdot l$ (Fig. 2.15). We have to see that $C' = \bar{C}'$.

Suppose, now, in order to see the main point of the proof, that there is nothing special about the configuration in question, for example, we do not have $R = \bar{R}$. Then we have to show that QRS and $\bar{Q}\bar{R}\bar{S}$ are axially perspective. To do this, it would be sufficient to show that they are centrally perspective, i.e., that $Q\bar{Q}$, $R\bar{R}$, $S\bar{S}$ are concurrent. Since triangles PRS and $\bar{P}\bar{R}\bar{S}$ are axially perspective, $P\bar{P}$, $R\bar{R}$, $S\bar{S}$ are concurrent. Similarly, from triangles PQS, $\bar{P}\bar{Q}\bar{S}$ we see that $P\bar{P}$, $Q\bar{Q}$, $S\bar{S}$ are concurrent. So $R\bar{R}$ and $Q\bar{Q}$ pass through $P\bar{P}\cdot S\bar{S}$, whence $Q\bar{Q}$, $R\bar{R}$, $S\bar{S}$ are concurrent.

Now for the strict proof. First, if there are just 3 points on a line and $Q\begin{pmatrix} A & B & C \\ A' & B' & X \end{pmatrix}$, then A, B, C are the 3 points on a line; X is on that line, and so X is one of A, B, or C. It is not A or B; therefore $X = C$. If there are just 4 points on a line, then the Fundamental Theorem holds:

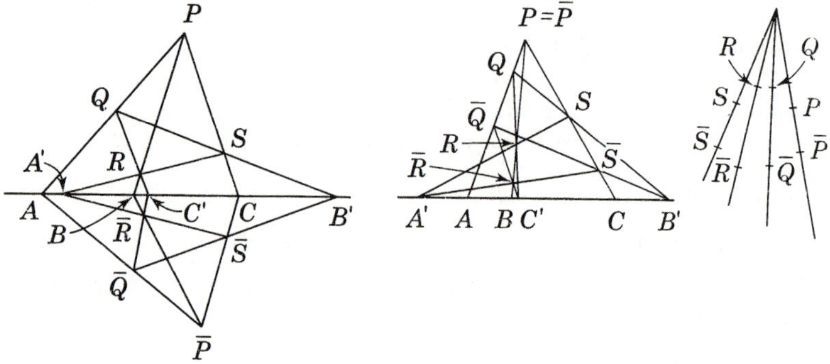

FIG. 2.15

there is just one projectivity sending $B \to B$, $B' \to B'$, $C \to A$, whence C' is uniquely determined as the image of A' in this projectivity (by the remark after the last theorem).

We may proceed, then, on the assumption that there are at least 5 points on a line. Consider the cases (a) $P = \bar{P}$, and (b) $P \neq \bar{P}$. If $P = \bar{P}$, then one sees that QRS and $\bar{Q}\bar{R}\bar{S}$ are perspective from P and draws the desired conclusion.

The case $P \neq \bar{P}$ will itself be divided into two cases: (I) $PA \neq \bar{P}A$ and $PC \neq \bar{P}C$ and $PB \neq \bar{P}B$; (II) either $PA = \bar{P}A$ or $PC = \bar{P}C$ or $PB = \bar{P}B$. Consider first case I. We want to argue as above that PRS and $\bar{P}\bar{R}\bar{S}$ are axially perspective: this can be done if and only if $PS \neq \bar{P}\bar{S}$, $PR \neq \bar{P}\bar{R}$, and $RS \neq \bar{R}\bar{S}$. Since we are in case I, we have the first two inequalities. Now $RS = \bar{R}\bar{S}$ is possible, but by case (a), the line $A'S$ may be varied: if $RS = \bar{R}\bar{S}$ we replace S by a different point on PC—we have to avoid P, C, and S, and so need at least 4 points on PC. Replacing S we would have $RS \neq \bar{R}\bar{S}$, and hence could conclude that $P\bar{P}$, $R\bar{R}$, $S\bar{S}$ are distinct and concurrent. Now we examine triangles PQS and $\bar{P}\bar{Q}\bar{S}$, and want to prove that $Q\bar{Q}$ passes through $P\bar{P}\cdot R\bar{R}$. We still have to contend with the possibility $QS = \bar{Q}\bar{S}$. Here again we vary S, but in order not to lose our previous arrangement, we need 5 points on PC. So we may assume $QS \neq \bar{Q}\bar{S}$ and complete the proof as previously (in case I).

For case II, take a line through P different from PB, PC, and PA. Let P^* be a point on this line that is not on l, $\bar{P}C$, $\bar{P}B$, or $\bar{P}A$ (since $\bar{P}B$, $\bar{P}C$, or $\bar{P}A$ is on P, in case II, we shall automatically have $P \neq P^*$). In choosing P^*, 4 points have to be avoided. Now construct a third complete 4-point $P^*Q^*R^*S^*$. We now have $P^*A \neq PA$, $P^*B \neq PB$, $P^*C \neq PC$, and so by case I conclude that $Q^*R^* \cdot l = QR \cdot l$. We also have $P^*B \neq \bar{P}B$, $P^*C \neq \bar{P}C$, $P^*A \neq \bar{P}A$, whence $Q^*R^* \cdot l = \bar{Q}\bar{R} \cdot l$, from which $QR \cdot l = \bar{Q}\bar{R} \cdot l$ follows. Q.E.D.

To prove that a quadrangular hexad remains quadrangular under a projectivity, we introduce the notion dual to a quadrangular hexad of

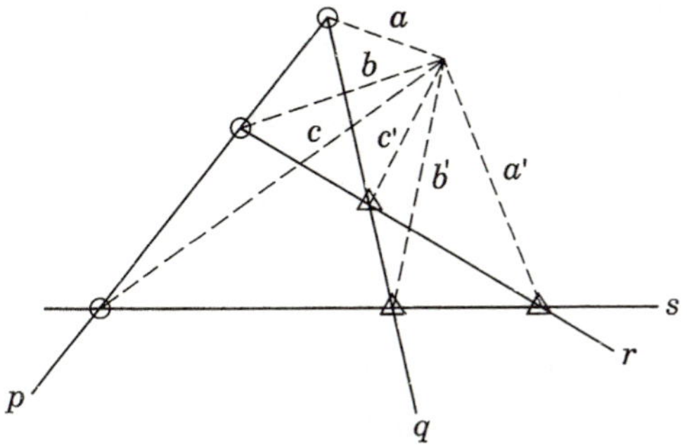

FIG. 2.16

points. Consider a complete 4-line. It has 6 vertices, which in several ways can be separated into a collinear triple and a triangle triple. Fig. 2.16 is dual to Fig. 2.14 and shows $Q\begin{pmatrix} a & b & c \\ a' & b' & c' \end{pmatrix}$. Examining Fig. 2.14, one sees the following immediately.

THEOREM. *If* $Q\begin{pmatrix} A & B & C \\ A' & B' & C' \end{pmatrix}$, *then* $Q\begin{pmatrix} PA' & PB' & PC' \\ PA & PB & PC \end{pmatrix}$.

Since P could be taken arbitrarily in exhibiting $Q\begin{pmatrix} A & B & C \\ A' & B' & C' \end{pmatrix}$, we see the following.

THEOREM. *If 6 concurrent lines* a, b, c, a', b', c' *lie, respectively, on* A, B, C, A', B', C', *and if*

$$Q\begin{pmatrix} A & B & C \\ A' & B' & C' \end{pmatrix}, \quad \textit{then} \quad Q\begin{pmatrix} a' & b' & c' \\ a & b & c \end{pmatrix}.$$

The dual of this, the following theorem, is also true.

THEOREM. *If 6 concurrent lines a, b, c, a', b', c' are cut by a secant in A, B, C, A', B', C', respectively, and if*

$$Q\begin{pmatrix} a & b & c \\ a' & b' & c' \end{pmatrix}, \quad \textit{then} \quad Q\begin{pmatrix} A' & B' & C' \\ A & B & C \end{pmatrix}.$$

Combining these two theorems we can prove the following.

THEOREM. *If $ABCA'B'C' \barwedge \bar{A}\bar{B}\bar{C}\bar{A}'\bar{B}'\bar{C}'$ and if*

$$Q\begin{pmatrix} A & B & C \\ A' & B' & C' \end{pmatrix}, \quad \textit{then} \quad Q\begin{pmatrix} \bar{A} & \bar{B} & \bar{C} \\ \bar{A}' & \bar{B}' & \bar{C}' \end{pmatrix}.$$

PROOF. We need only consider a perspectivity. So suppose $ABCA'B'C' \overset{E}{\bar{\wedge}} \bar{A}\bar{B}\bar{C}\bar{A}'\bar{B}'\bar{C}'$ and let $a = EA$, $b = EB$, etc. Then

$$Q\begin{pmatrix} A & B & C \\ A' & B' & C' \end{pmatrix} \quad \text{implies} \quad Q\begin{pmatrix} a' & b' & c' \\ a & b & c \end{pmatrix},$$

by the first of the last two theorems; and the second now implies

$$Q\begin{pmatrix} \bar{A} & \bar{B} & \bar{C} \\ \bar{A}' & \bar{B}' & \bar{C}' \end{pmatrix}. \quad \text{Q.E.D.}$$

Recall how, in an earlier part of the book, we proved that $ABCD \barwedge A'B'C'D'$ and $H(A, B; C, D)$ imply $H(A', B'; C', D')$. We gave two proofs. One used space and regarded the given perspectivity as a section of a projection. The configuration showing $H(A, B; C, D)$ then projected into a figure showing $H(A', B'; C', D')$. If we had space available, we could give the same proof for quadrangular hexads: a quadrangle showing

$$Q\begin{pmatrix} A & B & C \\ A' & B' & C' \end{pmatrix} \quad \text{would go into one showing} \quad Q\begin{pmatrix} \bar{A} & \bar{B} & \bar{C} \\ \bar{A}' & \bar{B}' & \bar{C}' \end{pmatrix}.$$

We do not have space, but the idea of that proof can be used. We consider transformations of the plane into itself that send collinear points into collinear points. Such transformations are called *collineations*. If distinct points go into distinct points, then the collineation is called *nonsingular*. Suppose, now, it were possible to regard a given perspectivity as part of some nonsingular collineation; i.e., given a perspectivity $T: l \barwedge l'$, suppose there were a collineation U such that $(X)T = (X)U$ for the points X on l. Then U would transform a 4-point showing $Q\begin{pmatrix} A & B & C \\ A' & B' & C' \end{pmatrix}$ (where $A, B, \cdots$ are on l) into a 4-point showing

$Q\begin{pmatrix}\bar{A} & \bar{B} & \bar{C} \\ \bar{A}', & \bar{B}', & \bar{C}'\end{pmatrix}$ (where $\bar{A}=(A)T$, $\bar{B}=(B)T,\cdots$). We leave the proof that U does exist as an exercise.

Exercise. Let E be a point and q a line not on E. Define a transformation of the (points and lines of the) plane as follows: E goes into E, every point of q goes into itself, every other point A goes into A' with $H(A, A'; E, EA\cdot q)$; q goes into q, every line l on E goes into itself, every other line l goes into l' with $H(l, l'; q, (q\cdot l)\cdot E)$. Show that this transformation is an isomorphism of the plane with itself.

11. Free planes. Previously we remarked that the axioms of alignment seem too weak to prove something like Desargues' Theorem. In what follows we bring this feeling to rigorous expression and show that it is correct.

By a *partial plane* one means the composite notion of a set of objects called points, a set of objects called lines, and a relation *on* between points and lines satisfying the following axiom:

AXIOM. *On any two (distinct) points there is at most one line.*

For partial planes the theorem holds that on any two lines there is at most one (common) point.

Any set of points together with any set of lines of a given projective plane, along with the notion *on* of the projective plane, furnishes an example of a partial plane.

By a *finite partial plane* one means a partial plane involving only a finite set of points and lines. A finite partial plane is also called a *configuration*.

A configuration is said to be *confined* if, in the configuration, there are at least three points on every line and at least three lines on every point. Roughly speaking, every theorem, or at any rate every theorem which concludes that some three points are collinear or some three lines are concurrent, must involve a confined configuration; for if in the configuration, there is a line on which there are only two points, the fact that this line is in the configuration does not tell us much, and so might be deleted from the configuration; similarly with any point on which there are at most two lines. The last two sentences need not be scrutinized for rigor, but one should check that the various theorems we have had, such as Desargues' Theorem, or Pappus' Theorem, or the theorem on the uniqueness of the fourth harmonic, do involve confined configurations.

Starting from a given partial plane π_0, we construct a partial plane π_1 as follows: for each pair of points in π_0 on which there is no line we introduce a new object, which we will call a line, and we will say that this line is on the two points; i.e., we define a new partial plane by keeping the points and lines of π_0 together with the incidence relations of π_0, but we enrich π_0 with new lines and new incidence relations. Of π_1 we can say: (a) on any two points there is at least one line. Now starting from π_1, we construct a partial plane π_2 as follows: for each pair of lines in π_1 on which there is no (common) point we introduce a new object, which we shall call a point, and we will say that this point is on the two lines. Of π_2 we can say: (b) on any two lines there is at least one point. The partial plane π_2, however, may not satisfy (a): so we proceed to patch it up by constructing π_3 from π_2 just as we constructed π_1 from π_0. Then π_3 satisfies (a); but (b) may be destroyed. So we patch π_3 up, by constructing π_4 from π_3 just as we did π_2 from π_1. And so we go, successively constructing $\pi_1, \pi_2, \pi_3, \pi_4, \cdots$. Each π_i with an odd subscript satisfies (a); each π_i with an even subscript satisfies (b). Let, now, π be the partial plane defined as follows: the points and lines of π shall be the points and lines of the π_i, for all i; as for incidence relations, letting P, l be a point and a line and letting π_j be the first of the partial planes containing P and l, the incidence relation between P and l in π_j is the same as the incidence relation of P and l in any succeeding π_i, and we define this to be the incidence relation between P and l in π. One checks now that π satisfies (a) and (b); for if P, Q are two points of π, then they occur in some π_i, i, even, and if in π_i there is no line on P and Q, then in π_{i+1}, and hence in π, there is such a line. Thus (a) is verified for π; and similarly (b) is verified.

If π_0 contains four points, no three of which are collinear, then π is, as one easily checks, a projective plane, called the *free extension* of π_0.

THEOREM. *π contains no confined configurations except those already contained in π_0.*

PROOF. To each point P and each line l of π assign a rank, namely, the least i such that π_i contains the point or line. Let C be a confined configuration in π: to see that all the lines and points of C belong to π_0. Since C contains only a finite number of points and lines, there is some element of C, either a point or a line, that has highest rank: say this is a line l, of rank i—a similar argument will hold if the element of highest rank is a point. We have to see that $i=0$. Suppose $i>0$. Since l is not in π_{i-1}, but is in π_i, it must have arisen in the above construction of π_i from π_{i-1}. Hence $i-1$ is even and l is on exactly two points of π_{i-1}. In passing from π_{i-1} to π_i we adjoined no points, so all points of π_i are in π_{i-1}; and since all points of C are in π_i, all points of C are in π_{i-1}. Hence on the line l of C there are only two points of C. This is a contradiction, since C is confined. Thus $i=0$; and all elements of C belong to π_0. Q.E.D.

Let now π_0 be the partial plane consisting of four points A, B, C, D and no lines (in particular, therefore, no three of A, B, C, D are collinear). The free extension π could not contain a nontrivial configuration of Desargues' Theorem, since such a configuration is confined, hence would be in π_0, whereas π_0 contains no confined configurations. Thus Desargues' Theorem fails in π unless perhaps it holds trivially. The remainder of the proof consists in showing that there are enough points and lines in π so that Desargues' Theorem could not hold trivially.

Proceeding to construct π as above, we introduce in π_1 the six lines AB, AC, etc.; then in π_2 the intersections of these lines—this introduces three new points F, G, H, the diagonal points of the complete 4-point $ABCD$. The points F, G, H are not collinear. Hence in π there must be at least four points on each line. Let O be a point of π. Through O take, in π, three lines l, m, n; on l, points A, A'; on m, points B, B'; and make this choice so that O, A, A', B, B' are distinct. Take C on n so that ABC is a triangle, $C \neq O$; take C' on n so that $A'B'C'$ is a triangle, $C' \neq O$, $C' \neq C$. Point C' can be chosen this way since on any line there are at least four points. Let $D = BC \cdot B'C'$, $E = CA \cdot C'A'$, $F = AB \cdot A'B'$. One verifies that O, A, B, C, A', B', C', D, E, F are ten distinct points. Were D, E, F collinear, then this figure for Desargues' Theorem would be a confined configuration (this involves checking certain lines to be distinct). By the above theorem, then, the figure would be in π_0, which is impossible, since π_0 contains no confined configurations.

CHAPTER III

ESTABLISHING COORDINATES IN A PLANE

1. Definition of a field. We are now almost in position to show how coordinates may be set up in the plane; but to understand just what this means, we have first to look at the finished product, i.e., what we want to attain. We have done this more or less in the first part of the book, but now we want to be more exact. Before, we constructed a projective plane analytically, i.e., using numbers. The numbers we had in mind were the so-called *real* numbers, but looking over what we have done, we see that, for example, all the considerations could equally have been made for *complex* numbers. More generally, let us treat numbers axiomatically, just as we have treated points and lines axiomatically. Here "numbers" will be the undefined terms: and there will be certain undefined operations, which we may call *plus* and *times*. Our axioms will be certain theorems, or rather, *like* certain theorems on real numbers with which we are familiar and which have proved fruitful. For example, consider the rule $a\cdot(b+c)=(a\cdot b)+(a\cdot c)$. This is a true theorem if a, b, c are real numbers and $+$ and $\cdot$ are the ordinary operations—this theorem can be proved, and its proof is not simple: we have just been taking it for granted. Nor are we going to prove it now. What we are going to do is to *take it as an axiom*, but now a, b, c are not understood to be real numbers, but merely objects having the stated property. Let us write down some of the familiar rules.

I. Rules for $+$:
- (a) $a+b=b+a$. (So-called *commutative* rule for *plus.*)
- (b) $(a+b)+c=a+(b+c)$. (So-called *associative* rule for *plus.*)
- (c) For every a and b, there is an x such that $a+x=b$. (So-called *solvability* rule for *plus.*)
- (d) There exists one and only one z such that $a+z=a$ for some a. (We usually write 0 for z, and call it *zero.*)

II. Rules for $\cdot$:
- (a) $a\cdot b=b\cdot a$. (*Commutative* rule for *times.*)

(b) $(a \cdot b) \cdot c = a \cdot (b \cdot c)$. (*Associative* rule for *times*.)
(c) For every a and b, with $a \neq 0$, there is an x such that $a \cdot x = b$. (*Right solvability* rule for *times*.)
(c′) For every a and b, with $a \neq 0$, there is a y such that $y \cdot a = b$. (*Left solvability* rule for *times*.)

III. RULES COMBINING + AND · :
(a) $a \cdot (b + c) = (a \cdot b) + (a \cdot c)$. (*Left distributive* rule.)
(b) $(b + c) \cdot a = (b \cdot a) + (c \cdot a)$. (*Right distributive* rule.)

We are going to take these rules as axioms: i.e., we have a collection of undefined objects, which we will call a field, in which there are two undefined relations + and · such that the above rules hold. Moreover, we suppose that $a + b$ and $a \cdot b$, for any elements a, b in the field, also represent elements of the field. We express this assumption by saying that the collection is *closed* under plus and times. We will also suppose the collection has at least two elements in it. We sum all this up in the following definition.

DEFINITION. A *commutative field* is a set of at least two elements that is closed under two operations, *plus* and *times*, that satisfy the ten axioms written above.

The word *commutative* in the definition refers specifically to the rule $a \cdot b = b \cdot a$. If we do not take this as an assumption we speak of a *not necessarily commutative field*; if $a \cdot b \neq b \cdot a$ for some a, b, then we speak of a *noncommutative field*.

The above rules, taken as axioms, are not independent: for example, the right distributive rule follows at once from the left distributive rule and commutativity of *times*; and IIc′ follows from IIc in virtue of IIa. In the case, though, that one would wish to speak of a not necessarily commutative field, i.e., if one dropped IIa, then it would be necessary to include IIc′ as well as IIc and IIIb as well as IIIa. Even then, however, the system is not independent, as Id follows from Ia, b, c: we proceed to prove this.

THEOREM 1. *If $a + x = a$ for some a, then $a + x = a$ for all a.*

PROOF. Given $a + x = a$ for some a, we have to show that also $b + x = b$ for any b. By Ic, there is a t such that $a + t = b$. Then $t + a = b$ also, from Ia. Now $t + (a + x) = t + a$, whence $(t + a) + x = t + a$, by Ib. Hence $b + x = b$.

DEFINITION. An element z such that $a + z = a$ for all a is called a *neutral element under plus*.

COROLLARY. *If $a + x = a$ for some a, then x is a neutral element under plus.*

THEOREM 2. *There exists at least one neutral element under* plus.

PROOF. Let a be any element (a field contains at least two elements). By Ic, there exists a z such that $a+z=a$, for this particular a. By Theorem 1, $a+z=a$ for all a.

THEOREM 3. *There exists at most one neutral element under plus: this element is also called* zero.

PROOF. Let z and w be neutral elements under plus. We have $z+w=w+z$ by Ia. Now $z+w=z$, because w is neutral, and $w+z=w$ because z is neutral; therefore $w=z$.

Rule Id follows at once from the above theorems, and since these were proved solely on the basis of Ia, b, c, we see that Id follows from Ia, b, c. Since this is the case, why did we include it in the definition of field? Because the zero element is used in IIc, and if we had not written down Id, we would have had to stop in the middle of the definition and establish it before proceeding. Incidentally, this shows one reason for not insisting upon independence.

THEOREM 4. *Given a, b, there exists at most one x such that $a+x=b$.*

PROOF. Let $a+x=b$, $a+y=b$. To prove $x=y$. We have $a+x=a+y$. Let a' be such that $a'+a=z$ (use Ic and Ia), where z is the zero element. Then $a'+(a+x)=a'+(a+y)$, whence $(a'+a)+x=(a'+a)+y$, $z+x=z+y$ and $x=y$. Q.E.D.

DEFINITION. *The* solution of $a+x=z$ is called $-a$. (There exists a solution by Ic and only one by Theorem 4.)

THEOREM 5. $-(-a)=a$.

PROOF. $-(-a)$ is the solution of $x+(-a)=z$. Now $a+(-a)=z$, since $-a$ is the solution of $a+t=z$. So a is the solution of $x+(-a)=z$.

THEOREM 6. $-(a+b)=(-a)+(-b)$.

Exercise. Prove Theorem 6.

DEFINITION. We write $a-b$ for $a+(-b)$ and $-a+b$ for $(-a)+b$.

Exercise. Prove that $-(a-b)=-a+b$.

THEOREM 7. $a\cdot z=z$ *for every* a.

PROOF. For some b, $b+z=b$. Therefore $a\cdot(b+z)=a\cdot b$, whence $(a\cdot b)+(a\cdot z)=a\cdot b$, whence $az=z$ by Theorem 1, Corollary, and Theorem 3.

THEOREM 8. $a(-b)=-(a\cdot b)$.

PROOF. $b+(-b)=z$, so $ab+a(-b)=a\cdot z$. Therefore $ab+a(-b)=z$, whence $a(-b)=-(a\cdot b)$. Q.E.D.

Exercise. Prove that $(-a)\cdot(-b)=a\cdot b$.

Using IIa (and other axioms) one can prove the existence of a unique neutral element under times, i.e., there is one and only one element e such that $a\cdot e=e\cdot a=a$ for all a (one first gets this for $a\neq z$, but $z\cdot e=e\cdot z=z$ by Theorem 7). Actually one can get this result, with a little extra effort, without using IIa. It would be very instructive to carry out all our considerations without using IIa, but this is not practicable in an introductory two-term course. Unfortunately, this will have the consequence that one cannot, in such a course, set up coordinates using Desargues' Theorem alone: another assumption will have to be made, and we shall make it at the point needed. (We do briefly consider the noncommutative case; see p. 86).

Having a unique neutral element under *times*, one can prove, just as in the corresponding situation with plus, that $ax=b$, with $a\neq 0$, has just one solution. This we write as $z=b/a$.

Exercise. Prove that $\frac{a}{b}+\frac{c}{d}=\frac{ad+bc}{bd}$.

2. Consistency of the field axioms. Just as with the axioms for a projective plane, so here for the field axioms, we raise the question of consistency. As before we establish consistency by means of a model.

Every field must contain at least two elements. Let us then take two objects a, b and try to make them into a field: this means that we must define *plus* and *times* in such way that the field axioms hold.

Every field contains a neutral element under *plus*, usually denoted 0. One of a, b is therefore neutral and will conveniently be denoted by 0; say $a=0$, so our two elements are denoted 0, b.

Every field contains a neutral element e under *times*. Note that $e\neq 0$, for if $e=0$, then for every a, $a=a\cdot e=a\cdot 0=0$, so the field would contain just one element, 0. Hence we denote our two elements by 0, e.

From theorems already established we must have (if we are to have a field at all) the following: $0\cdot 0=0$, $0\cdot e=e\cdot 0=0$, $e\cdot e=e$. These rules are conveniently summarized in a table:

	0	e
0	0	0
e	0	e

TABLE FOR TIMES

As for plus, we must have $0+0=0$, $0+e=e+0=e$. There remains the question of how to define $e+e$. In fact, we have no liberty, as $e+e=e$ implies e is a solution of $e+x=e$, hence (by a previous theorem) $e=0$, which is not so. So we must place $e+e=0$. These rules may be summarized as follows:

	0	e
0	0	e
e	e	0

TABLE FOR PLUS

The considerations so far can be summed up in the following theorem.

THEOREM. *There is at most one field containing exactly two elements.*

To show that there is at least one field containing exactly two elements, we have to verify that with plus and times as defined, the field axioms obtain. For example, we have to verify the associative rule: $a\cdot(b\cdot c)=(a\cdot b)\cdot c$. To do this we have to substitute 0 or e for each of a, b, c in this equation, apply the times table, and see that both sides are equal. To do this in detail would require verification of eight equalities (since there are two choices for each of a, b, c). In this way associativity of times can be checked: but one can argue more expeditiously as follows: if either a, b, or c is 0, then both sides are 0. Thus seven of the eight possibilities are taken care of; as for the eighth, here we must take $a=b=c=e$, in which case both sides are e.

In a similar way one verifies that all the field axioms obtain. Thus:

THEOREM. *There is one and only one field with exactly two elements.*

Exercise. Show that there is one and only one field with exactly three elements.

3. The Analytic Model. On the basis of the commutative field axioms, we can introduce subtraction and division, and so one computes in a field, as far as $+$, $\cdot$, $-$, and $\div$ are concerned, exactly in the familiar way. We can now form a projective plane over any (commutative) field F very much as we did in the first part of the book, but, because at that time we had not specified the axioms, let us go over this briefly again. By a *point* $P\begin{pmatrix}a\\b\\c\end{pmatrix}$ we shall mean a triple $\begin{pmatrix}a\\b\\c\end{pmatrix}$, not all a, b, $c=0$, *and* all its multiples $\begin{pmatrix}\lambda a\\\lambda b\\\lambda c\end{pmatrix}$, $\lambda\neq 0$; i.e., a point is a class of *column* triples. A *line* is defined in quite the same way as a class of *row* triples. We say $P\begin{pmatrix}a\\b\\c\end{pmatrix}$ is on $l(r, s, t)$ if $ra+sb+tc=0$. With these definitions one can easily verify, with the algebra set as a prerequisite for this book, the axioms of alignment: we will not go into the details for these axioms, but will pay some attention to Desargues' Theorem. Let us call the plane, so constructed, the projective number plane over F.

If $P\begin{pmatrix}a\\b\\c\end{pmatrix}$ is a point, then $\begin{pmatrix}a\\b\\c\end{pmatrix}$ are called coordinates of the point; we also say that the triple $\begin{pmatrix}a\\b\\c\end{pmatrix}$ *represents* the point, and, if no misunderstanding is to be feared, *is* the point. Similarly, we speak of coordinates of a line. If $l(a, b, c)$ is a line, then a point is on this line if and only if its coordinates satisfy the equation $ax+by+cz=0$. This equation is called an *equation of the line*. Given a line we can write down its equations; and conversely, given an equation $ax+by+cz=0$, not all $a, b, c=0$, we can write down a line one of whose equations is the given equation, viz., the line $l(a, b, c)$. If in $ax+by+cz=0$, we think of x, y, z as the coordinates of some *fixed* point $P\begin{pmatrix}x\\y\\z\end{pmatrix}$, and the equation as a condition on the *variable* line $l(a, b, c)$ to be on $P\begin{pmatrix}x\\y\\z\end{pmatrix}$, then the equation is called a (*line*) *equation of the point*.

THEOREM. *If* $P\begin{pmatrix}x_1\\x_2\\x_3\end{pmatrix}$, $P\begin{pmatrix}y_1\\y_2\\y_3\end{pmatrix}$ *are distinct points, then for any* λ *the numbers* $x_i+\lambda y_i$, $i=1, 2, 3$ *are not all* $=0$, *and* $P\begin{pmatrix}x_1+\lambda y_1\\x_2+\lambda y_2\\x_3+\lambda y_3\end{pmatrix}$ *is collinear with the first two points.*

PROOF. Were $x_1+\lambda y_1=0$, $x_2+\lambda y_2=0$, and $x_3+\lambda y_3=0$, then the x_i would be all the same multiple of the corresponding y_i and the points $P\begin{pmatrix}x_1\\x_2\\x_3\end{pmatrix}$, $P\begin{pmatrix}y_1\\y_2\\y_3\end{pmatrix}$ would be the same. Therefore the $x_i+\lambda y_i$ are not all $=0$, and we may speak of the point $P\begin{pmatrix}x_1+\lambda y_1\\x_2+\lambda y_2\\x_3+\lambda y_3\end{pmatrix}$. Let, now, $l(a_1, a_2, a_3)$ be a line on both $P\begin{pmatrix}x_1\\x_2\\x_3\end{pmatrix}$ and $P\begin{pmatrix}y_1\\y_2\\y_3\end{pmatrix}$.

Then

$$\begin{aligned}a_1x_1+a_2x_2+a_3x_3 &= 0\\a_1y_1+a_2y_2+a_3y_3 &= 0.\end{aligned}$$

Multiplying the second equation by λ and adding to the first, we get

$$a_1(x_1+\lambda y_1)+a_2(x_2+\lambda y_2)+a_3(x_3+\lambda y_3) = 0,$$

whence $P\begin{pmatrix}x_1+\lambda y_1\\x_2+\lambda y_2\\x_3+\lambda y_3\end{pmatrix}$ lies on $l(a_1, a_2, a_3)$. Q.E.D.

For simplicity in notation, we will write $\mathbf{x}$ for $\begin{pmatrix}x_1\\x_2\\x_3\end{pmatrix}$; and $\mathbf{x}+\lambda\mathbf{y}$ for $\begin{pmatrix}x_1+\lambda y_1\\x_2+\lambda y_2\\x_3+\lambda y_3\end{pmatrix}$.

COROLLARY (to last theorem). *With hypotheses as before, the same conclusion holds for* $\lambda\mathbf{x}+\mu\mathbf{y}$ *if not both* $\lambda=0$ *and* $\mu=0$. (PROOF: If $\lambda\neq 0$, say, apply the theorem to $\mathbf{x}+\frac{\mu}{\lambda}\mathbf{y}$.)

THEOREM. *If* $\mathbf{x}, \mathbf{y}, \mathbf{z}$ *are coordinates of three distinct collinear points, then there exists* λ, μ *such that* $\mathbf{z}=\lambda\mathbf{x}+\mu\mathbf{y}$.

PROOF. Let $l(a_1, a_2, a_3)$ be the line on these three points. Then we have:

$$\begin{aligned}a_1x_1+a_2x_2+a_3x_3 &= 0\\a_1y_1+a_2y_2+a_3y_3 &= 0\\a_1z_1+a_2z_2+a_3z_3 &= 0.\end{aligned}$$

Not all $x_1, x_2, x_3=0$. Say $x_1\neq 0$. Now

$$y_1 = \frac{y_1}{x_1}\cdot x_1, \qquad y_2 = \frac{y_1}{x_1}\cdot x_2, \qquad y_3 = \frac{y_1}{x_1}\cdot x_3$$

are not all possible since the points $P\mathbf{x}$, $P\mathbf{y}$ are distinct. The first of these equations holds, so the second or third does not. Say the second does not. We have, then, $y_2x_1-y_1x_2\neq 0$. Now solve the equations

$$\begin{aligned}\lambda x_1+\mu y_1 &= z_1\\\lambda x_2+\mu y_2 &= z_2.\end{aligned}$$

By elementary algebra it is known that these can be solved if $x_1y_2 - x_2y_1 \neq 0$, and this we have. Use the previous three equations now: multiply the first by λ, the second by μ, and subtract from the last. We get

$$a_3(z_3 - \lambda x_3 - \mu y_3) = 0.$$

If we knew that $a_3 \neq 0$, we would have $z_3 - \lambda_3 x_3 - \mu y_3 = 0$, and conclude that $\mathbf{z} = \lambda\mathbf{x} + \mu\mathbf{y}$. We proceed to prove that $a_3 \neq 0$. Suppose for a moment it were equal to zero. Then we would have

$$a_1x_1 + a_2x_2 = 0$$
$$a_1y_1 + a_2y_2 = 0.$$

Multiply the first by y_2, the second by x_2, and subtract. We get $a_1(x_1y_2 - x_2y_1) = 0$, whence $a_1 = 0$. Similarly we can get $a_2 = 0$. Then a_1, a_2, a_3 are all 0, and this is a contradiction, since these are the coordinates of a line. This completes the proof.

THEOREM. *Desargues' Theorem holds in the projective number plane over F (where F is any field).*

PROOF. Let A, B, C, A', B', C' be the vertices of two centrally perspective triangles, and let O be the center (Fig. 3.1). Let O have

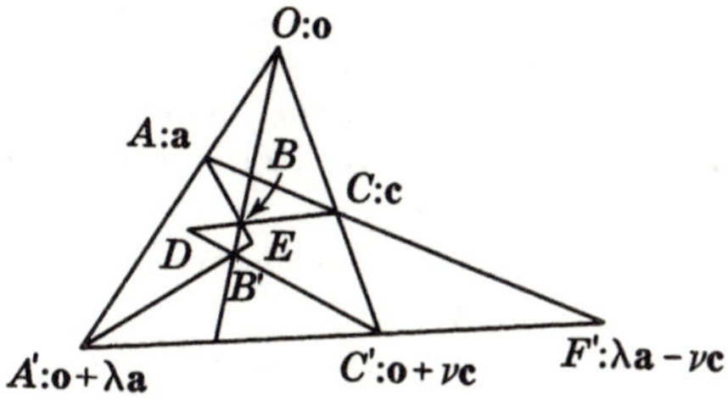

FIG. 3.1

coordinates $\mathbf{o}$; A, coordinates $\mathbf{a}$; B, coordinates $\mathbf{b}$; C, coordinates $\mathbf{c}$. By the last theorem, there are numbers λ, μ such that $\lambda\mathbf{a} + \mu\mathbf{o}$ are coordinates of A'. Here $\mu \neq 0$, as if $\mu = 0$, we see that $A = A'$, which is not so. Therefore, $\frac{\lambda}{\mu}\mathbf{a} + \mathbf{o}$ are also coordinates of A'. Changing the meaning of λ (i.e., designating by λ what was previously designated by $\frac{\lambda}{\mu}$), we can say that A has coordinates $\mathbf{o} + \lambda\mathbf{a}$. Similarly, B' has coordinates $\mathbf{o} + \mu\mathbf{b}$; and C', coordinates $\mathbf{o} + \nu\mathbf{c}$. Consider the point with coordinates $\lambda\mathbf{a} - \nu\mathbf{c}$. By a previous theorem it is on line AC: Since $\lambda\mathbf{a} - \nu\mathbf{c} = (\mathbf{o} + \lambda\mathbf{a}) - (\mathbf{o} + \nu\mathbf{c})$, it is, by the same theorem, on $A'C'$. So the point in question is $F = AC \cdot A'C'$; F has coordinates $\lambda\mathbf{a} - \nu\mathbf{c}$. Similarly one sees that D has coordinates $\nu\mathbf{c} - \mu\mathbf{b}$, and E has coordinates $\mu\mathbf{b} - \lambda\mathbf{a}$. Since $\lambda\mathbf{a} - \mu\mathbf{b} = (\lambda\mathbf{a} - \nu\mathbf{c}) + (\nu\mathbf{c} - \mu\mathbf{b})$, the points D, E, F are collinear. Q.E.D.

In setting up the above problem, we could have taken $\lambda\mathbf{a}$ instead of $\mathbf{a}$ as coordinates of A. With a change of notation, this makes A have coordinates $\mathbf{a}$, A', coordinates $\mathbf{o} + \mathbf{a}$. Similarly, we might have supposed B' to have coordinates $\mathbf{o} + \mathbf{b}$; C', coordinates $\mathbf{o} + \mathbf{c}$. Usually such simplifying methods, elimination of unnecessary letters, are helpful. The above proof was so simple, however, that to have introduced the idea at once would have obscured the proof.

Exercises

1. Let $\mathbf{a}$, $\mathbf{b}$, $\mathbf{a} + \lambda\mathbf{b}$ be the coordinates of three distinct collinear points A, B, C. Find coordinates of the point D such that $H(A, B; C, D)$. (*Answer*: $\mathbf{a} - \lambda\mathbf{b}$ are such coordinates.)

2. Establish Pappus' Theorem analytically.

4. Geometric description of the operations plus and times. Start with a commutative field F, and construct the projective number plane over F. We now ask whether the field makes itself manifest in any fairly direct manner in the geometry of this plane.

Let us introduce nonhomogeneous coordinates x_1/x_3, x_2/x_3 for the point $\mathbf{x}$, where $x_3 \neq 0$ is assumed. Every point, then, except those on the line $x_3 = 0$, has nonhomogeneous coordinates. Every point on the line $x_2 = 0$, except the point $P\begin{pmatrix}1\\0\\0\end{pmatrix}$, has nonhomogeneous coordinates of the form $(x, 0)$, where $x = x_1/x_3$. Conversely, for any number x in F, we see that $(x, 0)$ are the coordinates of some point on $x_2 = 0$, namely, the point $P\begin{pmatrix}x\\0\\1\end{pmatrix}$. Thus we see that the points of the line $x_2 = 0$ *minus one point* are in one-to-one correspondence with the elements of F. We can visualize to ourselves the field F as spread out on the line $x_2 = 0$, covering it completely except for the point $P\begin{pmatrix}1\\0\\0\end{pmatrix}$. Label this point $P\begin{pmatrix}1\\0\\0\end{pmatrix}$ with the letter Ω.

Let us take two points A, B on $x_2 = 0$, $A \neq \Omega$, $B \neq \Omega$. Let A have coordinates $(a, 0)$, B, coordinates $(b, 0)$. Now we can combine a and b by the field operations and form $a + b$. There is a point C such that C

has coordinates $(a+b, 0)$. We now ask whether the position of C can be described in any simple geometric fashion.

Go back to what is familiar, the ordinary Euclidean plane (Fig. 3.2). Here the coordinate a merely represents the distance from the point $O=(0, 0)$. To get $C=(a+b, 0)$, pick up the segment OA, place it down again with O on B, and A will fall on C. This description is a bit too

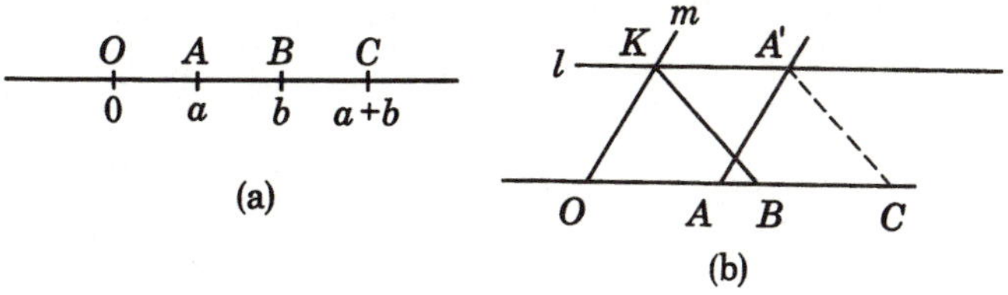

FIG. 3.2

informal (Fig. 3.2a). We can put it this way. Let l be a line parallel to OA. Through O take an arbitrary line $m(\neq OA)$, and let the line through A and parallel to m cut l in A'. Then the line through A' parallel to KB (where $K=l\cdot m$) cuts OA in the point C (Fig. 3.2b).

Let us carry out this same idea in the projective number plane over F (Fig. 3.3). Take a line l through Ω, say the line $x_2=x_3$; and take a line m through O, say the line $x_1=0$. Let m intersect $l: x_2=x_3$ in K and

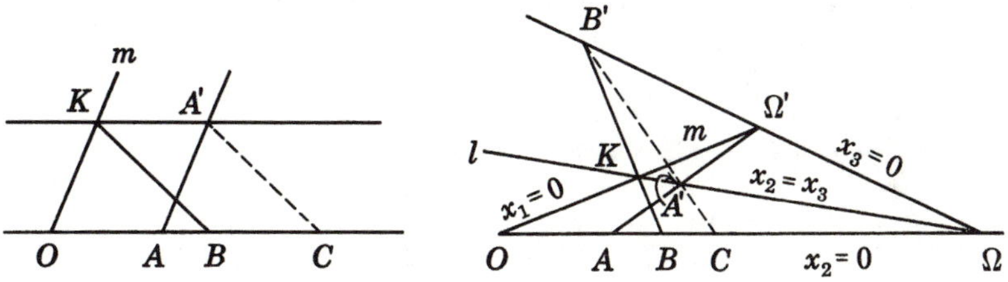

FIG. 3.3

$x_3=0$ in Ω'. Let $\Omega'A$ cut l in A'. Let KB cut $x_3=0$ in B'. If $A=(a, 0)$, $B=(b, 0)$, we claim that $C=B'A'\cdot OA=(a+b, 0)$. First,

$$K=\begin{pmatrix}0\\1\\1\end{pmatrix}, \quad \text{so} \quad B'=\begin{pmatrix}b\\-1\\0\end{pmatrix}; \quad \text{and} \quad A'=\begin{pmatrix}a\\1\\1\end{pmatrix}, \quad \text{so} \quad C=\begin{pmatrix}a+b\\0\\1\end{pmatrix},$$

or $C=(a+b, 0)$. Now note, using the 4-point $\Omega'KA'B'$, that $Q\begin{pmatrix}O & A & \Omega\\ C & B & \Omega\end{pmatrix}$. This is the geometric description we were looking for.

(Note, incidentally, that the above breaks down for $A=O$, or $B=O$.)

Let us do the same for times; the question is: if $A=(a, 0)$, $B=(b, 0)$, and $C=(ab, 0)$, then are A, B, C related in some simple geometric fashion? Again go back to what is familiar, and recall how one would construct $x=a\cdot b$, given a and b. Here one has to fix a unit length, and supposing this to have been done, we may write $1\cdot x=a\cdot b$ or $x/b=a/1$.

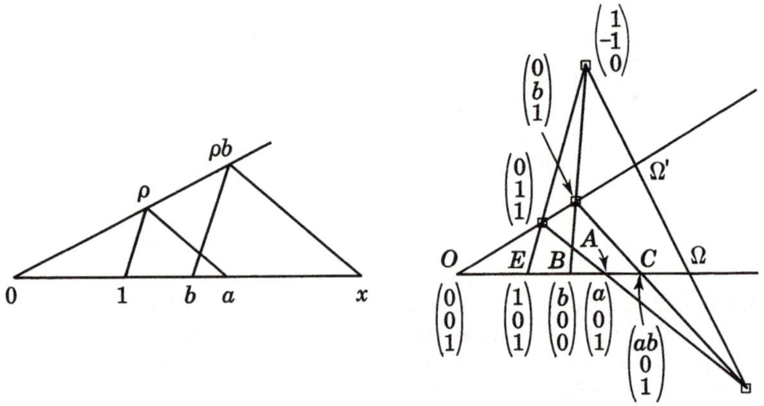

FIG. 3.4

We can solve for x geometrically using similar triangles. In Fig. 3.4, using similar triangles one sees that

$$b/1 = \rho b/\rho = x/a, \qquad \text{or} \qquad x = a\cdot b.$$

Carrying out this idea in the number plane over F, we see that if

$$O = \begin{pmatrix}0\\0\\1\end{pmatrix},\quad E = \begin{pmatrix}1\\0\\1\end{pmatrix},\quad A = \begin{pmatrix}a\\0\\1\end{pmatrix},\quad B = \begin{pmatrix}b\\0\\1\end{pmatrix},\quad C = \begin{pmatrix}ab\\0\\1\end{pmatrix},\quad \Omega = \begin{pmatrix}1\\0\\0\end{pmatrix},$$

then $Q\begin{pmatrix}E & A & O\\ C & B & \Omega\end{pmatrix}$.

Exercises

1. Establish the last statement in detail. (Note, incidentally, that the above breaks down for $A=E$ or $B=E$.)

2. How many points are there in the analytic model over the field of two elements? Over the field of three elements?

5. Setting up coordinates in the projective plane. We now take a quite general Desarguesian plane and ask whether we can associate coordinates to its points in such manner that the plane becomes a projective number plane over some field; or we can put the question as follows: Can we assign coordinates to the points in such manner that the lines have linear equations? Keep in mind that *we are not given any field*! If a field is to enter the discussion, we must construct it ourselves in terms of the given plane.

The work of the last section was by way of analysis. It in no way enters into the work of this section except as a guide. We proceed on the basis of Axioms 0–5 and Desargues' Theorem: at a certain point we shall add another axiom.

What, then, shall be our field F? Keeping in mind the analysis of the last section, we see that if we are to be successful in introducing coordinates, then the points of any line of our plane minus one point will be in one-to-one correspondence with the elements of F. We therefore take a line l, fix on it a point Ω, and try to make $l-\Omega$ into a field. We further select two points O and E on l, and try to make $l-\Omega$ into a field such that O be the zero, E the identity of the field.

Still being guided by the analysis of the last section, we write down the following definition.

DEFINITION. Let A, B be two points on l, $A\neq\Omega$, $B\neq\Omega$, $A\neq O$, $B\neq O$. Let C be the point such that $Q\begin{pmatrix} O & A & \Omega \\ C & B & \Omega \end{pmatrix}$. Then we place $C=A+B$ *by definition* (i.e., we are defining $+$). We also place $X+O=O+X=X$ for any X on l, $X\neq\Omega$.

Our problem now is to show that *plus* so defined satisfies the field axioms. Note first that $l-\Omega$ is closed under plus; i.e., $A+B$ is never Ω. This is so because of the necessary inequalities on quadrangular hexads.

We have to check that $A+B=B+A$. For $A=O$ or $B=O$, this is immediate, by definition. Consider, then, the other case. Let $A+B=C$, $B+A=D$. To prove $C=D$. We have $Q\begin{pmatrix} O & A & \Omega \\ C & B & \Omega \end{pmatrix}$ and $Q\begin{pmatrix} O & B & \Omega \\ D & A & \Omega \end{pmatrix}$. Why is $C=D$? By a theorem we can interchange a *pair* from above and below, so interchanging A with B and Ω with Ω we get $Q\begin{pmatrix} O & B & \Omega \\ C & A & \Omega \end{pmatrix}$. Now there is only one solution to $Q\begin{pmatrix} O & B & \Omega \\ X & A & \Omega \end{pmatrix}$, so $C=D$.

How about solvability? Can we solve $A+X=B$? This comes to asking: Can we solve $Q\begin{pmatrix} O & A & \Omega \\ B & X & \Omega \end{pmatrix}$? Yes, we certainly can, by our theorems on quadrangular hexads, at least if $A\neq O,\neq B$. If $A=O$, then $X = B$ is the required solution; if $A = B$, then $X = O$ is the required solution.

Now for the associative rule: to prove that $(A+B)+C = A+(B+C)$. Let us, to orient ourselves, look at how we would carry out these operations in the affine plane (Fig. 3.5).

We translate OA off onto l and then back to line $OABC$ in constructing $A+B$. Now we have to construct $B+C$, Let $K = l \cdot m$, where m is a line through O (in fact, the path of O in translating OA). We have to translate OB onto l. We could do this either in the direction of OK as in Fig. 3.5a or in the direction of BK as in Fig. 3.5b. Offhand it is not clear which of these would be better. One could try both ways and see which gives the simpler figure. We suppose this to have been done,

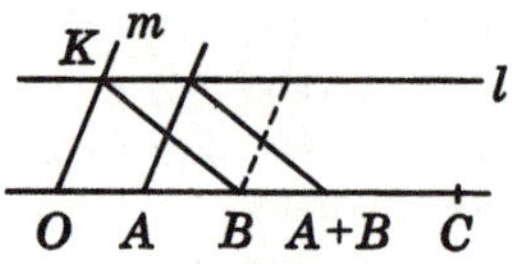

FIG. 3.5a

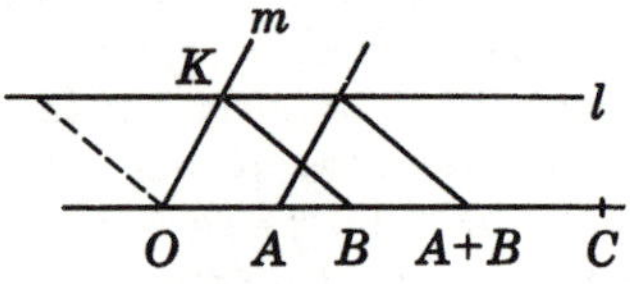

FIG. 3.5b

and find that the second figure is simpler. (These remarks are entirely informal and need not really be made in the proof.)

Proceeding then with the construction as in Fig. 3.6, one sees that the very line leading to $A+(B+C)$ is the same as the one leading to $(A+B)+C$. It is now entirely a straightforward matter to carry out these considerations in the projective plane. (One may ask whether

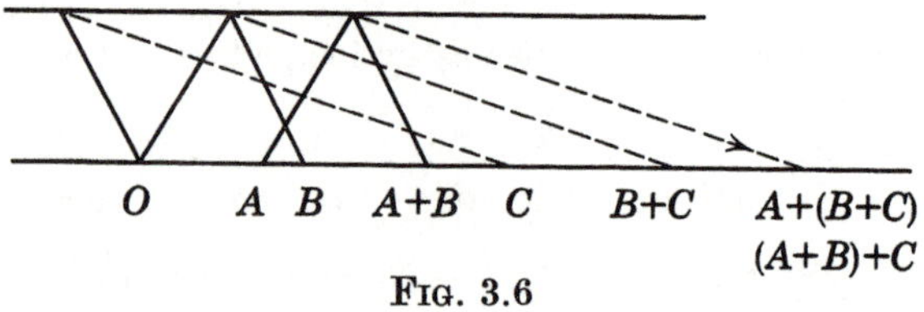

FIG. 3.6

special attention has to be given to special cases—for example, $C = O$ or $A + B = O$—since the definitions of $U + V$ depend on whether U and V are O or not. Actually one sees that the constructions used to get $A+B$, $B+C$, etc., all hold quite generally, even, for example, if $C = O$ or $A+B = O$.)

Exercises

1. Draw the figure in the projective plane corresponding to the proof that $(A+B)+C = A+(B+C)$. Explain the drawing briefly.

2. Prove that if $A+B = O$ (but $A \neq O$), then $H(O, \Omega; A, B)$.

A similar analysis leads to an appropriate definition of *times*. We recall the construction of $x=a\cdot b$ of two line segments a, b. Clearly, if the product is to be represented by a line segment, then it is necessary to fix a unit length: we suppose, then, that a length 1 is available, and write the equation $x=a\cdot b$ in the form $x/a=b/1$. Let O, A, B, E be collinear, $OA=a$, $OB=b$, $OE=1$; through E and B draw lines at 45° to OA and let these cut the perpendicular at O in E' and B', all as in Fig. 3.7. Through B' draw a line parallel to $E'A$, cutting OA in C. Then

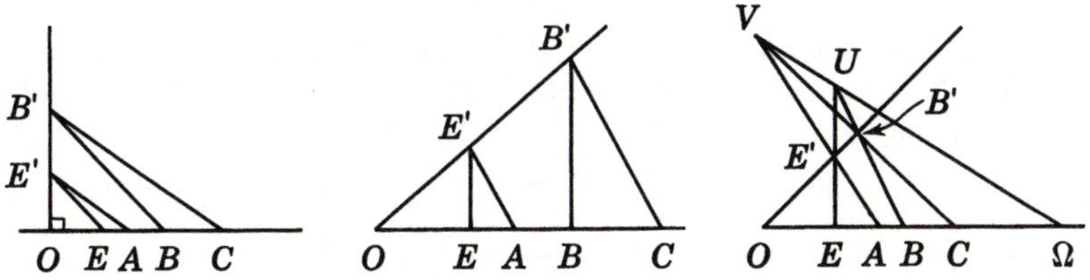

FIG. 3.7

$OC/OA=OB'/OE'$, or $OC/a=b/1$, so $OC=x$. One sees, however, that it is not necessary to have OE' perpendicular to OA, and the lines AE', CB' need not be at 45° to OA, but one needs only that they be parallel. In the projective plane, now, having selected a point Ω on line OA, we are led by analogy (and analysis) to the following definition.

DEFINITION. Having fixed, O, E, Ω on a line l, if A, B are points on l, different from O, E, Ω, we define $A\cdot B$ to be the point C, where C is the point such that $Q\begin{pmatrix} E & A & O \\ C & B & \Omega \end{pmatrix}$. We also define $A\cdot O=O\cdot A=O$, $A\cdot E=E\cdot A=A$.

With this definition, we now try to check the field rules for *times*. The discussion for solvability and associativity is parallel to the previous

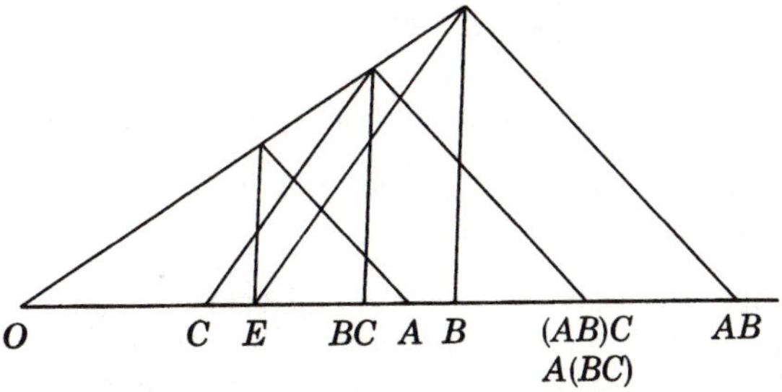

FIG. 3.8 FIGURE FOR ASSOCIATIVITY OF TIMES

discussion of these items for *plus* and shall not be written out here (Fig. 3.8). We postpone discussion of *commutativity* and consider next the *distributive* rules.

Let A be a point on our line l, $A \neq O$, $A \neq \Omega$. Consider the operation of passing from a point X on l to the point $A \cdot X$. We call such an operation a *left multiplication*. Similarly we define a *right multiplication*.

THEOREM. *The transformation $X \to A \cdot X$ is a projectivity.*
The transformation $X \to X \cdot A$ is a projectivity.

PROOF. In the construction of $A \cdot B$ (Fig. 3.7), think of A as fixed and $X = B$ as variable. The points E', U, V are also fixed. Then the passage from B to $C = A \cdot B = A \cdot X$ can be described as follows. Consider the projectivity $l \overset{U}{\underset{\wedge}{-}} m \overset{V}{\underset{\wedge}{-}} l$. Then B goes into C under this projectivity. That is, $X \to A \cdot X$ is a projectivity. The second statement of the theorem is proved similarly.

COROLLARY. *The above projectivities send O into O, Ω into Ω.*

THEOREM. $A \cdot (B+C) = (A \cdot B)+(A \cdot C)$
$(B+C) \cdot A = (B \cdot A)+(C \cdot A)$.

PROOF. We have $Q\begin{pmatrix} O & B & \Omega \\ B+C & C & \Omega \end{pmatrix}$, by definition of $B+C$. Now apply the left multiplication by A. This operation is a projectivity, and therefore sends quadrangular sets into quadrangular sets. Thus $Q\begin{pmatrix} O & AB & \Omega \\ A(B+C) & AC & \Omega \end{pmatrix}$, since $O \to O$, $\Omega \to \Omega$ under the projectivity in question. But by definition of *plus*, we then get $(A \cdot B)+(A \cdot C) = A \cdot (B+C)$. Q.E.D.

We proceed to the *commutative rule.* Here one comes upon a difficulty, and to overcome it we will assume Pappus' Theorem; i.e., it is sufficient to assume Pappus' Theorem in order to prove the commutative rule: it is also necessary, but this point is not under immediate discussion.

To our axioms, then, we at this point add Pappus' Theorem.

THEOREM. $A \cdot B = B \cdot A$; *multiplication satisfies the commutative rule in a Pappian plane.*

PROOF. It is important to keep in mind that A and B play essentially distinct roles in the construction of $A \cdot B$. In the construction of $B \cdot A$, the roles of the two letters are reversed. First, to orient ourselves, let us construct $B \cdot A$, in the Euclidean plane, reversing the roles of A and B. We proceed with Fig. 3·9a giving $A \cdot B = C$. Through A draw a line parallel to BB' cutting OE' in A'. Through A' draw a line parallel to $E'B$. Then this line cuts OE in $B \cdot A$. Fig. 3.9b gives the construction in the projective plane. The question is: Does WA' pass

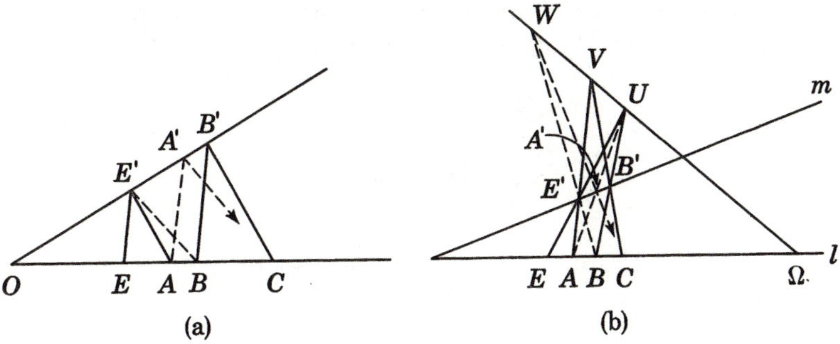

FIG. 3.9

through C? Applying Pappus' Theorem to the points (W, V, U; B', A', E'), we see that the answer is yes.

We see that Pappus' Theorem is sufficient for commutativity. It is also necessary, i.e., if commutativity holds, then so does the theorem. To see this, we first prove the following.

THEOREM. *Let l be a line, O, E, Ω three points thereon; let l' be a line, O', E', Ω' three points thereon. Let* times *be defined as above on l relative to O, E, Ω; and let* times *also be defined similarly on l' relative to O', E', Ω'. If one of these operations is commutative, then so is the other.*

PROOF. Suppose commutativity holds on l. Let A', B' be points on l'. Consider a projectivity sending O', E', Ω', respectively, into O, E, Ω. Let $A' \to A$, $B' \to B$; let $A'B' = C' \to C$, $B'A' = D' \to D$. Now $Q\begin{pmatrix} E' & A' & \Omega' \\ C' & B' & O' \end{pmatrix}$, and since quadrangular sets go into quadrangular sets under projectivity, we have $Q\begin{pmatrix} E & A & \Omega \\ C & B & O \end{pmatrix}$. Thus $C = A \cdot B$. Similarly $D = B \cdot A$. We are assuming commutativity on l. Therefore $C = D$. Hence $C' = D'$ (under a projectivity distinct points go into distinct points). So $A' \cdot B' = B' \cdot A'$, i.e., commutativity holds on l' also.

THEOREM. *If commutativity holds, then so does Pappus' Theorem.*

PROOF. In Fig. 3.9b, we started with the line ABC, and proceeded to construct the rest of the figure. Thus the points $W, V, U; B', A', E'$ conceivably are in some special position. But now consider W, V, U; B', A', E' as given, and complete the construction, so that now O, A, B, C, Ω are constructed, not given. On $O\Omega$, relative to O, E, Ω, define *times* in the standard manner. Commutativity holds by assumption. To prove Pappus' Theorem with respect to the points

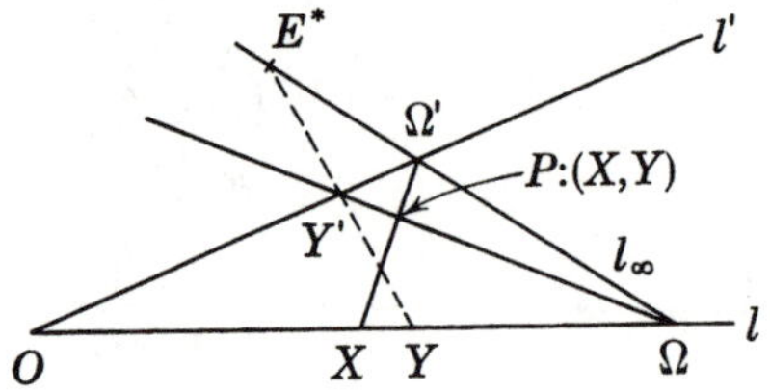

FIG. 3.10

$W, V, U; B', A', E'$ comes to showing that VB' and WA' meet on line AB. Since VB' cuts at $A \cdot B$ and WA' cuts at $B \cdot A$, we see that VB' and WA' do meet on line AB, and the proof is complete.

We now have a field, and proceed to set up coordinates in the plane. The line l will play a role analogous to the x-axis in the Euclidean plane. Let l' be another line through O and let l_∞ be a line through Ω (Fig. 3.10). Let $l' \cdot l_\infty = \Omega'$. The line l' will play a role analogous to the y-axis. On l_∞ take a point $E^* \neq \Omega, \neq \Omega'$. We think of all the points on l as having been named: A, B, C, and so on; and we designate $l' \cdot E^*X$,

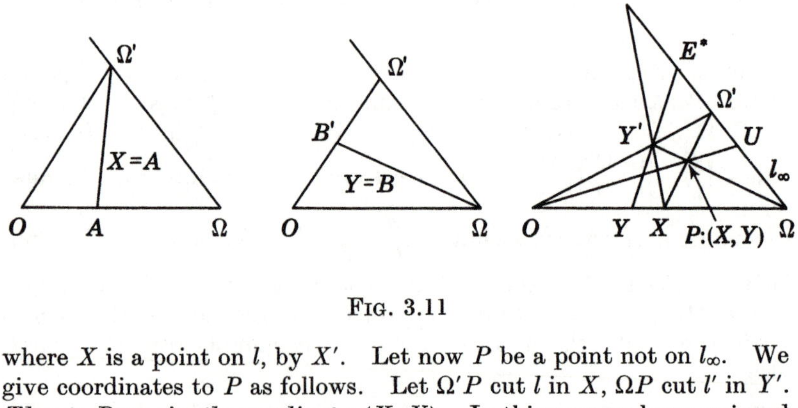

FIG. 3.11

where X is a point on l, by X'. Let now P be a point not on l_∞. We give coordinates to P as follows. Let $\Omega' P$ cut l in X, ΩP cut l' in Y'. *Then to P we give the coordinates* (X, Y). In this way we have assigned coordinates to all the points except those on l_∞. *The essential point to be proved is that with this definition, lines have linear equations.* We are here taking care of the plane minus l_∞.

Consider a line through Ω'. Let it cut l in the point A (Fig. 3.11). Then one sees that the equation of the line is $X = A$. Similarly consider a line through Ω and let it cut l' in B'. Then its equation is $Y = B$.

Now consider a line through O. Let it cut l_∞ in U (Fig. 3.11). Let P: (X, Y) be a point on it. Then ΩP cuts l' in Y', $\Omega' P$ cuts l in X.

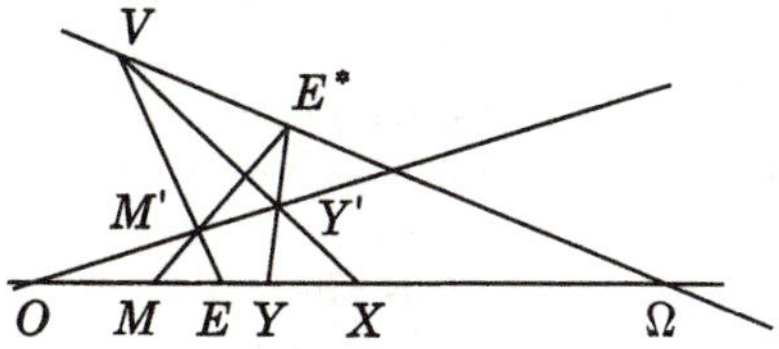

FIG. 3.12

Now note that if $V = XY' \cdot l_\infty$, then $H(\Omega\Omega'; UV)$, whence V is fixed as P varies on OU. We have E^*Y' cutting l in Y, and want the relation between X and Y. Let $Q = OU \cdot \Omega' E$. Then Q has coordinates (E, M). In Fig. 3.12 we have only drawn what is essential for the rest of the

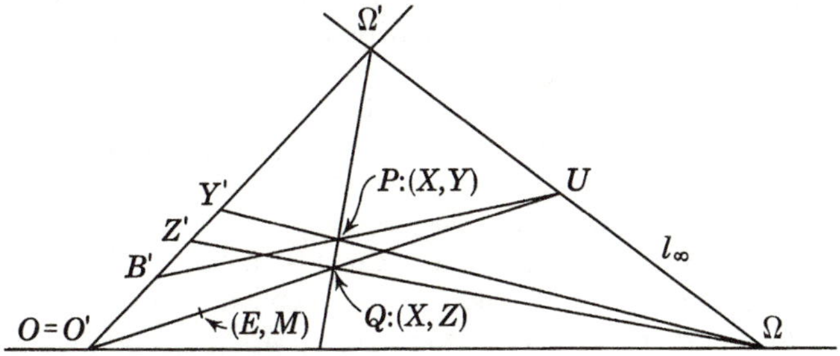

FIG. 3.13

argument: the fact that EM' passes through V expresses the condition that $Q:(E, M)$ is on OU. Now one sees that

$$Q\begin{pmatrix} E & M & O \\ Y & X & \Omega \end{pmatrix}, \quad \text{whence } Y = M \cdot X.$$

This is the equation of OU.

Now consider a line not through O, Ω, or Ω'. Let the given line cut l' in B' and l_∞ in U (Fig. 3.13). Let $P:(X, Y)$ be a point on $B'U$, and let $\Omega' P$ cut OU in $Q:(X, Z)$. Let (E, M) be on OU (i.e., there is a point on OU with first coordinate E; and let M be the second coordinate). Then $Z = M \cdot X$ by the previous case. From the complete 4-point ΩUPQ one sees that $Q\begin{pmatrix} O' & B' & \Omega' \\ Y' & Z' & \Omega' \end{pmatrix}$. Dropping the primes amounts to a perspectivity from l' to l, so $Q\begin{pmatrix} O & B & \Omega \\ Y & Z & \Omega \end{pmatrix}$. Therefore, $Y = Z + B$, $Y = M \cdot X + B$. This is the equation of the line $B'U$.

We have now set up nonhomogeneous coordinates in the affine plane consisting of the given plane minus l_∞. Much as has already been explained, one could now introduce homogeneous coordinates into the plane. In this way we see that *a plane in which Desargues' Theorem and Pappus' Theorem hold is a projective number plane over some field.* The importance of this theorem is that it shows that there is no real difference between so-called synthetic and so-called analytic geometry.

6. The noncommutative case. Toward the end of the preceding discussion we assumed Pappus' Theorem as an axiom, and we needed to do this in order to get commutativity of times. Pursuing the matter a bit further, one sees that one can carry out all our previous considerations without Pappus' Theorem and over a not necessarily commutative field. This is worth doing, as the deeper study shows us more exactly the relation between Desargues' Theorem and setting up coordinates, whereas the preceding treatment slightly obscures the relation by introducing Pappus' Theorem.

First we have to consider fields without laying down the commutativity of times as an axiom. Previously we got the neutral element of times in just the way we got the neutral element of plus. Now we have to get the neutral element of times by a slightly deeper argument.

By a right neutral element e_R under times we mean an element e_R such that $ae_R = a$ for every a in the field. We establish the existence of e_R as follows: for some $a \neq 0$ "solve" $ax = a$, i.e., assert the existence of an element x such that $ax = a$; then for any element b in the field, also $bx = b$; in fact, write $b = ya$; then $bx = yax = ya = b$; call this element x, e_R. Similarly obtain a left neutral element e_L; this has the property that $e_L a = a$ for every a. Now we claim that $e_L = e_R$. In fact, $e_L \cdot e_R = e_L$, since e_R is right neutral; and $e_L \cdot e_R = e_R$ since e_L is left neutral. Thus $e_L = e_R$, which we now designate as e, is a neutral element under times. There is only one such neutral element, since, if e' is another, then arguing as we just did, we find $e \cdot e' = e = e'$. Summing up:

THEOREM. *In a not necessarily commutative field there is one and only one element e such that $ae = ea = a$ for every a in the field.*

By a right inverse a_R^{-1} of an element $a(\neq 0)$ we mean an element x such that $ax=e$; by a field axiom, at least one such a_R^{-1} exists for every $a\neq 0$. Similarly we obtain for each $a\neq 0$ a left inverse a_L^{-1}. Now one sees that $a_L^{-1}=a_R^{-1}$; in fact $(a_L^{-1}a)a_R^{-1}=a_L^{-1}(aa_R^{-1})=a_L^{-1}=(a_L^{-1}a)a_R^{-1}=a_R^{-1}$. This shows, incidentally, that there is at most one right inverse a_R^{-1} and at most one left inverse a_L^{-1} (for any given $a\neq 0$) since every a_R^{-1} must equal any given a_L^{-1}, and every a_L^{-1} must equal any given a_R^{-1}. Summing up:

THEOREM *In a not necessarily commutative field there is for every $a\neq 0$ one and only one element a_R^{-1} such that $aa_R^{-1}=e$; one and only one element a_L^{-1} such that $a_L^{-1}a=e$; moreover $a_L^{-1}=a_R^{-1}$. This element is designated a^{-1} and is called the inverse of a.*

Now we get the uniqueness of solvability for times.

THEOREM. *The equation $ax=b$, $a\neq 0$, has one and only one solution. Similarly $ya=b$, $a\neq 0$, has one and only one solution.*

PROOF. Multiplying both sides of $ax=b$ on the left by a^{-1}, we obtain $a^{-1}ax=a^{-1}b$, i.e., $x=a^{-1}b$. This shows that the only possible solution is $a^{-1}b$, and hence there is at most one solution. By an axiom there is at least one solution, so it is not necessary to check that $a^{-1}b$ is a solution.

REMARK. For noncommutative fields one should not designate the solution of $ax=b$ by $x=b/a$. The symbol b/a would presumably equally stand for ba^{-1} or $a^{-1}b$, but in a noncommutative field, these may not be equal.

Now we will discuss the analytic model of a not necessarily commutative field. We may be tempted to define a point as a column triple $\begin{pmatrix}x_0\\x_1\\x_2\end{pmatrix}$, with not all entries equal to zero, and a line similarly as a row triple (a_0, a_1, a_2); and say that one is on the other if $a_0x_0+a_1x_1+a_2x_2=0$. If, however, just as before we allow $\begin{pmatrix}\lambda x_0\\\lambda x_1\\\lambda x_2\end{pmatrix}$, for $\lambda\neq 0$, also to be coordinates of the point $\begin{pmatrix}x_0\\x_1\\x_2\end{pmatrix}$ we run into difficulty, for we cannot deduce $a_0\lambda x_0+a_1\lambda x_1+a_2\lambda x_2=0$ from $a_0x_0+a_1x_1+a_2x_2=0$. This difficulty can, however, be overcome. Namely, one puts into the same class two column triples $\begin{pmatrix}x_0\\x_1\\x_2\end{pmatrix}$, $\begin{pmatrix}y_0\\y_1\\y_2\end{pmatrix}$, each different from $\begin{pmatrix}0\\0\\0\end{pmatrix}$ if, and only if, one is a *right* multiple of the other (in which case the other is a right multiple of the one): i.e., $\begin{pmatrix} x_0 \\ x_1 \\ x_2 \end{pmatrix}$, $\begin{pmatrix} y_0 \\ y_1 \\ y_2 \end{pmatrix}$ go into the same class if and only if there is a λ such that $x_0 = y_0\lambda$, $x_1 = y_1\lambda$, $x_2 = y_2\lambda$. Similarly we put two row triples (not equal (0, 0, 0)) into the same class if and only if one is a *left* multiple of the other. Now defining *point* as a class of column triples $\left\{\begin{pmatrix} x_0\lambda \\ x_1\lambda \\ x_2\lambda \end{pmatrix}\right\}$ and *line* as a class of row triples $\{(\mu a_0, \mu a_1, \mu a_2)\}$, and incidence by the condition that $\mu a_0 \cdot x_0\lambda + \mu a_1 \cdot x_1\lambda + \mu a_2 \cdot x_2\lambda = 0$, one can verify that the axioms of alignment and Desargues' Theorem obtain.

Exercises

1. Go over our previous considerations for homogeneous linear equations, but this time over a not necessarily commutative field.

2. Establish Desargues' Theorem analytically over a not necessarily commutative field. The technique is the same as before.

3. Show analytically that Pappus' Theorem does not hold over a noncommutative field.

The remainder of the considerations in the noncommutative case are quite like those already made in the commutative case: the axioms of alignment and Desargues' Theorem are sufficient for establishing coordinates with entries from a not necessarily commutative field.

CHAPTER IV

RELATIONS BETWEEN THE BASIC THEOREMS

In what follows, the axioms of alignment are always assumed. On the basis of them we can formulate many theorems, for example, Desargues' Theorem, Pappus' Theorem, Theorem A (see page 25), and The Fundamental Theorem.

We now ask which of these imply the others. For example, we have already seen that the Fundamental Theorem implies the others. We will now show that Desargues' Theorem and Pappus' Theorem, together, imply Theorem A. In the formulation of Theorem A one

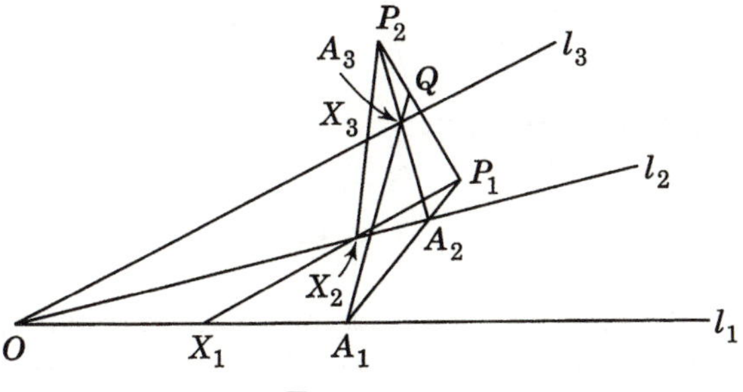

FIG. 4.1

speaks of a projectivity from a line l to a line l'. We first show that, on the basis of Desargues' Theorem alone, any projectivity from a line l to a line $l' \neq l$ can be written as a product of two perspectivities. First we need a lemma.

LEMMA. (*Assuming Desargues' Theorem*). *If* l_1, l_2, l_3 *are* distinct *concurrent lines, then any projectivity* $l_1 \overset{P_1}{\underset{\wedge}{-}} l_2 \overset{P_2}{\underset{\wedge}{-}} l_3$ *is a perspectivity.*

PROOF. Let A_1 be a point on l_1, $A_1 \neq O = l_1 \cdot l_2$, and let $A_1 \to A_2 \to A_3$ under the perspectivities in question (Fig. 4.1). Let $Q = A_1A_3 \cdot P_1P_2$.

We claim that the given projectivity is the perspectivity $l_1 \overset{Q}{\underset{\wedge}{-}} l_3$. To prove this, take a point X_1 on l_1, $X_1 \neq O$, and let $X_1 \to X_2 \to X_3$ under the perspectivities in question. We have to show that X_1X_3 passes through Q. This follows at once from Desargues' Theorem, since triangles $A_1A_2A_3$, $X_1X_2X_3$ are centrally perspective and the proof is complete.

Let now T be a projectivity from a line l to a line $l' \neq l$. T is a product of perspectivities, but the perspectivities entering into T are not uniquely determined, i.e., T can be written in many ways as a product of perspectivities. Let

$$(1) \qquad T: l = l_0 \overset{P_1}{\underset{\wedge}{-}} l_1 \overset{P_2}{\underset{\wedge}{-}} l_2 \overset{P_3}{\underset{\wedge}{-}} l_3 \overset{P_4}{\underset{\wedge}{-}} \cdots \overset{P_n}{\underset{\wedge}{-}} l_n = l'$$

be one such way. The question is: To what extent can we rewrite the above in a simpler way? By simpler, we mean with less perspectivities.

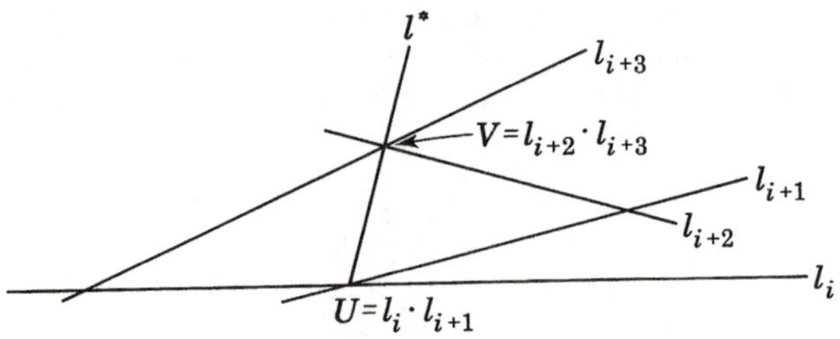

FIG. 4.2

If $n = 1$ or 2, the above would already be sufficiently simple; we suppose then that $n \geq 3$. Now suppose that four successive lines occurring in (1) are such that no three are concurrent: suppose these are l_i, l_{i+1}, l_{i+2}, l_{i+3} (Fig. 4.2). Also suppose each line has at least 4 points. Let l^* be the line on $l_i \cdot l_{i+1}$ and $l_{i+2} \cdot l_{i+3}$. By assumption, no three of the lines l_i, l_{i+1}, l_{i+2}, l_{i+3} are concurrent, and it follows that l^* differs from each of these lines. Now to get one main idea of the proof (leaving aside an exceptional possibility for the moment), we insert l^* into (1) as follows:

$$(2) \quad l = l_0 \overset{P_1}{\underset{\wedge}{-}} l_1 \overset{P_2}{\underset{\wedge}{-}} \cdots \overset{P_i}{\underset{\wedge}{-}} l_i \overset{P_{i+1}}{\underset{\wedge}{-}} l_{i+1} \overset{P_{i+2}}{\underset{\wedge}{-}}$$

$$l^* \overset{P_{i+2}}{\underset{\wedge}{-}} l_{i+2} \overset{P_{i+3}}{\underset{\wedge}{-}} l_{i+3} \overset{P_{i+4}}{\underset{\wedge}{-}} \cdots \overset{P_n}{\underset{\wedge}{-}} l_n = l'.$$

Note that (2) is a way of writing T. Now by the lemma, l_{i+1} and l_{i+2} can be dropped from (2), i.e.,

$$(3) \qquad l = l_0 \frac{P_1}{\wedge} l_1 \frac{}{\wedge} \cdots \frac{P_i}{\wedge} l_i \frac{}{\wedge} l^* \frac{}{\wedge} l_{i+3} \frac{P_{i+4}}{\wedge} \cdots \frac{P_n}{\wedge} l_n = l'$$

is also a way of writing T. Since we have inserted one line, but removed two, we have written T with one less perspectivity. This is the main point in the proof of the following theorem.

THEOREM. (*Assuming Desargues' Theorem.*) *Any projectivity from a line l to a line $l' \neq l$ can be written as a product of two perspectivities.*

There are two gaps in the above proof. First, there is the possibility that l^* lies on P_{i+2}, so that we do not have perspectivities as indicated. To overcome this difficulty (supposing P_{i+2} on l^*), we change l_{i+1} as follows: Let l_{i+1}' be a line through $l_{i+1} \cdot l_{i+2}$, $l_{i+1}' \neq l_{i+1}$, $\neq l_{i+2}$, and let l_{i+1}' be not on P_{i+1}. In the part of (2) that reads $l_i \frac{P_{i+1}}{\wedge} l_{i+1} \frac{P_{i+2}}{\wedge} l_{i+2}$ insert l_{i+1}' to obtain $l_i \frac{P_{i+1}}{\wedge} l_{i+1}' \frac{P_{i+1}}{\wedge} l_{i+1} \frac{P_{i+2}}{\wedge} l_{i+2}$; and then apply the previous lemma to remove l_{i+1}: in this way we change l_{i+1} to l_{i+1}'. At the same time we get a new P_{i+2} and a new l^*. Why does not the new P_{i+2} lie on the new l^*? To answer this, let $U = l_i \cdot l_{i+1}$, $V = l_{i+2} \cdot l_{i+3}$; then $U \to V$ under the given projectivity from l_i to l_{i+3}. Changing l_{i+1} changes U but not V, so the new U does not go into V.

The other thing that remains to be proved is that there are, or may be assumed to be, four lines in the position mentioned. Let us then look at l, l_1, l_2, l_3. It may well be that $l_3 = l$, but we will now show that T can be written with $l_3 \neq l$. First, if in (1) we have $n = 3$, then $l_3 = l'$ and we were given $l' \neq l$, so $l_3 \neq l$: suppose, then, that $n > 3$ and that $l_3 = l$. We may, of course, assume $l_2 \neq l_3$; were $l_2 = l_3$, then $l_2 \barwedge l_3$ would be the identity, and we could drop l_2, thus shortening the product (1). Let now l^* be a line on $l_2 \cdot l_3$, $l^* \neq l_2$, $l^* \neq l_3$, l^* not on P_4. Then $l_2 \frac{P_3}{\wedge} l_3 \frac{P_4}{\wedge} l_4$ can be rewritten $l_2 \frac{P_3}{\wedge} l_3 \frac{P_4}{\wedge} l^* \frac{P_4}{\wedge} l_4$, and then, by the lemma, rewritten as $l_2 \barwedge l^* \barwedge l_4$: thus l^* replaces l_3, and since $l_3 = l$, we have $l^* \neq l$. In short, the *new* l_3 (i.e., l^*) is different from l. We now proceed under the assumption $l_3 \neq l$. We already said that, for obvious reasons, we assume $l_2 \neq l_3$; also we assume $l \neq l_1$, $l_1 \neq l_2$, as otherwise we obviously can simplify (1). It is quite possible, however, that $l_1 = l_3$: but if that is so, then just as previously we changed l_3, so here we can change l_1. Thus we may assume $l_1 \neq l_3$. We may then still have $l_2 = l$, in which case we change l_2. So now we may assume that $l \neq l_1$, $l \neq l_2$, $l \neq l_3$, $l_1 \neq l_2$, $l_1 \neq l_3$, $l_2 \neq l_3$, i.e., l, l_1, l_2, l_3 are distinct. It remains to be seen why we may assume no three concurrent. Were l, l_1, l_2 concurrent, then by the lemma, we could drop l_1: similarly if l_1, l_2, l_3 are concurrent. So suppose l, l_1, l_2 not concurrent, and also l_1, l_2, l_3 not concurrent. It is still possible, however, that l, l_2, l_3 may be concurrent, or that l, l_1, l_3 be concurrent. Both cases are similar; consider then the case that l, l_2, l_3 are concurrent, in which case l, l_1, l_3 are not concurrent (Fig. 4.3). Let then l^* be a line on $l_1 \cdot l_2$, $l^* \neq l_1$, $l^* \neq l_2$, l^* not on P_3. We can write $l \underset{\wedge}{\overset{P_1}{-}} l_1 \underset{\wedge}{\overset{P_2}{-}} l_2 \underset{\wedge}{\overset{P_3}{-}} l_3$ as $l \underset{\wedge}{\overset{P_1}{-}} l_1 \underset{\wedge}{\overset{P_2}{-}} l_2 \underset{\wedge}{\overset{P_3}{-}} l^* \underset{\wedge}{\overset{P_3}{-}} l_3$, and drop l_2 by the lemma, so that we have $l \,\overline{\wedge}\, l_1 \,\overline{\wedge}\, l^* \,\overline{\wedge}\, l_3$. Now of

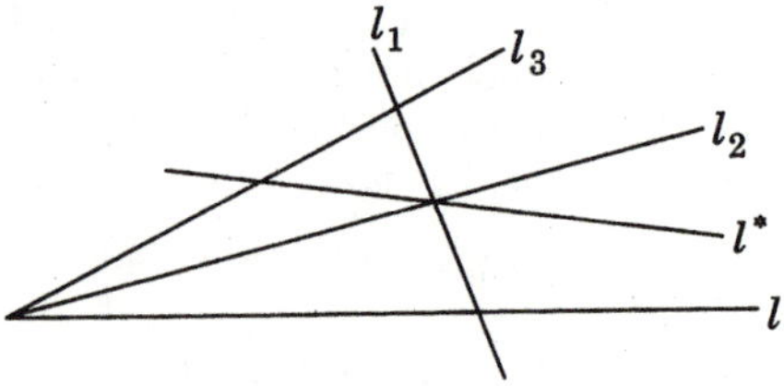

FIG. 4.3

these four lines no three are concurrent, so by the argument given above, (1) can be rewritten with fewer perspectivities. This concludes the proof.

Exercise. In the above proof it was tacitly assumed that on every line there are at least four points. Where? Was it assumed that on each line there are at least five points? Establish the theorem by a separate argument in the 7-point and 13-point planes.

THEOREM. *Desargues' Theorem and Pappus' Theorem together imply Theorem A.*

PROOF. Theorem A, recall, says that if T is a projectivity from a line l to a line $l' \neq l$, which leaves $l \cdot l'$ invariant, then T is a perspectivity. Let T be such a projectivity. By the last theorem, T can be written as a product of two perspectivities, say $T: l \underset{\wedge}{\overset{P_1}{-}} m \underset{\wedge}{\overset{P_2}{-}} l'$. (Here we have used Desargues' Theorem.) We now consider two cases: (i) m is concurrent with $l \cdot l'$, (ii) m is not concurrent with $l \cdot l'$. In the first case, by the lemma above, T is a perspectivity. Now for case (ii) (Fig. 4.4), let $O = l \cdot l'$; $U = l \cdot m$; $V = l' \cdot m$. Let $W = P_1 O \cdot m$. Since under the perspectivity $l \underset{\wedge}{\overset{P_1}{-}} m$, $O \to W$, and since $l \underset{\wedge}{\overset{P_1}{-}} m \underset{\wedge}{\overset{P_2}{-}} l'$ sends O into O, $m \underset{\wedge}{\overset{P_2}{-}} l'$ sends $W \to O$. This implies that P_1, O, P_2 are collinear. Let now X_1 be a (variable) point on l; and let $X_1 \rightarrow X_2 \rightarrow X_3$ under the perspectivities in question. Then T sends X_1 into X_3 and we have to see that X_1X_3 is on a fixed point, i.e., a point that is the same for all

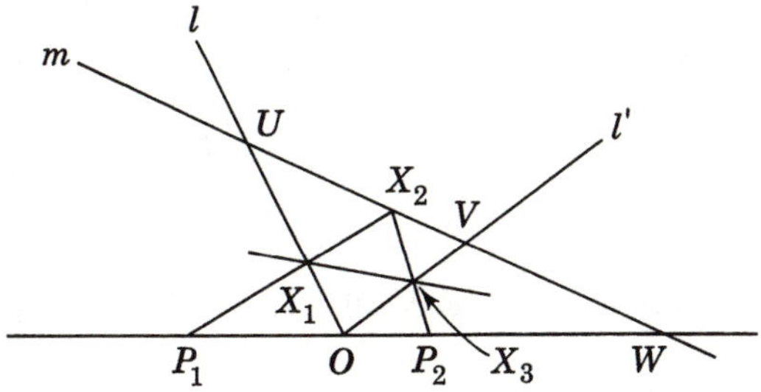

FIG. 4.4

positions of X_1 (and X_3). Let $Q = P_1V \cdot P_2U$. Then by Pappus' Theorem, X_1X_3 is on Q. Since Q is obviously fixed, the proof is complete.

We can formulate part of the proof as a separate theorem. We will call it a lemma.

LEMMA. *If* $l \overset{P_1}{\underset{\wedge}{-}} m \overset{P_2}{\underset{\wedge}{-}} l'$ *is a projectivity with* l, m, l' *not concurrent and if the projectivity leaves* $l \cdot l'$ *invariant, then the projectivity is a perspectivity: this depends only on Pappus' Theorem.*

The proof of the lemma is contained in the proof above. Note that the lemma is a special case of Theorem A. The point is that this

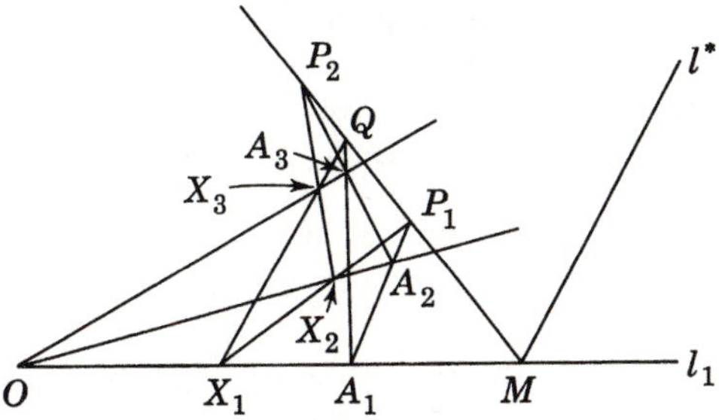

FIG. 4.5

special case depends only on Pappus' Theorem. Using it, we will prove the following.

THEOREM. *Pappus' Theorem implies Desargues' Theorem.*

PROOF. Look at Fig. 4.5. We are given triangles $A_1A_2A_3$, $X_1X_2X_3$

centrally perspective from O, and wish to prove P_1, Q, P_2 are collinear. We have to consider separately the possibilities (i) P_2P_1 passes through O, (ii) P_2P_1 does not pass through O. We first consider case (ii) and then come back to (i). Through $M = P_1P_2 \cdot l_1$ take a line l^*, $l^* \neq l_1$, $l^* \neq P_1P_2$, l^* not on P_2. Now T: $l_1 \overset{P_1}{\underset{\wedge}{-}} l_2 \overset{P_2}{\underset{\wedge}{-}} l_3$ can be written as $l_1 \overset{P_1}{\underset{\wedge}{-}} l_2 \overset{P_2}{\underset{\wedge}{-}} l^* \overset{P_2}{\underset{\wedge}{-}} l_3$. By the last lemma, $l_1 \overset{P_1}{\underset{\wedge}{-}} l_2 \overset{P_2}{\underset{\wedge}{-}} l^*$ is a perspectivity. So T can be written $l_1 \underset{\wedge}{-} l^* \overset{P_2}{\underset{\wedge}{-}} l_3$. Now T leaves $O = l_1 \cdot l_3$ invariant: this is given. So, again applying the last lemma, T is a perspectivity. This proves case (ii).

Now consider case (i). Observe that the points P_1, P_2, Q all play similar roles in the statement of Desargues' Theorem. If P_1P_2 goes through O, we can try P_1Q. If P_1Q does not pass through O, then we are through by case (ii). So suppose P_1Q also passes through O. But then certainly P_1, P_2, Q are collinear, and the proof is complete.

Since Pappus' Theorem implies Desargues' Theorem, we also have that *Pappus' Theorem implies Theorem A*.

We are now going to prove that Theorem A implies the Fundamental Theorem. We have already seen that the Fundamental Theorem implies Theorem A: in fact, Theorem A is a special case of the Fundamental Theorem.

THEOREM. *Any special case of the Fundamental Theorem implies the Fundamental Theorem.*

By a special case, we mean: given *some* lines l, l' with A, B, C distinct points on l, and A', B', C' distinct points on l', there is just one projectivity sending A, B, C, respectively, into A', B', C'. We have to prove that a similar situation obtains for *any* lines $\bar{l}$, $\bar{l}'$, points $\bar{A}$, $\bar{B}$, $\bar{C}$ on $\bar{l}$ and $\bar{A}'$, $\bar{B}'$, $\bar{C}'$ on $\bar{l}'$. Suppose then that there are lines $\bar{l}$, $\bar{l}'$, with $\bar{A}$, $\bar{B}$, $\bar{C}$ (distinct) on $\bar{l}$, $\bar{A}'$, $\bar{B}'$, $\bar{C}'$ on $\bar{l}'$, and two projectivities sending $\bar{A}$, $\bar{B}$, $\bar{C}$, respectively, into $\bar{A}'$. $\bar{B}'$, $\bar{C}'$. We will show that there also exist two projectivities sending A, B, C, respectively, into A', B', C' (Fig. 4.6).

To say that there are two projectivities V_1, V_2 sending $\bar{A}$, $\bar{B}$, $\bar{C}$ into $\bar{A}'$, $\bar{B}'$, $\bar{C}'$, respectively, is to say that for some $\bar{X}$, V_1 sends $\bar{X}$ into $\bar{X}_1'$, and V_2 sends $\bar{X}$ into $\bar{X}_2'$, where $\bar{X}_2' \neq \bar{X}_1'$ (and also, of course, that V_1, V_2 send $\bar{A}$, $\bar{B}$, $\bar{C}$, respectively, into $\bar{A}'$, $\bar{B}'$, $\bar{C}'$). Let now T be any projectivity sending A, B, C, respectively, into $\bar{A}$, $\bar{B}$, $\bar{C}$, and let U be any projectivity sending $\bar{A}'$, $\bar{B}'$, $\bar{C}'$, respectively, into A', B', C'. Let T^{-1} be the projectivity that carries points of $\bar{l}$ back into the points of l from which they came under T. If T^{-1} sends $\bar{X}$ into X, then T sends

X into $\bar{X}$. Note that if U sends $\bar{X}_1'$ into X_1' and $\bar{X}_2'$ into X_2', then $X_1' \neq X_2'$, since a projectivity is one to one. Then one sees that $T\ V_1\ U$, $T\ V_2\ U$ are projectivities sending A, B, C, respectively, into A', B', C'; but $T\ V_1\ U$ sends X into X_1' and $T\ V_2\ U$ sends X into X_2'. So $T\ V_1\ U \neq T\ V_2\ U$, which is a contradiction.

COROLLARY. *Theorem A implies the Fundamental Theorem.*

PROOF. Let l, l' be two distinct lines. Let $A = l \cdot l'$, and let B, C be two further points on l; B', C' two further points on l'. How many

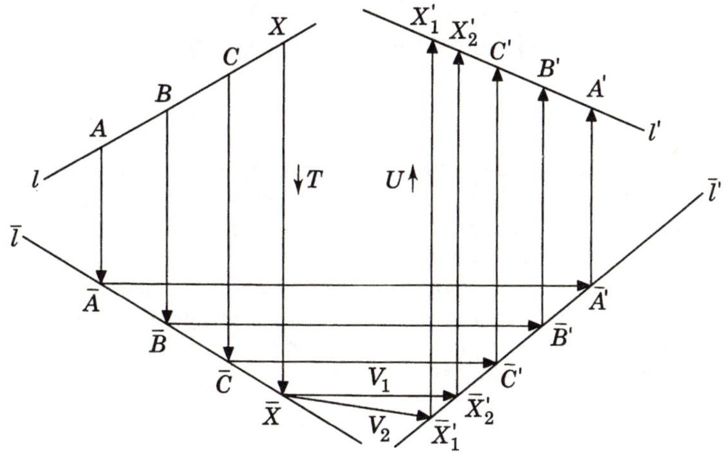

FIG. 4.6

projectivities are there sending A, B, C, respectively, into A', B', C'? If just one, then the corollary follows at once. Now any projectivity sending A, B, C, respectively, into A', B', C' must, by Theorem A, be a perspectivity. So now the question is: How many perspectivities can send A, B, C, respectively, into A', B', C'? Clearly only one, namely, the perspectivity of center $BC \cdot B'C'$. This completes the proof.

Previously we found that Pappus' Theorem implies Theorem A, and now we know that Pappus' Theorem implies the Fundamental Theorem. Also we know that the Fundamental Theorem implies Pappus' Theorem. Therefore we have:

Pappus' Theorem and the Fundamental Theorem are equivalent.

Exercise. Which of the basic theorems imply, and which are implied by, the statement that the plane is a number plane over a commutative field? Over a not necessarily commutative field?

CHAPTER V

AXIOMATIC INTRODUCTION OF HIGHER–DIMENSIONAL SPACE

1. Higher-dimensional, especially 3-dimensional projective space. So far we have considered only the plane, but now we want to set up axioms for space; i.e., we again take undefined terms, in fact, *point*, *line*, and *on*, as before, but now want to put down axioms that will give us an object akin to ordinary space. In ordinary space, a pair of lines may not meet for either of two reasons: one, they are parallel, or two, they are skew, that is, non-coplanar. In axiomatizing the plane, we wanted to get around this possibility of parallelism and put down as an axiom that on any two lines there should be at least one point. But if now we want to study higher-dimensional space, we will not want to retain that axiom, not because we want to get back the concept of parallelism, but because of the possibility of skew lines. The above indicates roughly the direction in which we want to go, but now let us put down exactly the axioms.

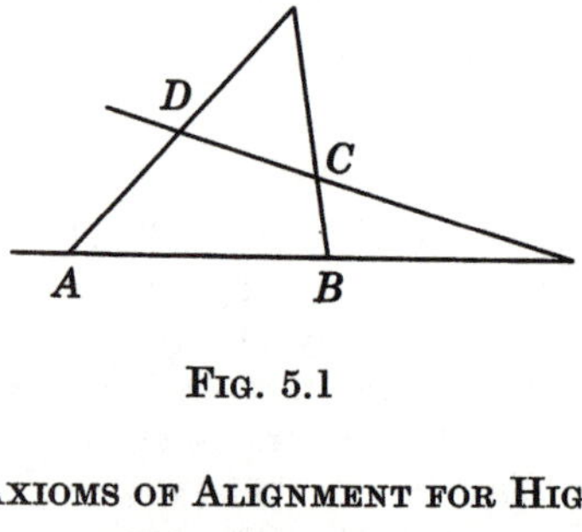

Fig. 5.1

Axioms of Alignment for Higher-Dimensional Projective Space.

0. *There exists at least one point and at least one line.*
1. *On any two points there is at least one line.*
2. *On any two points there is at most one line.*
3. Pasch's Axiom. *If A, B, C, D are distinct points and the lines on AB and CD are on a common point, then also the lines AD and BC are on a common point* (Fig. 5.1).
4. *There are at least three points on every line.*
5. *Not all points are on the same line.*

Higher-dimensional space is thus what we have called a system of type Σ, i.e., it is defined in terms of the undefined terms *point, line, on*. Let $\mathscr{S}$ be a system (of type Σ) satisfying the above axioms. Let $\mathscr{S}'$ be another system (of type Σ) constructed as follows: the points of $\mathscr{S}'$ shall be the same as those of $\mathscr{S}$; a line l' of $\mathscr{S}'$ shall consist of the set of points on a given line l of $\mathscr{S}$; incidence in $\mathscr{S}'$ is defined by membership. Then $\mathscr{S}$ and $\mathscr{S}'$ are isomorphic. Hence we may as well assume that a line consists of the points on it. In our previous work in the plane, it was better to put *point* and *line* on the same footing because of a duality between them that exists in the plane. This duality between point and line is lost in higher-dimensional spaces, and we shall find it convenient to regard a line as simply the set of points on it.

Before, the object defined by our axioms was a projective plane by definition. But now within our geometry we want to *define* objects

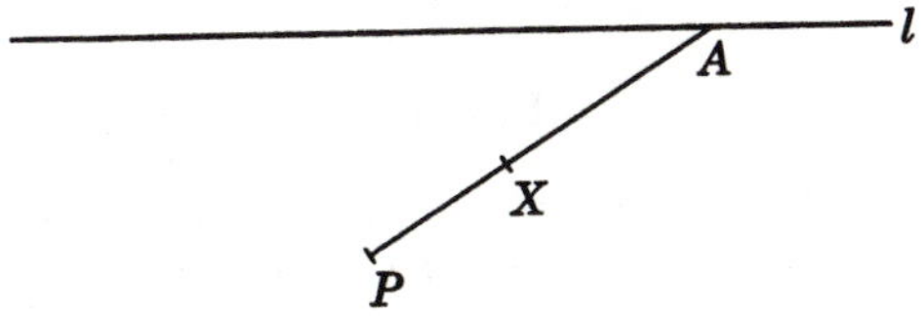

FIG. 5.2

that shall have the properties of a projective plane as previously axiomatized.

DEFINITION. Let l be a line, P a point not on l (Fig. 5.2). We say *the point X is in the plane* (P, l) if there is a point A on l such that X is on the line PA. (The set of points X satisfying this condition constitutes, by definition, the plane (P, l)).

The plane (P, l) consists of all the points on all the lines PA as A varies over l. Note in particular that P is a point of (P, l), as is every point of l.

THEOREM. *If A and B are two points of the plane* (P, l), *then every point on the line AB is a point of* (P, l). (*We will then say that line AB is in, or on, plane* (P, l).)

PROOF. Not both A and B are the same as P, since A and B are distinct, say $P \neq A$ (Fig. 5.3). Since A is in (P, l), there is a point A' on l such that P, A, A' are collinear. Every point of PA' (which is the same as PA) is in the plane (P, l), by definition. If B is on PA', in particular if $B = P$, then BA and PA' are the same line, and so every point on BA is in the plane (P, l). Consider then the case that B is not on PA'. By definition, we know there is a point B' on l such that P, B, B' are collinear. If $BA = l$, then every point of BA is in (P, l), which is what we want to prove; so we may suppose $BA \neq l$. We want to assert that AB and l meet. Were A, B, A', B' distinct, we could apply Pasch's axiom—BB' and AA' meet, hence so do AB and $A'B'$. Now $B \neq A$, by hypothesis; and $B \neq A'$, as we are supposing P, A, B

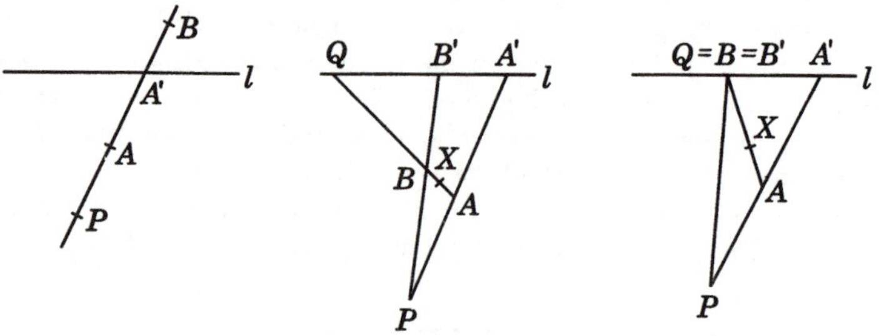

FIG. 5.3

not collinear; $B = B'$ is possible, but then certainly BA and l meet, namely in B'. Similarly we can dispose of any case in which A, B, A', B' are not all distinct, and in every case can assert that AB and l meet, say they meet in Q. We have $A' \neq B'$, as P, A, B are not collinear. Therefore, $Q \neq A'$ or $Q \neq B'$; say, $Q \neq A'$. Let X be a point on AB; if $X = A$, B, or Q, then X is in (P, l); suppose then that $X \neq A$, B, or Q. Then $X \neq A'$ as X is not on l. Also $P \neq X$, since P, A, B are not collinear; $P \neq Q$, $P \neq A'$, since P is not on l. So P, X, Q, A' are distinct. Now QX and PA' meet, so PX and QA' meet, say in X'. Thus X is collinear with P and a point of l; hence X is in (P, l). This completes the proof.

In the course of the proof, we have also established the following:

COROLLARY. *If the line m is in the plane (P, l), then m and l meet.*

We now want to see that the plane (P, l) as here defined is a projective plane in the sense previously considered, i.e., we should check our (previous) axioms of alignment for (P, l). (For the moment we are not concerned with Desargues' Theorem; we shall come back to that.) We have six axioms to check, but five of these, Axioms 0, 1, 2, 4, 5, are easily checked. It remains to check our previous Axiom 3.

Exercises

1. Axioms 0, 1, 2, 4, 5 for the projective plane are word for word the same as our present Axioms 0, 1, 2, 4, 5, which are given. Is it then necessary to check them in establishing that (P, l) is a plane?

2. Check Axioms 0, 1, 2, 4, 5 for (P, l).

THEOREM. *If m, m' are two lines in a plane (P, l), then there is a point on m and on m'.*

PROOF. If $m = l$ or $m' = l$, then m and m' meet by the last corollary. We may now suppose that $m \neq l$, $m' \neq l$. By the last corollary, m meets l, say in A, m' meets l, say in A' (Fig. 5.4). On m take another point $B(\neq A)$ and on m' a point $B'(\neq A')$. If $A = A'$ or $A = B'$ or $A' = B$ or $B' = B$, then certainly the lines m and m' meet. Therefore we may

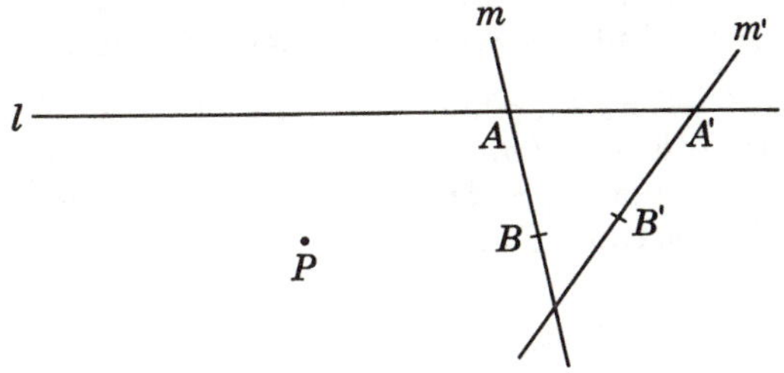

FIG. 5.4

suppose A, A', B, B' all distinct. Now AA' and BB' meet by the last corollary; so AB and $A'B'$ meet by Pasch's Axiom. Q.E.D.

From the above definition, it might seem that in a given plane a certain line and a certain point are specially important, or in some sense preferred. We want to show that this is not so. This is the object of the following theorem.

THEOREM. *If π is a plane (defined by P and l, say), and P' is a point of π, l' a line of π, and P' not on l', then $(P', l') = \pi$.*

PROOF. Let us designate (P', l') by π'. We have $\pi = (P, l)$. We want to prove $\pi = \pi'$. We will show that every point on π' is on π and every point on π is on π': this is the precise content of the statement $\pi = \pi'$. Let then A' be a point of π'. By definition there is a point Q'' on l' such that P', Q', Q'' are collinear. Now P' is in π (given), and l', and every point on it, in particular Q'', is in π (given): so every point on $P'Q''$ is in π, by the next to the last theorem; in particular Q' is in π. This proves that every point of π' is on π. Now let Q be a point of π. To see that Q is in π'; this is certainly the case if $Q = P'$, so suppose that $Q \neq P'$. Then the lines $P'Q$ and l' meet, since they both are in the plane π. But to say that $P'Q$ and l' meet is to say that Q is in (P', l'). This proves that every point of π is in π' and completes the proof.

COROLLARY. *There is one and only one plane containing three given noncollinear points.*

Our axioms so far do not exclude the possibility that our whole geometry consists of just one plane. We add an axiom definitely assuring us that we get something new.

AXIOM 6. *Not all points are in the same plane.*

(An entirely equivalent axiom is: Not every pair of lines meet. This last form has the advantage that it could immediately have been given with the first six axioms: the previous form required an intervening discussion. But there is no essential difference.)

DEFINITION. Let π be a plane, P a point not in π. By *the 3-space* (P, π) we mean the set of all points lying on lines on P and points of π (i.e., X is a point of (P, π) if and only if there is a point X' on π such that P, X, X' are collinear).

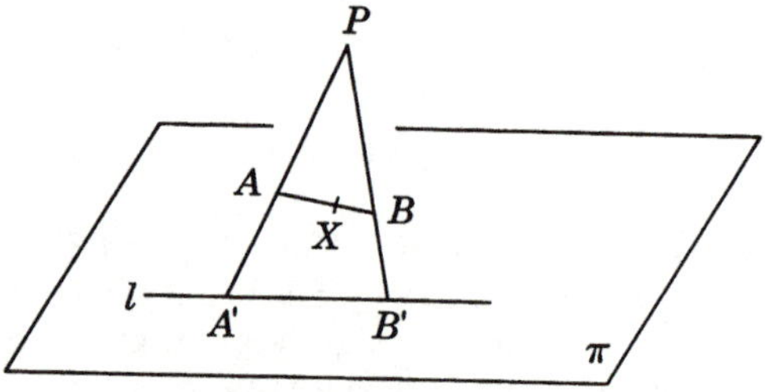

FIG. 5.5

THEOREM. *If A and B are two points in a 3-space (π, P), then every point on line AB is in the 3-space (π, P).*

PROOF. Either $A \neq P$ or $B \neq P$. Say, $A \neq P$. There is a point A' in π such that P. A, A' are collinear. Every point on this line is in the 3-space (π, P), by definition. If B is on this line, then AB coincides with this line, and there is nothing more to prove. We suppose then that P, A, B are not collinear. There is a point B' in π such that P, B, B' are collinear (Fig. 5.5). We have $B' \neq A'$, as we are supposing P, A, B noncollinear. Let $l = A'B'$. Then line AB is in plane (l, P), so if X is any point on AB, the lines PX and l meet, say in X'. Then X' is in π, and P, X, X' are collinear, so X is in the 3-space (π, P). Q.E.D.

THEOREM. *If A, B, C are three noncollinear points on the 3-space (π, P), then every point of the plane ABC is in the 3-space (π, P).*

PROOF. Let X be a point in the plane A, B, C. We may suppose $X \neq A$. Then AX and BC meet, say in Y. Now Y is in (π, P) by the previous theorem, and now one can assert that X is in (π, P), again by the previous theorem.

Just as no line or point is preferred in a plane, so in a 3-space there is no preferred plane or point, in spite of the definition, which appears to give such a preference. The following theorem shows this.

THEOREM. *Let V be a 3-space (defined by π and P, say), and P' a point of V, π' a plane of V, and P' not on π', then $(\pi', P') = V$.*

PROOF. Let us designate (π', P') by V'. Let Q' be a point of V'. Then there is a point Q'' on π' such that P', Q', Q'' are collinear. Now P', Q'' are in V, so Q' is in V by a previous theorem. This shows that every point of V' is in V. It does not follow "similarly," that every point of V is in V', i.e., V, V' do not enter into the argument symmetrically. If we knew that P and π were in V', then we *could* say, similarly, that every point of V is in V'. We proceed, then, to prove this, but first will consider two special cases: (a) $\pi = \pi'$, (b) $P = P'$.

Case (a). Here we are given $\pi = \pi'$, and would have nothing to prove if $P = P'$; we suppose, then, that $P \neq P'$. We want to show that P is in $V' = (\pi'; P')$. Since P' is in V, there is a point P'' in π such that P, P', P'' are collinear. But this also shows that P is in (π, P'), which is V'.

Case (b). Here we are given $P = P'$ and will suppose $\pi \neq \pi'$. We want to show that π is in $V' = (\pi', P')$. Let A', B', C' be three non-collinear points in π'. Let A'', B'', C'' be in π; P', A', A'' collinear; P', B', B'' collinear; P', C', C'' collinear: the points A'', B'', C'' exist since π' is in V. Now A'', B'', C'' could not be collinear, since if they were collinear, the plane containing them and P' would contain A', B', C'; this is impossible, since the one and only plane containing A', B', C' is π' and P' is not on π'. Now A'', B'', C'' are in V', so the plane on them, i.e., π, is also in V'.

Consider now any situation. We may, and will, suppose $\pi \neq \pi'$ and $P \neq P'$. It is possible that P is on π' and P' on π, but suppose that is not the case, at first, and we shall come back to this possibility. Say that P is not on π'. Then $(\pi, P) = (\pi', P)$, by case (b), and $(\pi', P) = (\pi', P')$ by case (a), so $(\pi, P) = (\pi', P')$. Similarly if P' is not on π. Finally consider the case that P is on π' and P' on π. The line PP' is not on π, since P is not on π, and it is not on π', since P' is not on π'. So PP' has no point in common with π but P, and no point in common with π' but P'. Let P'' be a point on PP', $P \neq P''$, $P' \neq P''$. Then $(\pi, P) = (\pi, P'') = (\pi', P'') = (\pi', P')$. This completes the proof.

THEOREM. *If π and l are a plane and a line in a 3-space V, then π and l meet.*

PROOF. If l is in π, there is nothing to prove, so suppose l is not in π. Let P be a point of l not in π. We may use π and P to define V, i.e., by the last theorem $V=(\pi, P)$. Let now P' be another point of l, i.e., $P'\neq P$. Since l is in V, so is P'; hence, by definition of (π, P), there is a point P'' in π such that P, P', P'' are collinear. Thus l meets π in P''. Q.E.D.

THEOREM. *If π, π' are distinct planes in a 3-space V, then π and π' meet in a line.*

PROOF. Let P be a point of π' not in π. Then $V=(\pi, P)$. Let A be a point of π', $A\neq P$. Then there exists a point A^* in π such that P, A, A^* are collinear. Let B be a point of π', B not on PA. Then there exists a point B^* in π such that P, B, B^* are collinear. Moreover, $B^*\neq A^*$, since B is not on PA. The line A^*B^* is on both π and π'. Q.E.D.

We may still point out explicitly that if not all points are in the same plane (Axiom 6), then every plane is Desarguesian; the proof is along the lines of one of our proofs at the beginning of the book, but now, of course, could be given in an entirely rigorous manner.

Later we will continue the study of higher-dimensional space. Meanwhile we mention that 4-space may be defined as a system of type Σ consisting of all the points on all the lines joining the points of a 3-space to a point not in the 3-space, together with the lines on the pairs of these points. Similarly 5-space can be defined; and, by induction, n-space.

Exercises

1. Develop the geometry of 4-dimensional space. In particular show that two planes meet in one and only one point if and only if they are not together in some 3-space.

2. *The analytic model.* Let F be a field, not necessarily commutative. Define a *point* to be a $(n+1)$-tuple $(x_0,\cdots, x_n)$, $n\geq 2$, of elements from F together with all of its right multiples, $(x_0\lambda,\cdots, x_n\lambda)$, λ in F, $\lambda\neq 0$; and define a *line* as all right linear combinations $(x_0\lambda+y_0\mu,\cdots, x_n\lambda+y_n\mu)\neq(0,\cdots, 0)$ of two $(n+1)$-tuples $(x_0,\cdots, x_n)$, $(y_0,\cdots, y_n)$, neither of which is a right multiple of the other. Show that with these definitions we get a model of our system. This standard model will be called PF^n (=projective number space over F of dimension n).

2. Desarguesian planes and higher-dimensional space. We have already observed that if space is more than 2-dimensional, i.e., if not all points lie in a plane, then Desargues' theorem holds, whence every plane of space is Desarguesian. Conversely, *every Desarguesian plane is part of a 3-space.* In fact, let π be a Desarguesian plane. Then we can coordinatize π by means of a not necessarily commutative field F. The points of π then become triples (x_0, x_1, x_2), not all $x_i = 0$, together with all multiples $(x_0\lambda, x_1\lambda, x_2\lambda)$, $\lambda \neq 0$. Using a notation introduced in the last exercise, we see that π is isomorphic to PF^2. Now consider the standard analytic model PF^3: this consists of classes of quadruples (x_0, x_1, x_2, x_3) as points, and of lines as previously defined. In PF^3 consider all the points in which the representatives have the fourth entry equal to zero. If one makes $P(x_0, x_1, x_2, 0)$ correspond to $P(x_0, x_1, x_2)$ (and similarly with lines), then one sees that PF^3 contains a plane isomorphic with PF^2 and hence with π. This shows that π is isomorphic with a plane π_1 of PF^3. By a technique into the details of which we will not enter, we remove π_1 from PF^3 putting π in its place: the result is a 3-space containing π as a part.

Later we will resume the study of higher-dimensional spaces, but there is one further point that we can conveniently take up here, and that is the introduction of coordinates. Since in the plane this depended on Desargues' Theorem, and here we have Desargues' Theorem, we should perhaps expect no essential difficulty, and this, in fact, with our previous work, is so. Still there is a technical detail or two worth examining.

It is not difficult to show that any two 3-spaces each of which has just three points per line are isomorphic, hence isomorphic to PF^3, where F is the field of two elements; hence such 3-spaces can be coordinatized. In what follows we suppose that every line contains at least four points.

By an *affine 3-space* one means a system of type Σ obtained from a projective 3-space by deletion of a 2-dimensional subspace: i.e., if S is 3-dimensional and H is a 2-dimensional subspace, an affine 3-space A is obtained by defining the points of A to be those of $S - H$; a line l of A to be a line l' of S, not in H, minus $l' \cdot H$; and P incident with l if and only if P is incident with the (corresponding) l'. Every affine 3-space can (by definition) be obtained in this way.

THEOREM. *Let A be an affine 3-space with each line containing at least three points. Then to A it is possible to adjoin "ideal" points, one per line, to obtain a projective 3-space S in which the adjoined points form a 2-dimensional subspace H of S and such that A can be obtained from S by deletion of H; and this can be done in only one way.*

PROOF. By definition, A is obtained from a 3-space S by deletion of a 2-space H, from which it is clear that the ideal points can be adjoined in at least one way: the main point is to see that this adjunction can be made in only one way.

Let A be obtained from S by deletion of H and from S' by deletion of H'. To each point P of S we will associate a point P' of S', each point of A being associated with itself, in such way that collinear points of S will correspond to collinear points of S', and vice versa; and we will show that this can be done in only one way.

Let P be a point of H, l a line through P, l not in H. To l corresponds a line $\bar{l}$ of A; and to l corresponds a line l' of S'; this line l' cuts H' in a point P'. If collinearity is to be preserved, one must associate P' with P. This gives the uniqueness of our procedure.

Let l_1 be a second line through P (giving rise via $\bar{l}_1$ to l_1' in S'). In the plane of l and l_1 take a point Q not on H, l, or l_1: here we use the fact that there are at least three points on every line of A, hence at least four on every line of S. On Q take two lines cutting l and l_1 in R, S, T, U, none in H. Using Pasch's axiom, one sees that l' and l_1' must meet, in fact, in P'. To P we associate P', and the present argument shows that P' depends only on P, not on a special line chosen through P.

Let now P, Q, R be three collinear points of S. If all of them are in A, then clearly collinearity is preserved. We suppose, then, that just one or all three of P, Q, R are in H. In the first case, we see by definition of P' that P', Q, R are collinear. In the second case, let O be a point of A and in the plane $OPQR$ take a line not through O, P, Q, R, cutting OP in P_1, OQ in Q_1, OR in R_1. Then P', Q', R' lie in H' and in plane $OP_1Q_1R_1$, whence P', Q', R' are collinear. The same argument shows that if P', Q', R' are collinear, then P, Q, R are collinear, hence if P, Q, R are not collinear, then P', Q', R' are not collinear. The proof is now complete.

REMARK. In the case that each line of A has just two points, the preceding theorem is false.

THEOREM. *Every projective 3-space is isomorphic to a* PF^3.

PROOF. Fix four points O, Ω, Ω', Ω'', no three of which are coplanar. Space minus the plane $\Omega\Omega'\Omega''$ will turn out to be an affine 3-space with $O\Omega$, $O\Omega'$, $O\Omega''$ as, respectively, x, y, z axes. On $O\Omega$, $O\Omega'$, $O\Omega''$ we fix three points E, E', E'' as "unit" points. Let the plane of $EE'E''$ meet $\Omega\Omega'\Omega''$ in the line m. On $OE\Omega$ fix coordinates as explained in our work on planes. Then using the pencil of planes through m, map the field of line $OE\Omega$ onto the lines $OE'\Omega'$ and $OE''\Omega''$ (in our work on planes, we mapped the field on $OE\Omega$ onto the field on $OE'\Omega'$ by means of the pencil of lines on $EE' \cdot \Omega\Omega'$). Since corresponding elements of the fields on the three axes are cut out by a single plane on m, the three axes enter symmetrically into our considerations. In analogy with the procedure in the plane case, we identify the fields on the three axes: one and the same field F is imagined as spread out on the three axes; and if C is the field element of a point P on $O\Omega$ (different from Ω), then C is also the field element for the points cut out on $O\Omega'$, $O\Omega''$ by the plane through m and P.

Let a plane through $\Omega'\Omega''$ cut $O\Omega$ in a point with associated field element C; then we give each point of this plane a first coordinate, namely, C. Similarly we give every point of space minus plane $\Omega\Omega'\Omega''$ a second and third coordinate.

Let l be a line not in plane $\Omega\Omega'\Omega''$. We seek conditions on the coordinates x, y, z of a point in order that the point lie on l. The line l cannot meet all three sides of $\Omega\Omega'\Omega''$, say it does not meet $\Omega\Omega'$. From Ω project l onto plane $O\Omega''\Omega'$, getting line l_1; and from Ω' project l onto plane $O\Omega''\Omega$, getting line l_2. The projection of (x, y, z) from Ω onto plane $O\Omega''\Omega'$ is $(0, y, z)$. Taking, as we may, the last two coordinates as coordinates in plane $O\Omega''\Omega'$, we see that equation of l_1 is:

$$(1) \qquad ay+bz+c = 0.$$

Similarly, in $O\Omega''\Omega$ we find as equation of l_2:

$$(2) \qquad dx+ez+f = 0.$$

Since l does not meet $\Omega\Omega'$, line l_1 does not pass through Ω', whence $a \neq 0$; similarly $d \neq 0$. Solving (1) and (2) for y and x (and writing t for z), we see that all the points (x, y, z) of l (not on plane $\Omega\Omega'\Omega''$) are obtained from equations

$$(3) \qquad \begin{aligned} x &= x_0+ut \\ y &= y_0+vt \\ z &= z_0+wt, \quad \text{not all } u, v, w = 0, \end{aligned}$$

by assigning values to t. Moreover, every value of t (i.e., of z) yields a point of l; (namely, fixing z arbitrarily, we solve for y and x from (1) and (2) and make use of the fact that l is the intersection of planes (1) and (2)). Introducing homogeneous coordinates we see that the points of l are given by all right linear combinations $(1, x_0, y_0, z_0)\lambda+(0, u, v, w)\mu$ of $(1, x_0, y_0, z_0)$ and $(0, u, v, w)$ with $\lambda \neq 0$. Conversely, any such set of right linear combinations of $(1, x_0, y_0, z_0)$ and $(0, u, v, w)$ arise from a line. In this way we see that space minus plane $\Omega\Omega'\Omega''$ is isomorphic to a PF^3 minus the plane $x_0=0$ (the locus $x_0=0$ *is* a plane). Applying the preceding theorem, we conclude that every 3-space is isomorphic to a PF^3.

CHAPTER VI

CONICS

1. Study of the conic on the basis of high school geometry. We have now laid a firm axiomatic foundation for projective geometry and want to pursue geometry as a strictly deductive discipline. To do this, each new object introduced must be defined in terms of things previously discussed. We want now to introduce conics, and will, in a little while, define them in terms of projectivities, but if we do this right away, the definition will appear as a mystery. We prefer to start with what is familiar, even if not rigorously conceived, and later make the considerations rigorous. This does not mean that we abandon our axiomatic foundation, but simply that we put it aside temporarily.

We take the circle as our starting point. Since we are interested only in projective properties, however, the projection of a circle Γ from one plane π to another plane π' yields a curve Γ' in π' that should claim as much of our attention as the circle Γ; similarly, the projection from π' to a third plane π'' yields a curve Γ'', and the properties of Γ that interest us are also properties of Γ''; and so on for further projections. We therefore lay down the following definition.

Definition. A *conic* is a figure obtained from a circle by a product of a finite number of projections.

Note that a figure obtained from a conic by projection is also a conic.

Let A be a point on a circle Γ. Every line through A with one exception cuts Γ in exactly one more point. Hence we get the following projective theorem.

Theorem. *Let Γ be a conic, A a point on Γ. Every line through A with one exception cuts Γ in exactly one more point.*

Definition. The exceptional line through A is called the *tangent to Γ at A.*

Let Γ be a circle, and A a point on Γ. Let l, m be lines through A cutting Γ in points B, C. The circle Γ, lines l, m, and points A, B, and

C are in the ordinary Euclidean plane. If this plane is π and B, C are taken on the vanishing line of a projection to a second plane π', then the images l', m' of l, m will pass through A', the image of A, but will not again cut Γ', the image of Γ. This is a type of difficulty we have already encountered at the beginning of the book and we know it is not serious: we can either introduce ideal points in both Γ and Γ' in order to get around the difficulty, or we can simply dismiss such difficulties on the ground that for the time being we are not concerned with rigor.

Let A, B be two points on a circle Γ. We recall the property of a circle that states that angle APB is constant as P varies on Γ: at least this is true if P remains on one of the arcs bounded by A and B; and if P passes to the other arc, then angle APB passes into its supplement. Hence we can say that the sine of angle APB remains constant. We can also allow $P=A$ if by PA we mean, as we shall, the tangent to Γ at A.

Let now A, B, C, D be four points on a circle Γ, and let P be a variable fifth point on Γ. Since

$$R(PA, PB; PC, PD) = \frac{\sin CPA}{\sin CPB} \Big/ \frac{\sin DPA}{\sin DPB},$$

we see that $R(PA, PB; PC, PD)$ remains constant as P varies over the circle. Since cross-ratio is invariant under projection, we have the following theorem.

THEOREM. *Let Γ be a conic, A, B, C, D four points thereon; and let P be a variable fifth point on Γ. Then $R(PA, PB; PC, PD)$ remains constant as P varies over the conic Γ.*

Let Γ be a circle, and A, B, C, D, P five points on Γ. Let

$$R(PA, PB; PC, PD) = k.$$

If Q is on Γ, then we know that $R(QA, QB; QC, QD)=k$. What if Q is not on Γ? Is it then true that $R(QA, QB; QC, QD)\neq k$? The answer is yes. To prove this, suppose otherwise and consider the lines QA, QB, QC, QD (Fig. 6.1). The point Q is on at most one of the sides of the complete quadrangle $ABCD$, so we may suppose (by changing the lettering if necessary) that Q is not on AB or CD. Since from Q one can draw at most two tangents to Γ, at least one of the lines QA, QB, QC, QD is not a tangent; say, QC is not a tangent. Since QC is not a tangent, QC cuts Γ in a second point R; and $R\neq D$ since Q is not on CD. Since R is on Γ we have

$$R(RA, RB; RC, RD) = R(PA, PB; PC, PD) = R(QA, QB; QC, QD).$$

Let $QC \cdot AB = C'$, $QD \cdot AB = D'$, $RD \cdot AB = D''$. Then $R(A, B; C', D') = R(A, B; C', D'')$, whence $D' = D''$. This is a contradiction if Q is not on Γ.

Hence we see that Γ is the locus of points Q such that

$$R(QA, QB; QC, QD) = k.$$

By projection we get the following theorem:

THEOREM. *Let Γ be a conic, A, B, C, D, P, five points on Γ, and let $R(PA, PB; PC, PD) = k$. Then Γ is the locus of points Q such that*

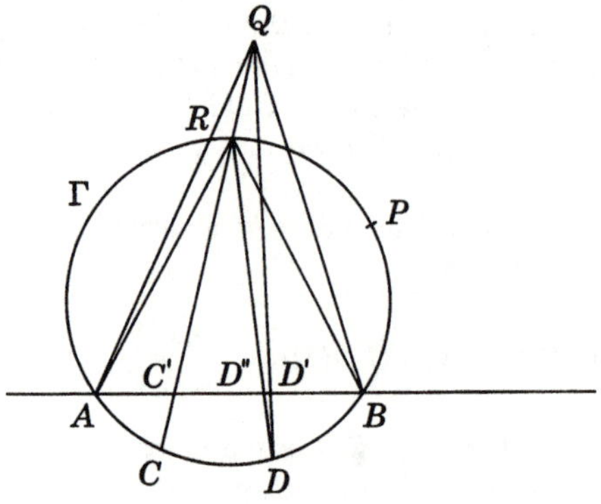

FIG. 6.1

$R(QA, QB; QC, QD) = k$. In other words, the point Q is on Γ if and only if $R(QA, QB; QC, QD) = k$.

COROLLARY. *Given five points A, B, C, D, P (no three of which are collinear), there is at most one conic through them.*

This is true because, if there is a conic Γ through A, B, C, D, P, it consists of the points Q such that

$$R(QA, AB; QC, QD) = R(PA, PB; PC, PD).$$

Incidentally, we do not have to require that no three of A, B, C, D, P are collinear, because if three are collinear, then certainly there is no conic through them.

LEMMA. *If A, B, C, D are four points in the plane π, no three of which are collinear, and similarly for the points A', B', C', D' in the plane π', then we can send π into π' by a sequence of projections so that A, B, C, D go into A', B', C', D', respectively.*

PROOF. Let $M = AB \cdot CD$, $M' = A'B' \cdot C'D'$. By a sequence of projections we can send π into a plane $\pi_1 \neq \pi'$ such that A, B, M go into

A', B', M', respectively. If C, D go into the points C_1, D_1 of π_1, then C_1, D_1, C', D' are coplanar, and a further projection from π_1 to π' will send C_1, D_1 into C', D', respectively, and, of course, leave A', B' fixed.

THEOREM. *Let A, B, C, D, P be five points, no three of which are collinear. Then there is one and only one conic through A, B, C, D, P.*

PROOF. We have seen that there is at most one, and now we have to see that there is at least one. Since no three of A, B, C, D, P are collinear, we may consider the cross-ratio $R(PA, PB; PC, PD)$. Let this number be k. In a plane π take a circle Γ and on Γ a point P'; and through P' draw four lines l_1, l_2, l_3, l_4 such that $R(l_1, l_2; l_3, l_4) = k$. Let these lines cut Γ in A', B', C', D'. By the lemma there is a product of projections sending A, B, C, D, respectively, into A', B', C', D'. We do not claim that P goes into P'; however, P goes into a point $\bar{P}$ such that $R(PA, PB; PC, PD) = R(\bar{P}A', \bar{P}B'; \bar{P}C', \bar{P}D')$. Hence,

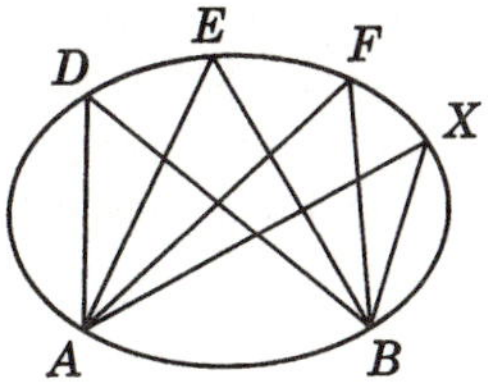

FIG. 6.2

$R(\bar{P}A', \bar{P}B'; \bar{P}C', \bar{P}D') = k$, and $\bar{P}$ is on the circle Γ. Let T be the transformation constructed above: T is a product of projections. Let T^{-1} be the product of these projections taken in the reverse order. Then T^{-1} sends A', B', C', D', $\bar{P}$ into A, B, C, D, P; and it takes the circle through A', B', C', D', $\bar{P}$ into a conic through A, B, C, D, P. This completes the proof.

Exercise. Let A, B, C, D be four points, no three of which are collinear; k, a number, $k \neq 0$, $k \neq 1$. Show that there is a conic through A, B, C, D consisting of the points for which $R(PA, PB; PC, PD) = k$. (This will follow if one can find even one point P such that $R(PA, PB; PC, PD) = k$.)

Let Γ be a conic, and A, B two points thereon. Define a transformation τ from the set of lines on A to the set of lines on B as follows: if l is a line on A, then (with one exception) l intersects Γ in a second point X; let τ associate BX with l, i.e., $\tau: l \to BX$. The one exceptional line l is the tangent at A, and to this we assign line BA.

THEOREM. *τ is a projectivity. τ is not a perspectivity.*

PROOF. Through A take three lines (different from the tangent at A) cutting Γ in points D, E, F, respectively (Fig. 6.2). We know there is one and only one projectivity sending lines AD, AE, AF, respectively, into lines BD, BE, BF; call this projectivity $\bar{\tau}$. (Since τ sends AD, AE, AF, respectively, into lines BD, BE, BF, if we can prove the theorem at all, we must have $\tau = \bar{\tau}$.) Let now l be any other line through A; l cuts Γ in a second point X. We claim $\bar{\tau}: AX \to BX$. To see this, let $\bar{\tau}: AX \to m$ and recall that

$$R(AD, AE; AF, AX) = R(BD, BE; BF, m),$$

since $\bar{\tau}$ is a projectivity and projectivities preserve cross-ratio. But we also have, by a proved property of conics, $R(AD, AE; AF, AX) = R(BD, BE; BF, BX)$. Hence, $m = BX$, i.e., $\bar{\tau}: AX \to BX$; but this is just the definition of τ, namely, $\tau: AX \to BX$. Hence, $\tau = \bar{\tau}$. Since $\bar{\tau}$ is a projectivity by construction, τ is a projectivity. To see that τ is not a perspectivity, recall explicitly the definition of a perspectivity between two pencils: a perspectivity between pencils at A and at B is determined by means of an axis of perspectivity a; and corresponding lines on A and B meet on a, hence the locus of intersections of the corresponding lines is a line. That is not the case with τ, since D, E, F are not collinear; hence, τ is not a perspectivity.

NOTATION. The above theorem determines for any conic Γ and distinct points A, B on Γ a projectivity τ. Designate this projectivity as follows: $\tau = \tau(\Gamma; A, B)$.

THEOREM. *Let A, B be two distinct points, τ a projectivity between the pencils at A and at B, τ not a perspectivity. Then there exists a conic Γ such that $\tau = \tau(\Gamma; A, B)$.*

PROOF. Consider the locus of intersections of corresponding lines on A and B (i.e., lines corresponding under τ). Of these, take three points D, E, F distinct from A, B. Let Γ be the conic through A, B, D, E, F. Let $\tau_1 = \tau(\Gamma; A, B)$. Since τ and τ_1 both send AD, AE, AF, respectively, into BD, BE, BF, we have $\tau = \tau_1$ and $\tau = \tau(\Gamma; A, B)$.

COROLLARY. *The conic Γ of the theorem is uniquely determined.*

The last two theorems will motivate the definition of *conic* in the axiomatic treatment to be taken up in a moment. Before doing that, we still want to prove a famous theorem of Pascal on hexagons inscribed in a conic; later we shall prove the theorem again from the axioms.

PASCAL'S THEOREM. *The intersections of the opposite sides of a hexagon inscribed in a conic are collinear* (Fig. 6.3).

PROOF. Let Γ be the given conic, A, B, C, D, E, F the successive vertices of the given inscribed hexagon. We are to prove that $AB \cdot DE$,

$BC \cdot EF$, $CD \cdot FA$ are collinear. The proof is very much like the proof of Pappus' Theorem from the Fundamental Theorem. Let $\tau = \tau(\Gamma; A, C)$; let σ_1 = perspectivity from range ED to pencil of center A, σ_2 = perspectivity from pencil of center C to range EF. Consider the projectivity $\sigma_1\tau\sigma_2$. Just as in the proof of Pappus' Theorem (p. 26 f.), we prove $\sigma_1\tau\sigma_2$ to be a perspectivity (since it sends E into E), that its

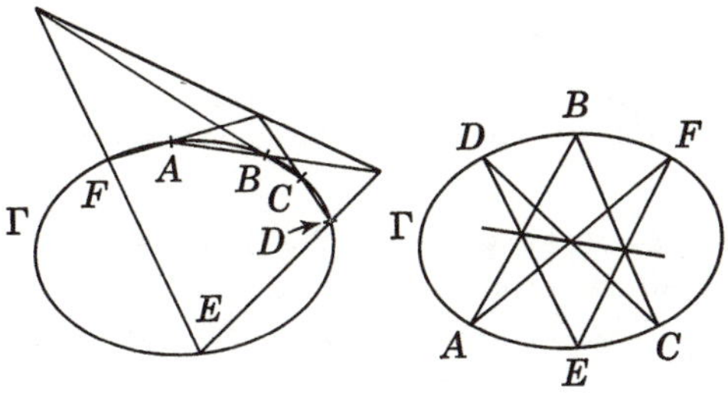

FIG. 6.3

center is $CD \cdot FA$ (since it sends D into $EF \cdot CD$; and $ED \cdot FA$ into F), and that it sends $AB \cdot DE$ into $BC \cdot EF$. Hence, $AB \cdot DE$, $BC \cdot EF$, $CD \cdot FA$ are collinear. Q.E.D.

Pascal called his theorem "the theorem on the Mystic Hexagon." Let us *define a Mystic Hexagon as a hexagon the opposite sides of which meet in three collinear points.* Then Pascal's Theorem could be stated as follows: *A hexagon inscribed in a conic is a Mystic Hexagon.* Pappus' Theorem could be stated as follows: *If three alternate vertices of a hexagon lie on a line and the other three vertices lie on another line, then the hexagon is mystic.*

THEOREM. *Let A, B, C, D, E be five points, no three of which are collinear, and let l be a line on E not on A, B, C or D. Then (with just one exception for l) there is one and only one point F on l such that $ABCDEF$ is a mystic hexagon.*

PROOF. Let $AB \cdot DE = U$, $l \cdot BC = W$, $UW \cdot DC = V$. Then F must be on AV and l, i.e., $F = l \cdot AV$. This fixes F, but one still has to see that $F \neq E$, otherwise one will not have a hexagon. The remainder of the proof is left as an exercise.

Exercise. Show that there is one and only one line l in the above construction such that $F = E$.

THEOREM. *If $ABCDEF$ is a mystic hexagon and no three of the vertices are collinear, then the points A, B, C, D, E, F, lie on a conic.*

PROOF. Let Γ be the conic through A, B, C, D, E. Every line through E, except one—namely, the tangent to Γ at E—cuts Γ in a second point F'; and $ABCDEF'$ is mystic. Therefore the one exceptional line spoken of in the last theorem is the tangent to Γ at E. Since AF is not that exception (since on AF there is a point F such that $ABCDEF$ is mystic), the line AF is not tangent to Γ. Hence AF cuts Γ in a second point F_1, and $ABCDEF_1$ is mystic. By the last theorem $F = F_1$, whence F is on Γ. Q.E.D.

Exercise. Given five points A, B, C, D, E, no three of which are collinear. Show how to construct the tangent at E to the conic that passes through the given points.

2. The conic, axiomatically treated. We now want to define *conic* using only our undefined terms and axioms (and terms and theorems derived from them). The theorems on pages 109 and 110 motivate the following definition.

DEFINITION. Let τ be a projectivity between pencils of centers A, B, $A \neq B$; τ not a perspectivity. Then the locus of intersections of the lines corresponding under τ is called a *conic*. NOTATION: $\Gamma = \Gamma(A, B; \tau)$.

Note that A and B are points of $\Gamma(A, B; \tau)$.

There is something good about this definition, and something bad, too. The good thing about it is that it *is* a definition, i.e., one accessible to our axiom system. (We recall briefly what this system contains; first, the axioms of alignment; second, Pappus' Theorem, from which we can deduce the Fundamental Theorem; projectivities can be defined as soon as we have the axioms of alignment.) The bad thing about the definition is that A and B appear to play a special role relative to the conic Γ, and that is actually not so: we shall see that if C, D are any points of $\Gamma = \Gamma(A, B; \tau)$, then there is also a projectivity σ such that $\Gamma = \Gamma(C, D; \sigma)$. Our first goal is to get this result. Curiously enough, we shall get Pascal's Theorem at the same time.

The transformation τ is considered to be *from* the pencil at A *to* that at B. The same correspondence, but considered as passing from lines through B to lines through A is written τ^{-1} (τ inverse). Clearly $\Gamma(A, B; \tau) = \Gamma(B, A; \tau^{-1})$. Now let C, D, E be three further points on Γ. Since τ is determined by the fact that it sends AC, AD, AE, respectively, into BC, BD, BE, we may also write $\Gamma = \Gamma(A, B; C, D, E)$. The three points C, D, E are not collinear; otherwise τ would be a perspectivity: also one checks immediately that no three of A, B, C, D, E are collinear. Conversely, given any five points A, B, C, D, E, no three of which are collinear, the points C, D, E determine a projectivity between the pencils at A and at B, and this projectivity, together with A, B, determines a conic $\Gamma(A, B; C, D, E)$. Note that in this symbol we can permute the first two letters, and we can permute the last three, to get the same conic; for example, $\Gamma(B, A; D, E, C) = \Gamma(A, B; C, D, E)$. But we do not yet know, for example, that $\Gamma(A, B; C, D, E) = \Gamma(C, A; B, D, E)$.

LEMMA. *Let $\Gamma = \Gamma(A, C; \tau)$ be a given conic. (Observe the strange notation!) Let B, D, E, F be four further points on Γ. Then $ABCDEF$ is a Mystic Hexagon.*

Note that this is a *special* case of Pascal's Theorem—the theorem is for the time being only asserted for inscribed hexagons of which the first and third points are the points used in defining Γ. As for the proof, recall the proof given before: there we used the fact that as a

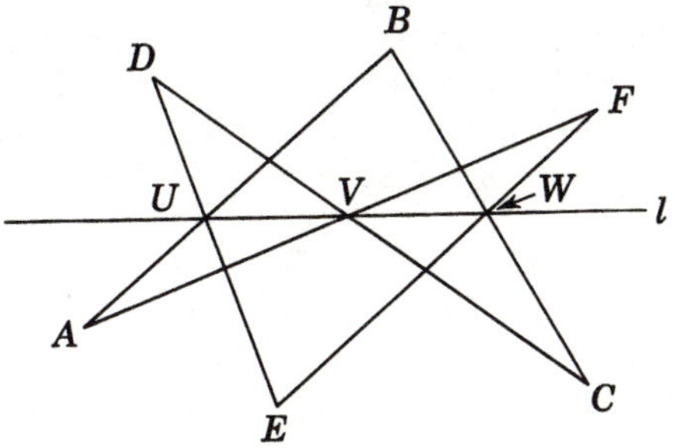

FIG. 6.4

point P varies over the conic, $AP \to CP$ is a projectivity. But here we have this from the definition, so the lemma follows.

Let $\Gamma = \Gamma(A, C; \tau)$ be a conic, B, D, E, three further fixed points. If F is a variable sixth point on Γ, then we have just seen that $ABCDEF$ is mystic. Our next goal is to see that if $ABCDEF$ is mystic, then F is on Γ.

THEOREM. *Let B, D, E be three points on a conic $\Gamma = \Gamma(A, C; \tau)$, A, B, C, D, E distinct. If F is a sixth point such that $ABCDEF$ is mystic, then F is on Γ.*

PROOF. The proof consists in a judicious rearrangement of the steps in the proof of Pascal's Theorem. Let $AB \cdot DE = U$, $CD \cdot FA = V$, $BC \cdot EF = W$ (Fig. 6.4). We are given that U, V, W are collinear. Let σ_2 be the perspectivity $(ED) \overset{V}{\underset{\wedge}{-}} (EF)$. This perspectivity sends U into W. Let σ_1 be the perspectivity from pencil of center A to range on line DE; and let σ_3 be the perspectivity from range on line EF to pencil at C. Then, by definition,

$$\begin{aligned}\sigma_1\sigma_2\sigma_3 : AD &\underset{\sigma_1}{\rightarrow} D \underset{\sigma_2}{\rightarrow} R(= DC \cdot EF) \underset{\sigma_3}{\rightarrow} CD;\\ AB &\rightarrow U \rightarrow W \rightarrow CB;\\ AE &\rightarrow E \rightarrow E \rightarrow CE.\end{aligned}$$

Therefore $\sigma_1\sigma_2\sigma_3$ has the same effect on AD, AB, AE as does τ; by the Fundamental Theorem, then, $\sigma_1\sigma_2\sigma_3 = \tau$. Now by definition $\sigma_1\sigma_2\sigma_3$: $AF \rightarrow S(=AF \cdot DE) \rightarrow F \rightarrow CF$, and therefore τ: $AF \rightarrow CF$; whence F is on Γ. Q.E.D.

By the above theorem, we can characterize $\Gamma(A, C; B, D, E)$ as follows: $\Gamma(A, C; B, D, E)$ = set of points F such that $ABCDEF$ is mystic; plus points A, B, C, D, E. Now the order of points in a hexagon is important; and, for example, hexagon $ABCDEF \neq$ hexagon $BACDEF$. But hexagon $ABCDEF$ = hexagon $EDCBAF$. Hence we can say: $\Gamma(A, C; B, D, E)$ = set of points F such that $EDCBAF$ is mystic; plus points A, B, C, D, E. But by the last theorem: $\Gamma(E, C; D, B, A)$ = set of points F such that $EDCBAF$ is mystic; plus points A, B, C, D, E.

Hence: $\Gamma(A, C; B, D, E) = \Gamma(E, C; D, B, A)$. In words: *One can interchange one of the last three letters in the symbol $\Gamma(A, B; C, D, E)$ with one of the first two to get a symbol for the same conic.* It is true we show only how to interchange the last of C, D, E with the first of A, B; but we know already that the order of C, D, E is immaterial, as is that of A, B. Moreover, *we can interchange both of the first two letters for two of the last three*: we only have to interchange the letters one at a time to get the desired result.

The above result shows that the points A, B used in defining a conic $\Gamma = \Gamma(A, B; \tau)$ are in no way special for Γ: if C, D are any other two points on Γ, we can write $\Gamma = \Gamma(A, B; C, D, E)$ for some E; hence,

$$\Gamma = \Gamma(C, D; A, B, E) = \Gamma(C, D; \sigma)$$

for some projectivity σ (σ not a perspectivity).

We conclude at the same time that *Pascal's Theorem holds quite generally*: the only reason the Lemma did not give us Pascal's Theorem in full generality is that at that stage A and C were special, but now we know that they are not really special, since any two points on Γ may play the roles of A and C.

THEOREM. *Let $\Gamma = \Gamma(A, B; \tau)$ be a conic. Every line through B, with one exception, cuts Γ in exactly one more point; the exceptional line does not cut Γ again. The exceptional line is $\tau(AB)$.*

PROOF. Let m be a line through B, l a line through A such that $\tau(l)=m$. Except for B, the intersection $l \cdot m$ is the only other point on m and Γ. Thus m cuts Γ in exactly two points except in the case that l passes through B. Q.E.D.

COROLLARY. *For any point P on a conic Γ, every line through P, with one exception, cuts Γ in exactly one more point. The exceptional line, by definition called the tangent at P to Γ, cuts Γ only in P.*

Note that the theorem itself could have been proved easily at the beginning of our discussion; not so the corollary.

In the proof of Pascal's Theorem, the main fact used was that the lines AB, AD, AE, AF could be sent projectively into CB, CD, CE, CF, respectively. This fact remains true if we let $B=C$ and if we mean by CB the tangent at B to Γ. Hence we have the following theorem.

THEOREM. *Let $ACDEF$ be a pentagon inscribed in a conic Γ; l, the tangent at C. Then $AC \cdot DE$, $l \cdot EF$, $CD \cdot FA$ are collinear.*

In exactly the same way we can get a theorem by letting ($B=C$ and) $F=A$. Note that the theorems we get this way are just what we might expect if we could use a *limit argument*: i.e., we let B approach C, finally falling on C; the line BC becomes the tangent at C; and the three points of Pascal's Theorem become the three points mentioned in the theorem. [Of course, the argument establishing the theorem (and given above) does not use limits because these have not been introduced axiomatically.] Similarly if F approaches A. Further, we should conjecture and prove a theorem corresponding to letting D approach E: we leave the discussion of this as an exercise.

Exercise. Write down a theorem conjectured from Pascal's Theorem by letting $B=C$, $F=A$, $D=E$. Prove the theorem (from the axioms, i.e., without a limit argument).

THEOREM. *Given five points A, B, C, D, E, no three of which are collinear, there is one and only one conic on them.*

PROOF. $\Gamma(A, B; C, D, E)$ is one such conic. Let $\Gamma=\Gamma(U, V; \tau)$ be a conic through A, B, C, D, E. Then $\Gamma=\Gamma(U, V; A, B, E)=\Gamma(A, B; U, V, E)=\Gamma(A, B; C, D, E)$. Q.E.D.

Exercise. Let A, B, C, D, be 4 points, no three of which are collinear, l a line through none of them. Call a pair of lines a *degenerate conic.* For any point X on l there is one and only one conic through A, B, C, D, X. This cuts l in a second point Y ($Y=X$ if l is tangent to the conic). Prove that $X \rightarrow Y$ is a projectivity.

3. The polar. We start with a motivation from elementary geometry. Let Γ be a circle, O its center, P a point not on Γ. Through P

take a variable secant line l, cutting Γ in R and S (Fig. 6.5). Let X be such that $H(R, S; P, X)$. Then X varies on a (fixed) line.

PROOF. Let PO cut the circle in A and B (we are assuming $P \neq O$) and let G be such that $H(A, B; P, G)$. A, B, P, and hence G, are fixed, as is also the line perpendicular to AB through G; we will show that X is on this line. Use the complete quadrangle $ABRS$ to construct G: let $RB \cdot SA = U$, $AR \cdot BS = T$, so that $TU \cdot AB = G$. Since angle ASB is inscribed in a semicircle, it is a right angle. Therefore AS is an altitude of triangle ABT. Similarly BR is an altitude. Since the altitudes of a triangle meet in a point, TU, which is TG, is the perpendicular through G to AB. Since harmonic sets go by a perspectivity into harmonic sets, if $TG \cdot RS = X$, we have $H(R, S; P, X)$. Hence XG, which is TG, is perpendicular to AB. Q.E.D.

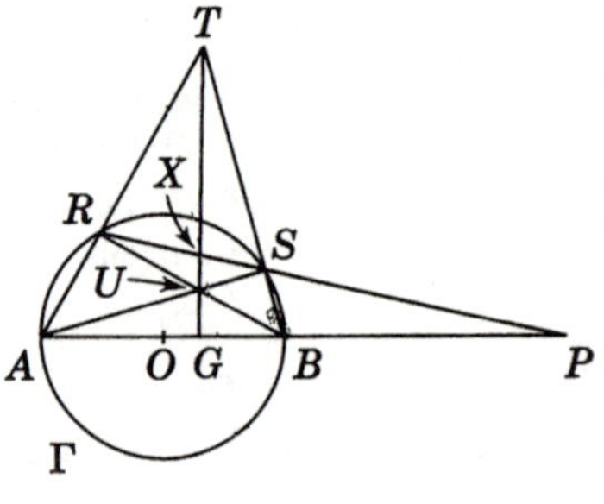

FIG. 6.5

Since *harmonic tetrad* is a projective notion, we have the following theorem for conics (where by a conic we mean a figure obtained by projections from a circle).

THEOREM. *Let Γ be a conic, P a point not on Γ, l a variable secant through P cutting Γ in R and S, X a point such that $H(R, S; P, X)$. Then X varies on a line, called the polar of P with respect to Γ.*

Going back to the circle and Fig. 6.5, we see, first, that from a point P not on Γ one can draw exactly two or no tangents to Γ according as P is outside or inside Γ (i.e., according as PO is greater or less than the radius; the concepts *inside* and *outside* will not occur in the axiomatic treatment). Moreover, if P is outside Γ, then G is inside, and the polar of P cuts Γ in two points, say M and N; then PM (as well as PN) is a tangent to Γ; for if not, and PM cuts Γ in a second point M', then the fourth harmonic of P with respect to M and M' would lie on the polar of P, and as M does also, this polar line would be PM; this implies $P = G$, which is not so. In this argument, we use the fact that $H(A, B; P, G)$ implies $P \neq G$. Finally, if P is inside Γ, then G is outside, and the polar of P does not cut Γ. Replacing the circle by a conic (by projections), and summing up, we have the following.

THEOREM. *If P is a point not on the conic Γ, then from P one can draw either exactly two tangents or no tangents to Γ. In the second case, the* *polar of P does not cut Γ; in the first case, it cuts Γ in two points M and N, and PM and PN are the tangents.*

4. The polar, axiomatically treated. To define *polar*, we will make use of the relation $H(A, B; C, D)$. Most of the properties of this relation that we shall want follow from our considerations on quadrangular sets, for example, the theorem that, given three distinct collinear points A, B, C, there is one and only one point D such that $H(A, B; C, D)$. However, $D \neq C$ does not follow: to see that it does not, just consider our 7-point geometry. Since we shall want $D \neq C$, we introduce an axiom precisely for this purpose.

AXIOM. *The diagonal points of a complete quadrangle are not collinear.*

The theorem that $H(A, B; C, D)$ implies $C \neq D$ is immediate. This theorem is, in fact, obviously equivalent to the axiom.

(In terms of a coordinate system over a field K, this axiom means that $1+1 \neq 0$ in K.)

Exercise. The dual of the above axiom is a true theorem.

THEOREM. *Let Γ be a conic (defined from our axiom system), and let P be a point not on Γ. If from P one can draw one tangent PM to Γ(M on Γ), then one can draw a second tangent from P to Γ.*

PROOF. Let R be a point on Γ, $R \neq M$. If PR is not a secant, then it is a second tangent. So suppose PR cuts Γ again in S. Let Q be such that $H(R, S; P, Q)$. Since $MQ \neq MP$, MQ is a secant, cutting Γ in a second point N. Let τ be the projectivity such that $\Gamma = \Gamma(N, M; \tau)$. On RS consider the transformation $\sigma_1: X \to Y$, where Y is the point such that $H(R, S; X, Y)$ (and $R \to R$, $S \to S$). One sees that σ_1 is a projectivity and sends $R \to R$, $S \to S$, $P \to Q$, and $Q \to P$. Now use τ to define a projectivity $\sigma_2: X \to Y$, where $Y = RS \cdot \tau(NX)$. Then σ_2 sends $R \to R$, $S \to S$, and $Q \to P$. By the Fundamental Theorem, $\sigma_1 = \sigma_2$; hence $\sigma_2: P \to Q$. For τ, this means that $\tau: NP \to MN$, or $\tau^{-1}: MN \to NP$. Hence, NP is tangent to Γ at N. Q.E.D.

THEOREM. *Let Γ be a conic, P a point not on Γ. From P one can draw at most two tangents to Γ.*

PROOF. The proof is a rearrangement of the steps of the last proof. Let PM, PN be two tangents to Γ, M, N on Γ (if two do not exist, then the theorem follows at once). Let R be a point on Γ, $R \neq M$, $R \neq N$. We want to prove that PR is a secant. Let $Q = PR \cdot MN$, and let S be such that $H(R, S; P, Q)$. Then using an argument like that occurring in the last proof, one proves that S is on Γ.

THEOREM. *Let* Γ *be a conic,* P *a point not on* Γ, l *a variable secant through* P *cutting* Γ *in* R *and* S, X *a point such that* $H(R, S; P, X)$. *Then* X *varies on a line, called the polar of* P *with respect to* Γ.

PROOF. Let PAB, with A, B on Γ, be a fixed secant through P. Let G be such that $H(A, B; P, G)$. Let a, b be the tangents at A, B, respectively. Let a, b meet at H. Consider the "hexagon" $AARBBS$. Then $a \cdot b = H$, $AR \cdot BS = U$, $RB \cdot SA = V$ are collinear; moreover UV passes through G. Hence, as secant PRS varies on P, the points U, V vary on the fixed line HG. Let $X = UV \cdot RS$. Then $H(R, S; P, X)$, and X varies on HG. Q.E.D.

DEFINITION. If P is on Γ, *the polar of P with respect to* Γ is the tangent at P to Γ.

THEOREM. *A point* P *is on* Γ *if and only if it is on the polar of* P.

PROOF. This follows immediately from the axiom on the diagonal points of a complete quadrangle.

THEOREM. *If* P *is not on the conic* Γ, *then the polar of* P *is not tangent to* Γ.

PROOF. Suppose the polar p of P touches Γ at Q. Since p is not on P, $p \neq PQ$. Therefore PQ is a secant, cutting Γ in a second point R. Let S be such that $H(Q, R; P, S)$. Then S is on p. This makes $p = QS = PQ$; contradiction.

THEOREM. *Let* Γ *be a conic,* P *a point not on* Γ. *If from* P *two tangents* PM, PN (M, N *on* Γ) *can be drawn, then* MN *is the polar of* P.

PROOF. In a previous theorem, given the tangent PM, we showed how to construct the tangent PN. A reconsideration of that construction gives the present theorem.

THEOREM. *Let* Γ *be a conic,* P *a point not on* Γ. *We know that the polar* p *of* P *either does not cut* Γ, *or cuts it in two points,* M, N. *In the latter case,* PM, PN *are tangents.*

PROOF. Immediate.

THEOREM. *If the polar of* P *passes through* Q, *then the polar of* Q *passes through* P.

PROOF. If P is on the conic Γ, then its polar is the tangent at P; QP is tangent at P; and the polar of Q passes through P. Similarly if Q is on Γ. So suppose neither P nor Q is on Γ. Take a secant PRS, with R, S on Γ. We assume that Q is not on RS, as in that case the proof is immediate. If QR and QS are tangent to Γ, then RS is the polar of Q

and passes through P. Suppose QR or QS, say QR, is not a tangent; QR cuts Γ again in T. We claim that PT is not a tangent; for if it were, then TQ would be the polar of P; but the polar of P does not pass through R, since it passes, rather, through the point X on RS such that $H(R, S; P, X)$. Let PT cut Γ again in U. Let $Q_1 = RT \cdot SU$. Using the quadrangle $RSTU$, we see that the polar of P passes through Q_1. Were $Q_1 \neq Q$, this polar would be QQ_1; not so. Therefore, $Q = Q_1$. Using the quadrangle $RSTU$ again, we see that the polar of $Q(=Q_1)$ passes through P. Q.E.D.

THEOREM. *Every line l is the polar of some point (with respect to a given conic Γ).*

PROOF. Let P, Q be two points on l, p, q their polars, R a point on p and q (actually, $p \neq q$, so $R = p \cdot q$). Then the polar of R passes through P and through Q, hence $PQ = l$ is the polar of R. Q.E.D.

DEFINITION. R is called the *pole of l*. I.e., R is pole of l if l is polar of R.

Exercise. Distinct points have distinct polars.

THEOREM. *Let Γ be a conic, P a point, p its polar. Let X be a variable point on p and consider the transformation $\tau: X \to$ polar of X. Then τ is a projectivity (from the range on p to the pencil on P).*

PROOF. First assume p is not a tangent, or, stated otherwise, P is not on p or Γ. The polar of X passes through P. To construct this polar, take a fixed point Q on Γ, Q not on p. Let QX meet Γ again in Y. (Generally, if Z is such that $H(Q, Y; X, Z)$, then PZ is the required polar.) Since P is not on Γ, $Q \neq P$; let $QP \cdot p = S$; $S \neq Q$, $S \neq P$, since neither P nor Q is on p. Since the polar of P is not on Q, the polar of Q, i.e., the tangent at Q, is not on P; so QP is a secant; let it cut Γ again in R. Then $H(Q, R; S, P)$. Let $YR \cdot p = W$. Then (unless $W = S$)

$$H(WQ, WR; WS, WP)$$

and therefore WP is the polar of X (also if $W = S$). $PW \to RY$ is a perspectivity. $RY \to QY$ is a projectivity; $QY \to X$ is a perspectivity. Hence, $PW \to X$ is a projectivity. This completes the proof for the case that p is not tangent to Γ.

Exercise. Give the proof for the case that p is tangent to Γ.

By the exercise, the proof is complete.

THEOREM. *Let Γ be a conic, a, b, two tangents to Γ (at A and B). Let X be a point on a. Through X (if $X \neq A$) there is a second tangent* *to* Γ; *let this second tangent* (*if* $\neq b$) *cut* b *in* Y. *Then* $X \to Y$ *is a projectivity, not a perspectivity.* (*For* $X = A$, *we define* $Y = a \cdot b$; *and for* $X = a \cdot b$ *we define* $Y = B$.)

PROOF. Let XY touch Γ in Z. Then AZ is the polar of X and BZ is the polar of Y. $X \to AZ$, $Y \to BZ$, and $AZ \to BZ$ are projectivities. Hence, $X \leftrightarrow AZ \to BZ \to Y$ is a projectivity. Since $a \cdot b \to AB \to b \to B \neq a \cdot b$, the projectivity is not a perspectivity. This completes the proof.

Let us dualize our considerations. What will be the dual of a conic? A conic is a set of points; so its dual will be a set of lines. To have a convenient term, let us call the conics so far considered "point-conics"; the dual notion will be that of a "line-conic." As for the definition, let a, b be two lines, τ a projectivity from a to b, τ not a perspectivity. For any point X on a, consider the line on X and $\tau(X)$. This set of lines forms a *line-conic*.

THEOREM. *The set of tangents to a point-conic is a line-conic.*

PROOF. This is a reformulation of the last theorem.

What is the dual of "tangent to a point-conic"? If Γ is a point-conic, a line either cuts Γ in two points, no points, or exactly one point if it is a tangent. Let us speak of "no-cut, one-cut, two-cut lines." Let Γ^* be a line-conic. The dual of a "line cutting Γ" is a "point lying on a line of Γ^*"; let us use the term "subtend" as dual of "cut." We then can speak of "no-subtend, one-subtend, two-subtend points." The dual of "tangent," i.e., "one-cut line" is "one-subtend point." A one-subtend point is a point on which there is just one line of Γ^*.

Let Γ be a point-conic. Its set of one-cut lines is a line-conic Γ^*. Dually, the set of one-subtend points of Γ^* is a point-conic Γ^{**}. It does not follow from duality that $\Gamma^{**} = \Gamma$; but this is in fact so, as one easily sees.

BRIANCHON'S THEOREM. *Let* Γ *be a point-conic,* a, b, c, d, e, f *six tangents thereto. Then the lines* $ab \cdot de$, $bc \cdot ef$, $cd \cdot fa$ *are concurrent.*

PROOF. This theorem sounds a little like the dual of Pascal's, but it could not be the dual: Pascal's Theorem is about a point-conic, so that its dual is about a line-conic, whereas Brianchon's Theorem is not about a line-conic, but about a point-conic. Still, the lines a, b, c, d, e, f are six lines of a line-conic, and applying the dual of Pascal's Theorem we get Brianchon's.

If we define *conic* to mean the set of points of a point-conic *plus* its set of tangents, then *conic* becomes a self-dual concept; and deleting the word *point* from the statement of *Brianchon's Theorem* we do get a theorem of the same content and one that *is the dual of Pascal's.*

5. Polarities. Let π, π' be two planes; each consists of a set of points and a set of lines. An *isomorphism* of π onto π' is a one-to-one mapping of the set of points of π onto the set of points of π' and of the set of lines of π onto the set of lines of π' such that incidence is preserved, i.e., if P is on l in π, then the image of P is on the image of l in π'. (See p. 55.) If $\pi = \pi'$, the isomorphism is called an *automorphism*. Let σ be an automorphism, l any line of π, $l' = \sigma(l)$. Then σ induces a transformation from the range on l to that on l'. If this is a projectivity for each l, then σ is called a *projective collineation*. A *correlation* is much like a projective collineation, the only difference being that it sends points into lines and lines into points. Let σ be a correlation. Let P be a point, $\sigma(P) = l$, $\sigma(l) = Q$. In general, we need not have $Q = P$. Were $\sigma^2 (= \sigma \cdot \sigma)$ the identity transformation, then we would have $P = Q$. A transformation whose square is the identity is called an *involution*. A correlation that is an involution is called a *polarity.*

Let Γ be a conic. Let σ be the transformation of π that sends each point into its polar and each line into its pole. By theorems established previously one sees that σ is a polarity.

In a polarity σ, if X and Y are points and Y lies on $\sigma(X)$, then X lies on $\sigma(Y)$, as one easily proves. In this situation one calls X and Y *conjugate* to each other. A point X is *self-conjugate* if it is conjugate to X, or, what is the same thing, if X is on $\sigma(X)$.

Let Γ be a conic, σ the associated polarity described in the next to the last paragraph. Then clearly Γ *consists of the self-conjugate elements of* σ. Conversely, start with a polarity σ (no conic given): we want, if possible, to associate a conic with σ. It may be that σ has no self-conjugate elements. but if it has at least one, we can show that the *set of self-conjugate elements is a conic* Γ; moreover, for any point X, $\sigma(X)$ will be the polar of X with respect to Γ. Due to lack of space, we will not write out a synthetic proof, though we will later give an analytical proof (see Chapter X, §5). One consequence of the facts just stated, however, is the following. At the very beginning of our discussion on conics, we could have defined the term *polarity*, and then could have defined *conic* as the set of self-conjugate elements in a polarity that has at least one self-conjugate element. The advantage of such a definition is that no point (or line) of the conic appears, from the definition, to be special. The disadvantages are obvious.

CHAPTER VII

HIGHER-DIMENSIONAL SPACES RESUMED

1. Theory of dependence. Preparatory to the further development of higher-dimensional projective spaces, we consider the so-called theory of dependence. This theory, if given axiomatically, has a great number of applications, the application to projective geometry being only one.

We are given a *set* S. Using the set S we build *strings* of elements from S: a string is simply a finite sequence $\langle a_1, \cdots, a_n \rangle$, where the a_i are elements of S. The order of the elements is of importance; for example, if $a_1 \neq a_2$, then the string $\langle a_1, a_2 \rangle$ is different from the string $\langle a_2, a_1 \rangle$. Thus *string* and *set* (or *subset*) do not mean the same thing: the set $\{a_1, a_2\}$ is the same as the set $\{a_2, a_1\}$; two sets are called equal if they contain the same items. Note also that the string $\langle a_1, a_1 \rangle$ is different from the string $\langle a_1, a_1, a_1 \rangle$, but that the sets $\{a_1, a_1\}$ and $\{a_1, a_1, a_1\}$ are the same.

Given an element a of S and a string $\langle a_1, \cdots, a_n \rangle$, we will sometimes say that *a depends on $\langle a_1, \cdots, a_n \rangle$*. The term *depends on* shall be undefined: it is a primitive term. The axioms concern this term.

The axioms are as follows:

I. THE IDENTITY AXIOM. *Given a string $\langle a_1, \cdots, a_n \rangle$, each $a_i (i = 1, \cdots, n)$ depends on $\langle a_1, \cdots, a_n \rangle$.*

II. THE EXCHANGE AXIOM. *If a depends on $\langle a_1, \cdots, a_n \rangle$ but not on $\langle a_1, \cdots, a_{n-1} \rangle$, then a_n depends on $\langle a_1, \cdots, a_{n-1}, a \rangle$.*

III. THE TRANSITIVITY AXIOM. If *c depends on $\langle b_1, \cdots, b_n \rangle$ and each $b_i (i = 1, \cdots, n)$ depends on $\langle a_1, \cdots, a_m \rangle$, then c depends on $\langle a_1, \cdots, a_m \rangle$.*

THEOREM 1. *Let $\langle a_1, \cdots, a_m \rangle$, $\langle b_1, \cdots, b_n \rangle$ be two strings and assume that $\{a_1, \cdots, a_m\} = \{b_1, \cdots, b_n\}$. Then c depends on $\langle a_1, \cdots, a_m \rangle$ if and only if c depends on $\langle b_1, \cdots, b_n \rangle$.*

In other words, whether c depends on $\langle a_1, \cdots, a_m \rangle$ depends solely on the set $\{a_1, \cdots, a_m\}$ and not on the order in which the a_i occur in the string $\langle a_1,\cdots, a_m\rangle$. Because of this, we sometimes speak of c being dependent on the *set* $\{a_1,\cdots, a_m\}$, as this can usually be done without confusion.

PROOF OF THEOREM 1. Assume c is dependent on $\langle a_1,\cdots, a_m\rangle$. Each a_i is dependent on $\langle b_1,\cdots, b_n\rangle$ by Axiom I. Hence by transitivity, c is dependent on $\langle b_1,\cdots, b_n\rangle$. Similarly, if c is dependent on $\langle b_1,\cdots, b_n\rangle$, then it is dependent on $\langle a_1,\cdots, a_m\rangle$.

DEFINITION. A string $\langle a_1,\cdots, a_m\rangle$ is said to be *dependent* if, for some i, a_i is dependent on $\langle a_1,\cdots, a_{i-1}, a_{i+1},\cdots, a_m\rangle$ (or more briefly: if one of the a_i is dependent on the others). A string is called *independent* if it is not dependent.

One also sometimes speaks of a *set* being independent, but here there are good chances for confusion. For example, suppose the string $\langle a_1,\cdots, a_m\rangle$ is independent: then one would also say that the set $\{a_1,\cdots, a_m\}$ is independent. Now $\langle a_1, a_1, a_2,\cdots, a_m\rangle$ is dependent (any string with replicas is dependent, by Axiom I). Presumably then one would say that $\{a_1, a_1, a_2,\cdots, a_m\}$ is dependent. But the set $\{a_1, a_2,\cdots, a_m\}$ is the same as the set $\{a_1, a_1, a_2,\cdots, a_m\}$. Still, with a little caution, one can avoid such ambiguities. (The best way to avoid such ambiguities is to use the word *string* when one means *string*.)

The *empty* string, i.e., the string with no elements, is independent, as one sees from the definition. A string $\langle a\rangle$ with one element a is dependent if and only if a is dependent on the empty string. We need not, in fact should not, puzzle our heads as to what could be meant by an element depending on the empty string, since quite generally what is meant by an element being dependent on a string is undefined.

THEOREM 2. *If $\langle a_1,\cdots, a_{n-1}\rangle$ is independent, but $\langle a_1,\cdots, a_n\rangle$ is not independent, then a_n is dependent on $\langle a_1,\cdots, a_{n-1}\rangle$.*

PROOF. Some a_i are dependent on the others: if one of these is a_n, then the proof is complete. So suppose that one of $a_1,\cdots, a_{n-1}$ is dependent on the other a_i. By Theorem 1, since order is immaterial, we may suppose this to be a_1: notationally, then, a_1 is dependent on $\langle a_2,\cdots, a_{n-1}, a_n\rangle$. Now a_1 is not dependent on $\langle a_2,\cdots, a_{n-1}\rangle$. Hence, by the exchange axiom a_n is dependent on $\langle a_2,\cdots, a_{n-1}, a_1\rangle$. By Theorem 1, then, a_n is dependent on $\langle a_1,\cdots, a_{n-1}\rangle$. Q.E.D.

THEOREM 3. *Given a set $\{a_1,\cdots, a_m\}$, there is a string built from the a_i which is independent and on which each a_i is dependent.*

PROOF. Using the a_i, one can build at least one independent string: for example, the empty string. Consider all such independent strings:

an independent string cannot contain replicas, so every such string has at most m elements; take one such string that is longest. Notationally, say this is $\langle a_1, \cdots, a_s \rangle$. Then each a_i is dependent on $\langle a_1, \cdots, a_s \rangle$. For $\langle a_1, \cdots, a_s, a_i \rangle$ is dependent; hence by Theorem 2, a_i is dependent on $\langle a_1, \cdots, a_s \rangle$. Q.E.D.

DEFINITION. Two strings $\langle a_1, \cdots, a_m \rangle$, $\langle b_1, \cdots, b_n \rangle$ are called *equivalent* if each a_i depends on $\langle b_1, \cdots, b_n \rangle$, and each b_i depends on $\langle a_1, \cdots, a_m \rangle$.

Note that any string is equivalent to itself and that strings equivalent to the same string are equivalent to each other.

THEOREM 4. (THE EXCHANGE THEOREM). *Let $\langle a_1, \cdots, a_r \rangle$ be an independent string, $\langle b_1, \cdots, b_s \rangle$ a string (not necessarily independent), with each a_i dependent on $\langle b_1, \cdots, b_s \rangle$. Then there exists a string $\langle c_1, \cdots, c_s \rangle$ equivalent to $\langle b_1, \cdots, b_s \rangle$ and such that each a_i occurs in the string $\langle c_1, \cdots, c_s \rangle$.*

PROOF. Suppose we have a string $\langle c_1, \cdots, c_s \rangle$ equivalent to $\langle b_1, \cdots, b_s \rangle$ and containing t of the a_i. (Note that the a_i are distinct.) If $t = r$, the proof is complete, so suppose $t < r$. We now show how to get a string equivalent to $\langle c_1, \cdots, c_s \rangle$ and containing one more a_i. Repeating the argument $r - t$ times, the proof will be complete.

The string $\langle c_1, \cdots, c_s \rangle$ contains t of the a_i; since the order of the a_i is immaterial, we may suppose these are $a_1, \cdots, a_t$; and since the order of the c_j is immaterial, we may suppose $c_1 = a_1$, $c_2 = a_2, \cdots, c_t = a_t$. Thus a_{t+1} is not dependent on $\langle c_1, \cdots, c_t \rangle$, but it is dependent on $\langle c_1, \cdots, c_t, \cdots, c_s \rangle$. Consider the strings $\langle c_1, \cdots, c_t \rangle$, $\langle c_1, \cdots, c_{t+1} \rangle$, $\langle c_1, \cdots, c_{t+2} \rangle, \cdots, \langle c_1, \cdots, c_s \rangle$; and take the shortest one on which a_{t+1} depends: say this is $\langle c_1, \cdots, c_t, \cdots, c_{t+n} \rangle$; we know that $n \geq 1$. Then a_{t+1} does not depend on $\langle c_1, \cdots, c_{t+n-1} \rangle$, so by the exchange axiom, c_{t+n} depends on $\langle c_1, \cdots, c_{t+n-1}, a_{t+1} \rangle$. Now consider the strings $\langle c_1, \cdots, c_{t+n-1}, a_{t+1}, c_{t+n+1}, \cdots, c_s \rangle$ and $\langle c_1, \cdots, c_{t+n-1}, c_{t+n}, c_{t+n+1}, \cdots, c_s \rangle$. Then one sees that these two strings are equivalent; and the first contains $t+1$ of the a_i. The proof is complete.

COROLLARY. *With the assumption of the theorem, $s \geq r$.*

THEOREM 5. *If $\langle a_1, \cdots, a_r \rangle$ and $\langle b_1, \cdots, b_s \rangle$ are independent and equivalent, then $r = s$.*

PROOF. By the corollary, $r \geq s$ and $s \geq r$, so $r = s$.

The strings that we consider will always be finite (by definition), but the set S may be infinite, and we will also consider various, possibly infinite, subsets of S. Let M be a subset of S. By a *basis* of M we mean an independent string on which each element of M depends.

A set M may not have a basis; but if it does, then by Theorem 5, any two bases of M have the same length. This common length is called the *rank* of M.

THEOREM 6. *If M has a basis, then any independent string*

$$\langle a_1, \cdots, a_m \rangle$$

with the a_i in M can be completed to a basis of M; i.e., there is a basis of the form $\langle a_1, \cdots, a_m, a_{m+1}, \cdots \rangle$, with the a_i in M.

PROOF. Let $r = \text{rank}\ (M)$. Then by Theorem 4, Corollary, no independent string of elements from M can be of length greater than r. Consider all independent strings of the form $\langle a_1, \cdots, a_m, a_{m+1}, \cdots, a_n \rangle$, with the a_i in M. Then $n \leq r$. So among these strings there is a longest. By the argument of the proof of Theorem 3, such longest string will be a basis.

COROLLARY. *If rank $(M) = r$, then any independent string of elements from M is of length at most r.*

COROLLARY. *If N is a subset of M and M has a basis, then N has a basis, and rank $N \leq$ rank M.*

Exercises

1. Show that a model of the axiom system we are considering can be obtained as follows: Let S be an arbitrary set, and let "a depends on $<a_1, \cdots, a_m>$" be taken to mean "a is one of the a_i".

2. Show that a model can be obtained as follows. Let F be a commutative field. Consider all ordered n-tuples of elements from F, where n is a fixed (positive) integer: this shall be the set S. Let "a depends on $<a_1, \cdots, a_m>$" be taken to mean that the n-tuple a is a linear combination of the n-tuples $a_1, \cdots, a_m$, if $m > 0$; for $m = 0$, "a depends on the empty string" shall mean that a consists of n zeros. (See remark below.)

3. Same as the last exercise except that F is to be a not necessarily commutative field, and the linear combinations are restricted to be *right* linear combinations.

4. Write out the proof of the last corollary.

5. In Exercise 2 above, take $n = 3$ and answer the following questions. Are $\begin{pmatrix}2\\3\\4\end{pmatrix}$, $\begin{pmatrix}5\\6\\7\end{pmatrix}$ dependent? Or, more exactly, is $\left\langle \begin{pmatrix}2\\3\\4\end{pmatrix}, \begin{pmatrix}5\\6\\7\end{pmatrix} \right\rangle$ dependent? Is $\left\langle \begin{pmatrix}2\\3\\4\end{pmatrix}, \begin{pmatrix}5\\6\\7\end{pmatrix} \begin{pmatrix}9\\12\\15\end{pmatrix} \right\rangle$ dependent? Is $\left\langle \begin{pmatrix}2\\3\\4\end{pmatrix}, \begin{pmatrix}5\\6\\7\end{pmatrix}, \begin{pmatrix}7\\8\\5\end{pmatrix} \right\rangle$ dependent?

REMARK ON THE SECOND EXERCISE. In building a model we must say what "a depends on $<a_1, \cdots, a_m>$" shall mean, whatever the string $<a_1, \cdots, a_m>$ be, in particular if it is the empty string. Suppose we have suggested to ourselves that "a depends on $<a_1, \cdots, a_m>$, $m > 0$" shall mean a is a linear combination of $a_1, \cdots, a_m$. Still, what could "a dependent on the empty string" mean? It makes no sense to take a linear combination of nothing. Here we argue to ourselves as follows. Let b be any n-tuple. If a is to depend on the empty string (and we are to come out with a model), then, for every b, a depends on $<b>$, since by Axioms I and III, if a depends on a string $<a_1, \cdots, a_m>$, it depends on any longer string $<a_1, \cdots, a_m, \cdots>$. Hence if a depends on the empty string, then a is a multiple of every n-tuple b. The only n-tuple with this property is the n-tuple consisting of n zeros: call this z. We thus have two possibilities: (1) either say z depends on the empty string, or (2) say no n-tuple depends on the empty string. Actually (2) is also excluded; because let b be any n-tuple, $b \neq z$, and suppose we have said (2). Then $<z>$ is independent, but $<z, b>$ is dependent: this makes b a multiple of z, hence equal to z, which is not so. Hence we are left with (1) as the only possibility. By the exercise, (1) does work.

2. Application of the dependency theory to geometry. We take *point*, *line*, and *on* as primitive terms and take as axioms the ones given on p. 96, in particular Axiom 3, which is Pasch's Axiom. We want to build a model of the dependency theory using geometric notions. The set S shall be simply the set of points. The first thing is to define *depends on*, i.e., given a point P and a string of points $\langle P_1, \cdots, P_n \rangle$, we must say what it shall mean to say that P depends on $\langle P_1, \cdots, P_n \rangle$. To facilitate this, we give a preparatory definition.

DEFINITION. We say Q *depends directly* on $\langle P_1, \cdots, P_m \rangle$ if Q is one of the P_i or if Q is collinear with some pair P_i, P_j, $P_i \neq P_j$.

DEFINITION. We say Q *depends on* $\langle P_1, \cdots, P_n \rangle$ if there are points $Q_1, Q_2, \cdots, Q_s$ such that Q_i depends directly on

$$\langle P_1, \cdots, P_n, Q_1, \cdots, Q_{i-1} \rangle$$

for $i = 1, \cdots, s$ and $Q = Q_s$. No point is to depend on the empty string.

Now to check the axioms. Axioms I and III are immediately verified. Axiom II gives some difficulty. The following lemma will be useful. (This lemma is rephrased in Theorem 5, p. 130).

LEMMA. *If Q is dependent on $\langle P_1, \cdots, P_n \rangle$, and $Q \neq P_n$, then there is a point Z on the line QP_n that is dependent on $\langle P_1, \cdots, P_{n-1} \rangle$.*

PROOF. Let $Q_1, \cdots, Q_s$ be such that Q_i is directly dependent on $\langle P_1, \cdots, P_n, Q_1, \cdots, Q_{i-1} \rangle$ for $i = 1, \cdots, s$ and $Q = Q_s$. There may be many choices for $Q_1, \cdots, Q_s$ but we suppose them taken in such way that s is as small as possible. Then we make an induction on s. Consider first the case $s = 1$. Then Q depends directly on $\langle P_1, \cdots, P_n \rangle$, hence is either one of the P_i or is collinear with a distinct pair P_i, P_j. In the first case, $Q = P_i$, we have $i \neq n$, since $Q \neq P_n$, whence Q is dependent on $\langle P_1, \cdots, P_{n-1} \rangle$ and we take $Q = Z$. In the second case, with Q collinear with P_i, P_j, if $i \neq n$, $j \neq n$, then Q is dependent $\langle P_1, \cdots, P_{n-1} \rangle$ and we take $Z = Q$; if $j = n$, then we take $Z = P_i$.

In the case $s=1$, the proof thus comes to an examination of a numbei of alternatives each of which is disposed of in a simple manner. The number of alternatives could have been reduced as follows. We first note that if Q depends on $\langle P_1,\cdots, P_{n-1}\rangle$, then the lemma follows at once by taking $Q=Z$. Thus we may suppose that Q is not dependent on $\langle P_1,\cdots, P_{n-1}\rangle$. The only alternative now is that QP_n is on one of the points $P_1,\cdots, P_{n-1}$, say P_i, and we take $Z=P_i$.

Proceeding to the inductive step, we first make some remarks of the kind occurring in the last paragraph. As there, we may assume that $Q(=Q_s)$ is not dependent on $\langle P_1,\cdots, P_{n-1}\rangle$; but also we may assume that no Q_i is dependent on $\langle P_1,\cdots, P_{n-1}\rangle$. For, assuming $s>1$ suppose, say, that $Q_i(i<s)$ were dependent on $\langle P_1,\cdots, P_{n-1}\rangle$. The we can get from $\langle Q_i, P_1,\cdots, P_{n-1}, P_n\rangle$ to Q in $s-1$ stages. By induc-

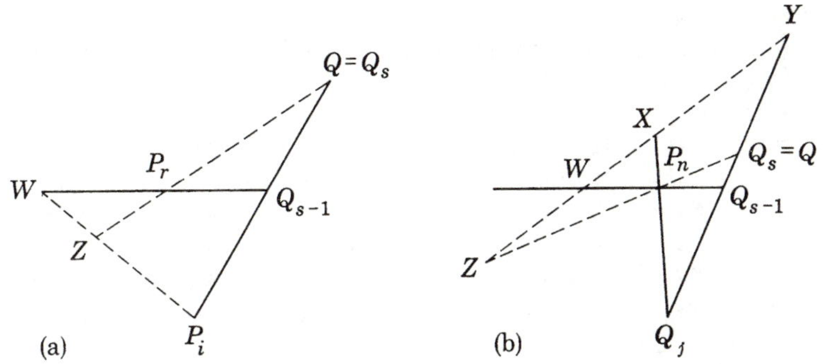

FIG. 7.1

tion, there is a point Z collinear with Q and P_n, and dependent on $\langle Q_i, P_1,\cdots, P_{n-1}\rangle$. By transitivity Z is dependent on $\langle P_1,\cdots, P_{n-1}\rangle$ and is the point sought. Likewise we may suppose, or, rather, here we can assert, that no Q_i is P_n: for if $Q_i=P_n$, we do not need Q_i among the Q's, and so we could get to Q in $s-1$ stages from $\langle P_1,\cdots, P_n\rangle$. Similarly we suppose that P_n is not dependent on $\langle P_1,\cdots, P_{n-1}\rangle$ and that the Q_i are distinct.

Now we come to the proof proper. By induction on s, there is a point W on P_nQ_{s-1} that is dependent on $\langle P_1,\cdots, P_{n-1}\rangle$. If P_n, Q_{s-1}, Q_s are collinear, we take $Z=W$; so we may suppose these points noncollinear. Since Q_s depends directly on $\langle P_1,\cdots, P_n, Q_1,\cdots, Q_{s-1}\rangle$ Q_s is collinear with two of the points $P_1,\cdots, Q_{s-1}$ (it is not equal to any of them by our preliminary remarks). Moreover one of the two points is Q_{s-1}; otherwise we can reduce s (i.e., we can get to Q in less than s stages). Thus Q_s, Q_{s-1} are collinear with another Q_j or with a P_i (Fig. 7.1).

Consider the two cases separately. The second one mentioned (Fig. 7.1a) is easier and will be considered first. The points W, Q_{s-1}, P_t determine a triangle (i.e., these points are distinct and noncollinear—this follows from the preliminary remarks). We also have $Q \neq Q_{s-1}$ and $P_n \neq Q_{s-1}$, so by Pasch's Axiom, QP_n and WP_t meet in a point Z. The point Z is dependent on $\langle W, P_1, \cdots, P_{n-1} \rangle$, hence, by transitivity, also on $\langle P_1, \cdots, P_{n-1} \rangle$. Therefore Z is the point sought.

In the second case (Fig. 7.1b), we first note that there is a point X on Q_jP_n that is dependent on $\langle P_1, \cdots, P_{n-1} \rangle$: this follows because of the induction assumption on s. From the preliminary remarks, $W \neq P_n$, $X \neq P_n$; applying Pasch's Axiom to triangle $P_nQ_jQ_{s-1}$, we get a point Y collinear with W and X and on Q_jQ_{s-1}. Since $W \neq X$, Y depends on $\langle W, X \rangle$, and since W, X each depend on $\langle P_1, \cdots, P_{n-1} \rangle$, by transitivity Y depends on $\langle P_1, \cdots, P_{n-1} \rangle$. Now applying Pasch's Axiom to triangle YWQ_{s-1}—note that Y, W, Q_{s-1} are not collinear—we get a point Z on P_nQ_s that is collinear with W and Y. Again using transitivity, we conclude that Z is dependent on $\langle P_1, \cdots, P_{n-1} \rangle$. The proof of the lemma is complete.

To check Axiom II of the dependency theory, let P depend on $\langle P_1, \cdots, P_{n-1}, P_n \rangle$ but not on $\langle P_1, \cdots, P_{n-1} \rangle$: to see that P_n depends on $\langle P_1, \cdots, P_{n-1}, P \rangle$. If $P = P_n$, this is immediate; so suppose $P \neq P_n$. By the lemma, there is a point Z on PP_n that is dependent on $\langle P_1, \cdots, P_{n-1} \rangle$. Then there are points $Q_1, \cdots, Q_s$ such that Q_i depends directly on $\langle P_1, \cdots, P_{n-1}, Q_1, \cdots, Q_{i-1} \rangle$, for each i, and $Q_s = Z$. Moreover P_n depends directly on $\langle P, P_1, \cdots, P_{n-1}, Q_1, \cdots, Q_s \rangle$. Hence P_n depends on $\langle P, P_1, \cdots, P_{n-1} \rangle$. Axiom II is thus verified and the theory of dependency established in the geometric situation.

DEFINITION. A subset M of S shall be called a *space* if M has a (finite) basis and if it contains together with any $P_1, \cdots, P_n$ in M any P in S that depends on $\langle P_1, \cdots, P_n \rangle$. (Note that this definition could be given in the abstract theory, but we shall be concerned only with the geometric model.)

THEOREM 1. *A subset M of S is a space if and only if it has a basis and contains with any two of its points A, B, $A \neq B$, all the points X collinear with A, B.*

The proof is immediate.

THEOREM 2. *The intersection of two spaces is also a space.*

PROOF. Let N be the intersection of spaces M_1, M_2. Since M_1 has finite rank, also N has finite rank (see p. 125, Theorem 6, second corollary). Thus N has one of the two required properties. Let A, B be points in N, $A \neq B$, and let X be collinear with A, B. Since A, B are in M_1, also X is in M_1; since A, B are in M_2, also X is in M_2; so X is in the intersection N. Q.E.D.

COROLLARY. *The intersection of any number of spaces, finitely or infinitely many, is also a space.*

Notation. $N = M_1 \cap M_2 =$ intersection of M_1 and M_2.

If $\{P_1, \cdots, P_n\}$ is any finite set of points, the set of points dependent on $\langle P_1, \cdots, P_n \rangle$ is a space, which we designate by $[P_1, \cdots, P_n]$. Here $\langle P_1, \cdots, P_n \rangle$ need not be independent. Any space M is of the form $[P_1, \cdots, P_n]$, with $\langle P_1, \cdots, P_n \rangle$ even independent: in fact, let $\langle P_1, \cdots, P_n \rangle$ be a basis of M, then $M = [P_1, \cdots, P_n]$, as one readily checks.

Any two spaces are contained together in a space. In fact, if $M = [P_1, \cdots, P_m]$, $N = [Q_1, \cdots, Q_n]$, then M and N are both in $[P_1, \cdots, P_m, Q_1, \cdots, Q_n]$.

DEFINITION. The intersection of all spaces containing the spaces M and N is called the *join* of M and N.

If $M = [P_1, \cdots, P_m]$ and $N = [Q_1, \cdots, Q_n]$, then any space containing M and N must contain the points $P_1, \cdots, P_m, Q_1, \cdots, Q_n$, hence also contain $[P_1, \cdots, P_m, Q_1, \cdots, Q_n]$. Thus $[P_1, \cdots, P_m, Q_1, \cdots, Q_n]$ is the join of M and N. This join is also denoted $[M, N]$.

With a definition like the one above, one can easily define the join of a finite number of spaces, and the join will be a space. Given infinitely many spaces, there may not be a space containing all of them, so we will not define the join in this case.

If $M = [P_1, \cdots, P_m]$, we say that $P_1, \cdots, P_m$ *generate* M, that M is generated by $P_1, \cdots, P_m$, or *spans* $P_1, \cdots, P_m$. Thus if $P_1, \cdots, P_m$ generate M and $Q_1, \cdots, Q_n$ generate N, then $P_1, \cdots, Q_n$ generate the join of M and N.

Any point P generates a space, namely, the space $[P]$: this space consists of just the one point P. If P, Q are distinct points, then the space they generate, $[P, Q]$, consists of the points on the line PQ. Any line spans any pair of its points; any line is the join of any two of its points.

DEFINITION. The *dimension* of a space is one less than its rank, i.e., one less than the number of elements in a basis.

Thus a point is a 0-dimensional space, a line is a 1-dimensional space. The empty set is a space; its dimension is -1.

THEOREM 3. *Let M be a space of dimension 2. Then any two lines in*

M meet. (We say that a line l is in M if every point on l is in M; note that if A, B are points of M, $A \neq B$, then the line AB is in M.)

PROOF. Let l, m be two given distinct lines. Let P be a point on m that is not on l. Let P_1, P_2 be two distinct points on l. $\langle P_1 \rangle$ is independent, and since P_2 does not depend on P_1, the string $\langle P_1, P_2 \rangle$ is independent (Theorem 2, p. 123); since P is not dependent on $\langle P_1, P_2 \rangle$, $\langle P_1, P_2, P \rangle$ is independent. The string $\langle P_1, P_2, P \rangle$ can be completed to a basis of M (Theorem 6, p. 125); but no independent string in M can be longer than 3 (Theorem 6, Cor.), so $\langle P_1, P_2, P \rangle$ is itself a basis. Now let Q be any point of m other than P. Then Q is dependent on $\langle P_1, P_2, P \rangle$. Hence by the lemma on p. 126, QP has on it a point Z that is dependent on $\langle P_1, P_2 \rangle$, i.e., is on P_1P_2. Thus P_1P_2 and QP meet in Z. Q.E.D.

THEOREM 4. *Let M be a space of dimension ≥ 3. Then some pair of lines in M do not meet.*

PROOF. Let $\langle A, B, C, D \rangle$ be independent. Then AB and CD do not meet. In fact, suppose they meet in X. Then D depends on C and X; and C and X each depend on $\langle A, B, C \rangle$. Thus D depends on $\langle A, B, C \rangle$, a contradiction.

We have a definition of *plane*, but using present notions, let us define *plane* as a 2-dimensional space. We should see, however, that the two definitions describe the same objects. Let π, then, be a plane according to the old definition. We have to see that π is 2-dimensional. Let, then, A, B, C be three noncollinear points: they are independent, and we have to see that any fourth point D depends on them. Let CD meet AB in X. We have: X depends on $\langle A, B \rangle$; D depends on $\langle C, X \rangle$; whence D depends on $\langle A, B, C \rangle$. Q.E.D.

Conversely, let π be a 2-dimensional space. We have to see that any two lines of π meet. This was just proved as Theorem 3.

Exercise. In a 3-space, every line meets every plane.

The lemma on p. 126 has the following important formulation.

THEOREM 5. *If Q is in the space $[P_1, \cdots, P_n]$ and $Q \neq P_n$, then the line QP_n meets the space $[P_1, \cdots, P_{n-1}]$.*

Exercises

1. In a 4-space, every line meets every 3-space.
2. In a 4-space, every pair of planes meet.
3. In a 5-space, there are planes that do not meet.
4. Let π_1, π_2 be distinct planes in a 4-space. $[\pi_1, \pi_2]$ is either 3-dimensional

or 4-dimensional. Show that $\pi_1 \cap \pi_2$ is a line or a point according as $[\pi_1, \pi_2]$ is 3- or 4-dimensional.

THEOREM 6. *Let l, m be two skew lines (i.e., lines that do not meet). The join of l and m is a 3-space. Every point of the join [l, m] is collinear with a point of l and with a point of m.*

PROOF. Let A, B be points of l; C, D points of m. Then $[A, B, C, D]$ is the join, and is therefore 3-dimensional or less; the dimension cannot be less, since AB and CD do not meet. For the second part of the theorem, let P be a point of $[l, m]$. If P is on l, then clearly it is collinear with a point of l and a point of m. Assume, then, that P is not on l. Then $[P, l]$ is a plane; and $[P, l]$ meets m in a point Q. We may assume $Q \neq P$, as otherwise P is on m. Assuming $Q \neq P$, PQ meets l in a point X. Then P is collinear with the point X on l and the point Q on m.

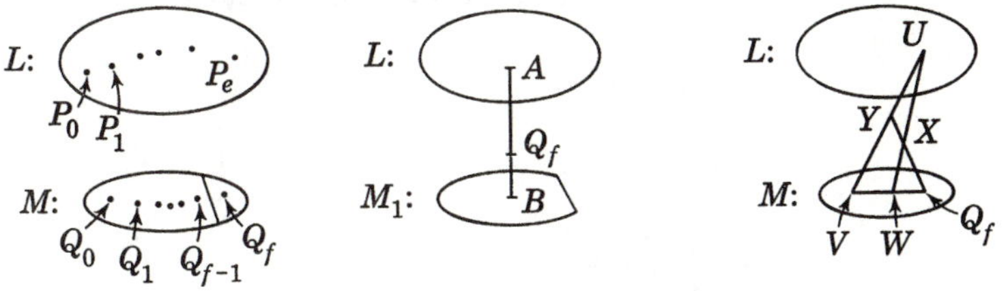

FIG. 7.2

Theorem 6 can be generalized as follows.

THEOREM 7. *Let L, M be two skew spaces (i.e., spaces that do not meet). Then* $\dim [L, M] = \dim L + \dim M + 1$. *Every point of the join is collinear with a point of L and a point of M.*

PROOF. Let $\dim L = e$, so that L has a basis $P_0, \cdots, P_e$; and let $\dim M = f$, so that M has a basis $Q_0, \cdots, Q_f$. The join $[L, M]$ is $[P_0, \cdots, P_e, Q_0, \cdots, Q_f]$. Note that there are $e+1+f+1$ points here. Hence, $\dim [L, M] \leq e+1+f+1-1 = e+f+1$. To see that equality holds, we must see that $\langle P_0, \cdots, Q_f \rangle$ is independent. To prove this we make an induction on f. To save space, we skip the verification for $f=0$. Assuming $f>0$, our induction hypothesis tells us that $\langle P_0, \cdots, P_e, Q_0, \cdots, Q_{f-1} \rangle$ is independent. We have yet to see that Q_f is not dependent on $\langle P_0, \cdots, Q_{f-1} \rangle$ (see Theorem 2 of the dependency theory, p. 123). Let $M_1 = [Q_0, \cdots, Q_{f-1}]$. If Q_f is dependent on $\langle P_0, \cdots, Q_{f-1} \rangle$, then Q_f is in the space $[L, M_1]$. Now let us make the second part of the present theorem a part of our induction. Then Q_f is collinear with a point A of L and a point B of M_1 (Fig. 7.2). But

BQ_f lies in M; hence, L and M meet at A. This is a contradiction. This completes the proof that $\dim [L, M] = e + f + 1$, *provided that we also prove the second part of the theorem.* Let, then, X be a point of $[L, M]$. By Theorem 5, Q_fX meets $[L, M_1]$ in a point Y. By induction hypothesis, Y is collinear with a point U of L and a point V of M_1 (Fig. 7.2). Aside from trifling special cases, we apply Pasch's Axiom to conclude that X is collinear with U and W, a point of Q_fV; since the whole line Q_fV is in M, the proof is complete.

The next theorem generalizes Theorem 7 to the case that L and M are not necessarily skew.

THEOREM 8. $\dim L + \dim M = \dim [L, M] + \dim L \cap M$.

Exercise. Apply this formula to the exercises on p. 130 f.

PROOF OF THEOREM 8. Let $\dim L = e$, $\dim M = f$, $\dim L \cap M = s$. Then $L \cap M$ has a basis $G_0, \cdots, G_s$; this can be completed to a basis

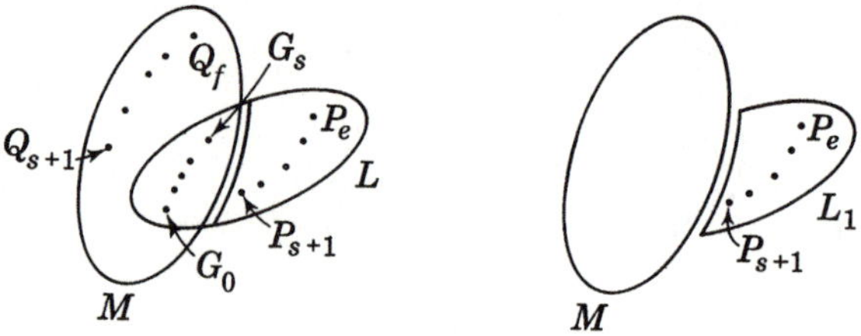

FIG. 7.3

$G_0, \cdots, G_s, P_{s+1}, \cdots, P_e$ of L; and a basis $G_0, \cdots, G_s, Q_{s+1}, \cdots, Q_f$ of M (Fig. 7.3). Then $[L, M] = [G_0, \cdots, G_s, P_{s+1}, \cdots, P_e, Q_{s+1}, \cdots, Q_f]$. Note that there are $(s+1) + (e-s) + (f-s)$, or $\dim L + \dim M - \dim (L \cap M) + 1$, points written here. Thus $\dim [L, M] \leq \dim L + \dim M - \dim (L \cap M)$. To obtain equality, we have to see that the points are independent. If $L_1 = [P_{s+1}, \cdots, P_e]$, then this will be seen by Theorem 7 if we can show that M and L_1 are skew. To see this, suppose X is a point in L_1 and M. Then X is in $L \cap M = [G_0, \cdots, G_s]$. Since X can be completed to a basis of $L \cap M$, and since $G_0, \cdots, G_s$ was an arbitrary basis, we may assume $X = G_0$. Since X is in L_1, G_0 depends on $P_{s+1}, \cdots, P_e$. This contradicts the fact that $G_0, \cdots, G_s, P_{s+1}, \cdots, P_e$ are independent.

A theorem we have actually used many times without stating it explicitly as a theorem is the following.

THEOREM 9. *Let L, M be spaces. If L is contained in* (*or equals*) *M, and* $\dim L = \dim M$, *then* $L = M$.

PROOF. Let $L=[P_0,\cdots, P_e]$, where $e=\dim L=\dim M$. The string $\langle P_0,\cdots, P_e\rangle$ is rank M long, so every point in M depends on $\langle P_0,\cdots, P_e\rangle$, whence $M=[P_0,\cdots, P_e]=L$. Q.E.D.

3. Hyperplanes. Let S be an n-dimensional space. For the next few theorems we shall be working in S.

By a *hyperplane*, we mean an $(n-1)$-dimensional subspace of S. A plane, as has already been said, is a 2-dimensional space. In 3-space, the two concepts fall together, but not in n-space if $n>3$ (or even if $n\neq 3$).

THEOREM 10. *Let A be an r-dimensional space in S, H a hyperplane. Then either H contains A or H intersects A in an $(r-1)$-dimensional space.*

PROOF. Assume that H does not contain A. Then the join $[A, H]$ contains H, and since it does not equal H, it must be of dimension $>\dim H=n-1$. Therefore, $\dim [A, H]=n$ (it cannot be $>n$, since we are in n-space). Hence, $\dim A\cap H=\dim A+\dim H-\dim [A, H]$ $=r+n-1-n=r-1$. Q.E.D.

Conversely, we have the following.

THEOREM 11. *Let A be an r-space, B an $(r-1)$-space in A. Then there is a hyperplane H such that $B=A\cap H$.*

PROOF. Let $P_0,\cdots, P_{r-1}$ be a basis for B. Complete this to a basis $P_0,\cdots, P_{r-1}, P_r$ for A, and to a basis $P_0, P_1.\cdots, P_r, P_{r+1},\cdots, P_n$ for S. The space $[P_0,\cdots, P_{r-1}, P_{r+1},\cdots, P_n]$ is a hyperplane H. It does not contain A, since P_r is not dependent on $P_0,\cdots, P_{r-1}, P_{r+1},\cdots, P_n$. Hence, $H\cap A$ is an $(r-1)$-space. $H\cap A$ also contains B, since $H\cap A$ contains $P_0,\cdots, P_{r-1}$. By Theorem 9, $H\cap A=B$.

THEOREM 12. *Every r-space is the intersection of $n-r$, but no fewer, hyperplanes.*

PROOF. Let A be a given r-space, $P_0,\cdots, P_r$ a basis for A. Let this basis be completed to a basis $P_0,\cdots, P_r, P_{r+1},\cdots, P_n$ for S. Consider the spaces $A=A_0=[P_0,\cdots, P_r]$, $A_1=[P_0,\cdots, P_r, P_{r+1}]$, $A_2=[P_0,\cdots, P_{r+1}, P_{r+2}],\cdots, A_{n-r-1}=[P_0,\cdots, P_{n-1}]$. The last of these is a hyperplane, and every other one can be obtained from the next one by intersecting with a hyperplane. In this way we get A as the intersection of $n-r$ hyperplanes. Moreover, A cannot be the intersection of fewer than $n-r$ hyperplanes. For, suppose $A=H_1\cap H_2\cap\cdots\cap H_s$, where $H_1, H_2,\cdots$ are hyperplanes. Consider the spaces H_1, $H_1\cap H_2$, $H_1\cap H_2\cap H_3,\cdots$. Each of these spaces (after the first) has dimension at most one less than the preceding one (if $H_1\cap\cdots\cap H_t=H_1\cap\cdots\cap H_{t+1}$, however, one does not need H_{t+1} in the representation

$A = H_1 \cap \cdots \cap H_s$). Hence, $H_1 \cap \cdots \cap H_t \geq n-t$, and for $t=s$, we get $r \geq n-s$. Hence, $s \geq n-r$. Q.E.D.

Exercise. What familiar fact does the preceding theorem give for $n=3$, $r=1$?

4. The dual space. Starting from the n-space S, we will show how to construct from it another projective space S'. This other space is a set of objects satisfying certain properties, and these objects we would ordinarily call "points." But if the word "point" is already reserved for denoting the objects of S, we need another word to denote the objects of S'. To get around this difficulty, we will refer to the objects of S' as *Points* (with a capital P).

Now to state what these objects, these Points, are. They are the hyperplanes of S. *A Point of S' is a hyperplane of S.*

To have a projective space, we need not only points, but also lines. Again, to avoid confusion, we will write *Line* (with a capital L) when referring to "lines" of S'.

Let A, B, C be three Points, i.e., hyperplanes of S. We will say *they are on a Line if the hyperplanes A, B, C are on an $(n-2)$-space.*

Exercise. What is being said for $n=3$? What for $n=4$? What for $n=2$?

Thus, a *Line* will consist of all the Points, i.e., hyperplanes, on a given $(n-2)$-dimensional space in S. If S_{n-2} is a given $(n-2)$-space in S, then it is convenient to refer to the Line S_{n-2}, although strictly speaking that is not correct, because S_{n-2} is a set of points, whereas a Line is a set of hyperplanes. What we really mean when we speak of the Line S_{n-2} is the set of hyperplanes through S_{n-2}.

The problem now is to show that for Points and Lines thus defined, S' becomes a projective space.

Let us check Axiom 2, that on any two distinct Points there is at most one Line. Let us make clear to ourselves what property of hyperplanes is comprised in this statement. Let A, B be Points, i.e., hyperplanes. The space $A \cap B$ is an $(n-2)$-space, and so A and B are on the Line which consists of all hyperplanes on this $(n-2)$-space: this is the content of Axiom 1, however, not of Axiom 2. Axiom 2 comes to saying that there is at most one $(n-2)$-space on which both A and B are. If S_{n-2} is an $(n-2)$-space and S_{n-2} is contained in A and in B, then it is contained in $A \cap B$, and since $A \cap B$ is itself an $(n-2)$-space, we have $S_{n-2} = A \cap B$. In other words, the set of hyperplanes on $A \cap B$ is the Line in question.

Let us check Pasch's Axiom. In Fig. 7.4, let A, B represent hyperplanes of S, or, in other words, Points of S'. The part connecting A

and B represents the $(n-2)$-space $A \cap B$. Let C be a Point not on Line AB, i.e., a hyperplane not on $A \cap B$. Let D be a hyperplane on $B \cap C$ and E a hyperplane on $A \cap C$. From this we have to see that there is a hyperplane F on $D \cap E$ and on $A \cap B$. This we would have if $[A \cap B, D \cap E]$ were at most $(n-1)$-dimensional. Use the formula (Theorem 8, p. 132) to get $\dim [A \cap B, D \cap E] = \dim (A \cap B) + \dim (D \cap E) - \dim (A \cap B) \cap (D \cap E) = (n-2) + (n-2) - \dim (A \cap B) \cap (D \cap E)$. We would come to the desired conclusion if we had $\dim (A \cap B) \cap (D \cap E) \geq n-3$. Now note that $(A \cap B) \cap (D \cap E) = A \cap B \cap D \cap E = (A \cap E) \cap (B \cap D)$. Again use the formula to

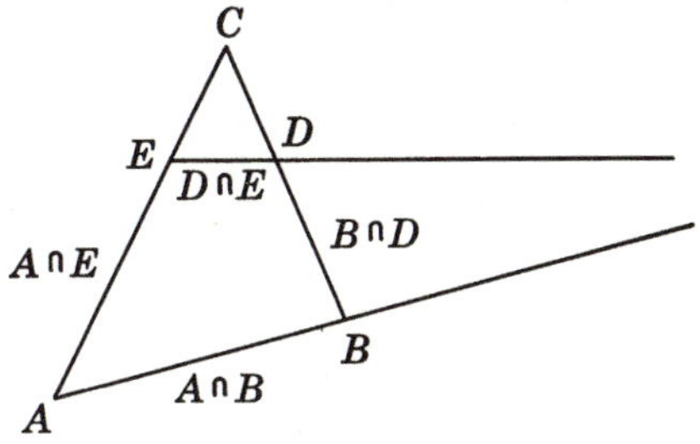

FIG. 7.4

get $\dim (A \cap E) \cap (B \cap D) = \dim (A \cap E) + \dim (B \cap D) - \dim [A \cap E, B \cap D] = (n-2) + (n-2) - \dim [A \cap E, B \cap D] \geq (n-2) + (n-2) - (n-1) = n-3$. This last inequality follows from the fact that $A \cap E$ and $B \cap D$ are both in C, and hence $[A \cap E, B \cap D]$ is at most $(n-1)$-dimensional.

Exercise. Verify the other axioms.

We next wish to see what dependency in S' means in terms of the space S.

THEOREM 13. *Let $p_1, \cdots, p_m$ be Points of S'* (i.e., *hyperplanes of S*). *The Point q of S' is dependent on $p_1, \cdots, p_m$ if and only if the hyperplane q contains the intersection of the hyperplanes $p_1, \cdots, p_m$.*

PROOF. Let q_1 depend directly on $p_1, \cdots, p_m$; then q_1 is either one of the p_i or lies on a Line with two of them p_i, p_j. In terms of the space S, the latter possibility means that q_1 contains the $(n-2)$-dimensional space $p_i \cap p_j$. Thus, in either event, q_1 contains $p_1 \cap p_2 \cap \cdots \cap p_m$. Now let q_2 depend directly on $p_1, p_2, \cdots, p_m, q_1$: then, as we have just seen, q_2 will contain $p_1 \cap p_2 \cap \cdots \cap p_m \cap q_1$, which is the same as

$p_1 \cap \cdots \cap p_m$. Proceeding in this way in several steps to q, we see that if q is dependent on $p_1, \cdots, p_m$, then q contains $p_1 \cap p_2 \cap \cdots \cap p_m$.

Conversely, suppose q contains $p_1 \cap \cdots \cap p_m = T$. The proof is by induction on m. If q contains $p_1 \cap \cdots \cap p_{m-1} = R$, then we are through by induction. We suppose, then, that q does not contain R; but q does contain $R \cap p_m$. Thus, if R is r-dimensional, then T is $(r-1)$-dimensional. Let $U = p_m \cap q$. We seek a hyperplane q' that will contain R and U. To see why, suppose we had such a hyperplane. Then q' depends on p_m and q, but not on p_m—otherwise it would be p_m and p_m would contain R, which is not so; so by dependency theory in S', q depends on p_m and q'. By induction q' depends on $p_1, \cdots, p_{m-1}$, hence by transitivity of dependence in S', q depends on $p_1, \cdots, p_m$. Thus it remains to find q'. Now

$$\dim\,[R, U] = r+n-2-\dim\,(R \cap U) = r+n-2-(r-1) = n-1,$$

so $[R, U]$ is the desired hyperplane q'.

THEOREM 14. *Every subspace of S' consists of all the hyperplanes containing a subspace of S; and conversely, all the hyperplanes containing a given subspace of S constitute a subspace of S'. All the hyperplanes on an r-dimensional subspace of S' form an $(n-r-1)$-dimensional subspace of S'. In particular, S' is n-dimensional.*

PROOF. Every subspace of S' consists of the Points Q dependent on some Points $p_1, \cdots, p_m$, i.e., consists of all hyperplanes q containing $p_1 \cap \cdots \cap p_m$. This proves the first point and the second point is proved similarly. If L is an r-dimensional subspace of S, then by Theorem 12 it is the intersection of $n-r$ but no fewer hyperplanes, say of $p_1, \cdots, p_{n-r}$. Thus the Points q that are the hyperplanes containing L are the Points depending on $p_1, \cdots, p_{n-r}$. The Points $p_1, \cdots, p_{n-r}$ are independent, as otherwise the Points dependent on these would also be dependent on some proper subset of them, so that L would be the intersection of fewer than $n-r$ hyperplanes. Hence the space of Points q is $(n-r-1)$-dimensional. Since the empty set is (-1)-dimensional, S' is $n-(-1)-1 = n$-dimensional.

Starting from S', we can build its dual space S''. The points of S'' are the hyperplanes of S'; we will write Hyperplanes when referring to the $(n-1)$-dimensional subspaces of S'. Given a subspace L of S we associate with it a subspace L' of S', namely, the set of hyperplanes on L. Similarly we associate with L' a space L'' of S''. Our next object is to prove the following theorem.

THEOREM 15. *The dual of the dual of S is isomorphic to S. Moreover, if L is a point of S and L'' is the corresponding point of S'' described* *above, then the mapping $L \to L''$ is an isomorphism of S and S'', called the canonical isomorphism (of S with S'').*

The main point of the proof is contained in the following lemma.

LEMMA. *Let H_1, H_2, H_3 be three points (of S) and consider the corresponding three hyperplanes H_1', H_2', H_3' (of S'). Then H_1', H_2', H_3' are on a common Line (of S'') if and only if H_1, H_2, H_3 are on a common line.*

PROOF OF THE LEMMA. H_1', H_2', H_3' are on a common Line if and only if $H_1' \cap H_2' \cap H_3'$ is $(n-2)$-dimensional. Here $H_1' \cap H_2' \cap H_3'$ means the set of hyperplanes on H_1 and on H_2 and on H_3. By Theorem 14, the dimension of this set (space) is $n-1-\dim [H_1, H_2, H_3]$; and this is $n-2$ if and only if $\dim [H_1, H_2, H_3]$ is 1, i.e., if and only if H_1, H_2, H_3 are on a line. Q.E.D.

5. The analytic case. Let K be a (commutative) field. By a point (in the analytic case) we will mean a class of $(n+1)$-tuples

$$\{(\lambda x_0, \lambda x_1, \cdots, \lambda x_n)\},$$

where the x_i are numbers from K not all zero and λ varies over K except that $\lambda \neq 0$: any one of the n-tuples $(\lambda x_0, \lambda x_1, \cdots, \lambda x_n)$ is called a representative of the point. We abbreviate, writing $\mathbf{x}$ for $(x_0, \cdots, x_n)$ and $\lambda\mathbf{x}$ for $(\lambda x_0, \cdots, \lambda x_n)$. We write $P\mathbf{x}$ or $P_{\mathbf{x}}$ for the point having $\mathbf{x}$ as representative; where no confusion is likely, we refer to the point $\mathbf{x}$ (rather than $P\mathbf{x}$). We say that $P\mathbf{x}$, $P\mathbf{y}$, $P\mathbf{z}$ are collinear if one of $\mathbf{x}$, $\mathbf{y}$, $\mathbf{z}$ is a linear combination of the others. If $n \geq 2$, as we shall assume throughout, then there exists noncollinear points. The other axioms for projective space can be easily verified (for arbitrary n) by techniques we have already studied. This model of projective space is designated PNK^n (=projective number space over K of dimension n). (Also for $n=0$, 1 one refers to the spaces PNK^0, PNK^1.)

Dependence is defined in terms of collinearity, but for PNK^n we want an analytic expression of dependency. If $P\mathbf{x}$ is directly dependent on $P\mathbf{y}$, $P\mathbf{z}$, $P\mathbf{w}, \cdots$, then $\mathbf{x}$ is a linear combination of (one or two of) $\mathbf{y}$, $\mathbf{z}$, $\mathbf{w}, \cdots$; and since a linear combination of linear combinations of $\mathbf{y}$, $\mathbf{z}$, $\mathbf{w}, \cdots$ is a linear combination of $\mathbf{y}$, $\mathbf{z}$, $\mathbf{w}, \cdots$, one sees that if $P\mathbf{x}$ is dependent on $P\mathbf{y}$, $P\mathbf{z}$, $P\mathbf{w}, \cdots$, then $\mathbf{x}$ is a linear combination of $\mathbf{y}$, $\mathbf{z}$, $\mathbf{w}, \cdots$. Conversely, if $\mathbf{x} \neq (0, \cdots, 0)$, and if $\mathbf{x}$ is a linear combination of $\mathbf{y}$, $\mathbf{z}$, $\mathbf{w}, \cdots$, then one sees that $P\mathbf{x}$ is a dependent on $P\mathbf{y}$, $P\mathbf{z}$, $P\mathbf{w}, \cdots$.

Of the points

$$\mathbf{e}_0 = (1, 0, \cdots, 0), \quad \mathbf{e}_1 = (0, 1, 0, \cdots, 0), \quad \mathbf{e}_2 = (0, 0, 1, 0, \cdots, 0), \cdots,$$

one sees that none is a linear combination of the others; hence these

$n+1$ points are independent. Since every $(n+1)$-tuple is clearly a linear combination of $\mathbf{e}_0, \cdots, \mathbf{e}_n$, these form a basis of PNK^n. *Thus PNK^n is an n-dimensional space.*

The next main point is incorporated in the following two theorems.

THEOREM. *The set of solutions (different from* $(0, 0, \cdots, 0)$*) of a linear equation* $a_0x_0 + \cdots + a_nx_n = 0$*, where not all* $a_i = 0$*, is an* $(n-1)$*-dimensional space.*

PROOF. If $\mathbf{y}$, $\mathbf{z}$ are two solutions of the equation, then so is $\lambda\mathbf{y} + \mu\mathbf{z}$, from which one sees that the set of solutions yields a space (see p. 128). Next we find a basis for this space. Not all $a_i = 0$, say $a_0 \neq 0$. If we fix the values of $x_1, \cdots, x_n$, the equation $a_0x_0 + \cdots + a_nx_n = 0$ determines uniquely a solution $x_0, x_1, \cdots, x_n$. Let us fix successively $x_1 = 1$, $x_2 = 0, \cdots, x_n = 0$; then $x_1 = 0$, $x_2 = 1$, $x_3 = 0, \cdots$; etc. We get solutions $\mathbf{u}_1 = (c_1, 1, 0, \cdots, 0)$, $\mathbf{u}_2 = (c_2, 0, 1, 0, \cdots, 0)$, $\mathbf{u}_3 = (c_3, 0, 0, 1, 0, \cdots, 0)$, etc. These are n independent points. Thus the space in question is either $(n-1)$-dimensional or n-dimensional; but as not every point yields a solution to $a_0x_0 + \cdots + a_nx_n = 0$, the possibility of being n-dimensional is excluded.

THEOREM. *Every* $(n-1)$*-dimensional subspace* H *consists of the solutions (not counting* $(0, 0, \cdots, 0)$*) of a linear equation* $a_0x_0 + \cdots + a_nx_n = 0$ *with not all* $a_i = 0$.

PROOF. Let $\mathbf{x}^{(1)}, \cdots, \mathbf{x}^{(n)}$ be a basis of H. We propose to find a linear equation $a_0x_0 + \cdots + a_nx_n = 0$, with not all $a_i = 0$, having $\mathbf{x}^{(1)}, \cdots, \mathbf{x}^{(n)}$ as solutions. In that event, the $(n-1)$-dimensional space given by the equation contains the $(n-1)$-dimensional space H, and hence, by a previous theorem, must coincide with H.

To complete the proof, we need the following lemma.

LEMMA. *A system*

$$\begin{aligned} a_{00}x_0 + a_{01}x_1 + \cdots + a_{0n}x_n &= 0 \\ &\vdots \\ a_{m0}x_0 + a_{m1}x_1 + \cdots + a_{mn}x_n &= 0 \end{aligned}$$

of homogeneous linear equations in which the number of equations is less than the number of unknowns always has a nontrivial solution.

The proof is by induction on the number of unknowns and follows easily by eliminating one of the unknowns. To save space, we omit writing out the details.

To continue with the proof of the theorem, we are looking for numbers $a_0, \cdots, a_n$ such that

$$a_0x_0^{(1)} + a_1x_1^{(1)} + \cdots + a_nx_n^{(1)} = 0$$
$$a_0x_0^{(2)} + a_1x_1^{(2)} + \cdots + a_nx_n^{(2)} = 0$$
$$\vdots$$
$$a_0x_0^{(n)} + a_1x_1^{(n)} + \cdots + a_nx_n^{(n)} = 0.$$

By the lemma, this system has a nontrivial solution $(a_0, \cdots, a_n)$. Q.E.D.

Just as in the case $n = 2$, one can show for arbitrary n that $a_0x_0 + \cdots + a_nx_n = 0$ and $b_0x_0 + \cdots + b_nx_n = 0$ have the same solutions if and only if $(a_0, \cdots, a_n)$ and $(b_0, \cdots, b_n)$ are multiples of each other. Thus the hyperplanes of PNK^n can be represented, in one-to-one fashion, as the points of a second PNK^n. This second PNK^n is the dual of the first; but there is one thing that remains to be proved for this to be established. Dependency is defined in a standard way in the dual space and is also defined in a standard way in the second PNK^n referred to. It remains to be seen that the two ways coincide. Since dependence is defined in terms of collinearity, it is sufficient to see that the notion of collinearity in the dual space coincides with that notion in the second PNK^n. This comes to establishing the following theorem.

THEOREM. *Let* $H_\mathbf{a}$: $a_0x_0 + \cdots + a_nx_n = 0$; $H_\mathbf{b}$: $b_0x_0 + \cdots + b_nx_n = 0$; $H_\mathbf{c}$: $c_0x_0 + \cdots + c_nx_n = 0$ *be three distinct hyperplanes. Then* $H_\mathbf{c}$ *contains* $H_\mathbf{a} \cap H_\mathbf{b}$ *if and only if* $\mathbf{c}$ *is a linear combination of* $\mathbf{a}$ *and* $\mathbf{b}$.

PROOF. If $\mathbf{c} = \lambda\mathbf{a} + \mu\mathbf{b}$, then clearly every common solution of $a_0x_0 + \cdots + a_nx_n = 0$ and $b_0x_0 + \cdots + b_nx_n = 0$ is a solution of $c_0x_0 + \cdots + c_nx_n = 0$, so $H_\mathbf{c}$ contains $H_\mathbf{a} \cap H_\mathbf{b}$. Conversely, let $H_\mathbf{c}$ contain $H_\mathbf{a} \cap H_\mathbf{b}$. The space $H_\mathbf{a} \cap H_\mathbf{b}$ is $(n-2)$-dimensional. Let $P_\mathbf{v}$ be a point in $H_\mathbf{c}$ not in $H_\mathbf{a} \cap H_\mathbf{b}$. Write down the condition that some linear combination of $a_0x_0 + \cdots + a_nx_n$ and $b_0x_0 + \cdots + b_nx_n$ vanish at $\mathbf{v}$: namely, write $\lambda(a_0v_0 + \cdots + a_nv_n) + \mu(b_0v_0 + \cdots + b_nv_n) = 0$. This is one equation in two unknowns (λ, μ), which therefore has a nontrivial solution: let us designate one such as (λ, μ). Moreover, $\mathbf{d} = \lambda\mathbf{a} + \mu\mathbf{b} \neq 0$; otherwise $H_\mathbf{a}$ would equal $H_\mathbf{b}$. Consider now the hyperplane $H_\mathbf{d}$: it is spanned by a basis of $H_\mathbf{a} \cap H_\mathbf{b}$ and $P_\mathbf{v}$; hence $H_\mathbf{d}$ is contained in $H_\mathbf{c}$, because the mentioned basis for $H_\mathbf{d}$ belongs to $H_\mathbf{c}$. Hence $H_\mathbf{d} = H_\mathbf{c}$ since $H_\mathbf{d}$, $H_\mathbf{c}$ have the same dimension. Therefore $\mathbf{c}$ is a multiple of $\mathbf{d}$ and so is a linear combination of $\mathbf{a}$ and $\mathbf{b}$. Q.E.D.

We leave it as a not difficult exercise to show that all the above considerations hold also for K not necessarily commutative. As a precaution, we emphasize that one will speak of a *right* PNK^n and a *left* PNK^n: in a right PNK^n a point consists of a $(n+1)$-tuple $(\neq (0, \cdots, 0))$ and its nonzero right multiples; similarly for a left PNK^n. The dual of a right PNK^n turns out to be a left PNK^n. (In the case that K is not commutative, a given PNK^n and its dual do not look alike; in case

K is commutative it would appear that the given PNK^n and its dual PNK^n are alike. Well, they are alike, but they are not the same, so it is convenient to regard a point as a class of column $(n+1)$-tuples and a hyperplane as a class of row $(n+1)$-tuples; this is a convenient notational device.)

Exercises

1. Let $P_{\mathbf{a}}$ be a point of a PNK^n. The hyperplanes on $P_{\mathbf{a}}$ constitute a hyperplane of the dual space; and this hyperplane is represented as a (column) $(n+1)$-tuple $\mathbf{b}$ (and its non-zero multiples) in the dual of the dual of the given PNK^n. Show that $\mathbf{b}$ is a multiple of $\mathbf{a}$.

2. Let $P_{\mathbf{a}(0)}, P_{\mathbf{a}(1)}, \cdots, P_{\mathbf{a}(n)}$ be points in a PNK^m, and let them generate a c-dimensional space. Show that the points $P_{\mathbf{x}}$ for which $\mathbf{a}^{(0)}x_0 + \cdots + \mathbf{a}^{(n)}x_n = 0$ form an $(n-c-1)$-dimensional subspace of PNK^n. Hence, exhibiting $\mathbf{a}^{(0)}, \cdots, \mathbf{a}^{(n)}$ in the form

$$\begin{pmatrix} a_0^{(0)}, & \cdots, & a_0^{(n)} \\ & \vdots & \\ a_m^{(0)}, & \cdots, & a_m^{(n)} \end{pmatrix},$$

show that the rows of this array generate a c-dimensional space in PNK^n.

By the *column rank* of a matrix one means the dimension of the space generated by the columns; similarly for *row rank*. The last exercise shows that column rank = row rank; and this number is called the *rank* of the matrix.

Exercise. Let U, V be matrices such that UV is defined. The columns of UV are linear combinations of the columns of U, hence rank $UV \leq$ rank U. By a similar argument on the rows of V, rank $UV \leq$ rank V.

CHAPTER VIII

COORDINATE SYSTEMS AND LINEAR TRANSFORMATIONS

1. Coordinate systems. Let K be a commutative field and consider the analytic model of the plane which was constructed in Chapter III, Section 3. We recall that a point P is a column $\begin{pmatrix} x_0 \\ x_1 \\ x_2 \end{pmatrix}$ of numbers, not all zero, together with all multiples $\begin{pmatrix} \lambda x_0 \\ \lambda x_1 \\ \lambda x_2 \end{pmatrix}$, $\lambda \neq 0$. It is incorrect to refer to $\begin{pmatrix} x_0 \\ x_1 \\ x_2 \end{pmatrix}$ as a point, since a point is not just one triple, but a collection of triples; it is correct, however, to refer to the point $P\begin{pmatrix} x_0 \\ x_1 \\ x_2 \end{pmatrix}$. Similarly a line is a triple (a_0, a_1, a_2), not all zero, together with its multiples $(\mu a_0, \mu a_1, \mu a_2)$, $\mu \neq 0$. We use the abbreviated notation $\mathbf{x}$ to stand for the triple $\begin{pmatrix} x_0 \\ x_1 \\ x_2 \end{pmatrix}$; and similarly $\mathbf{a}$ to stand for the triple (a_0, a_1, a_2). One will have to tell from the context whether a triple $\mathbf{c}$ is a row triple or a column triple. We say that $\mathbf{x}$ *represents* the point $P_{\mathbf{x}}$, or is a *representative* of $P_{\mathbf{x}}$.

Two triples $\mathbf{x}$, $\mathbf{y}$ are called *equivalent* if $\mathbf{x} = \lambda \mathbf{y}$, $\lambda \neq 0$. In mathematics, whenever the word equivalent is used, say in the expression $\mathbf{x}$ is equivalent to $\mathbf{y}$, or notationally $\mathbf{x} \sim \mathbf{y}$, it is important to check the following three properties:

r: $\mathbf{x} \sim \mathbf{x}$ for every $\mathbf{x}$ (reflexivity)
s: if $\mathbf{x} \sim \mathbf{y}$, then $\mathbf{y} \sim \mathbf{x}$ (symmetry)
t: if $\mathbf{x} \sim \mathbf{y}$ and $\mathbf{y} \sim \mathbf{z}$, then $\mathbf{x} \sim \mathbf{z}$ (transitivity).

Obviously if these properties do not hold, then using the word equivalent in saying $\mathbf{x} \sim \mathbf{y}$ will lead to confusion, since the word equivalent has connotations which suggest the properties r, s, t. Conversely, though we will not stop over the proof, if these three conditions obtain, then one can use the word equivalent without confusion. One checks without difficulty that the equivalency of triples as just defined has the three properties.

All the triples $\mathbf{x}'$ equivalent to a given triple $\mathbf{x}$ form a so-called *equivalence class* of triples. One equivalence class consists just of $\mathbf{0} = \begin{pmatrix} 0 \\ 0 \\ 0 \end{pmatrix}$.

Otherwise every equivalence class is a point; and every point is an equivalence class. This is true also for lines, of course.

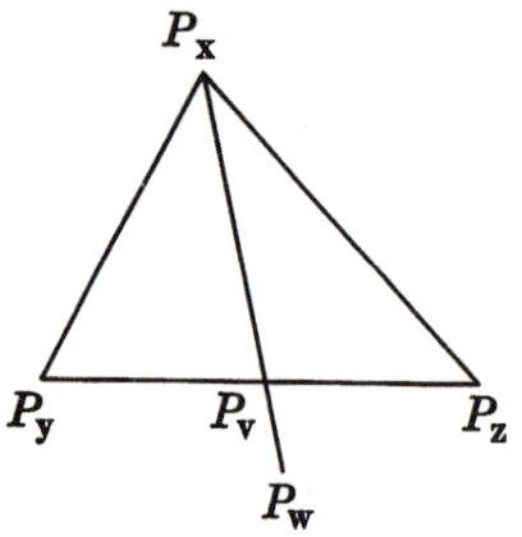

FIG. 8.1

We recall further (from Chapter III, Section 3) that, if $\mathbf{x}$ and $\mathbf{y}$ are representatives of distinct points, then $\mathbf{z} = \lambda\mathbf{x} + \mu\mathbf{y}$ represents a point collinear with them for every λ, μ with λ, μ not both zero. Conversely, every point on the line $P_\mathbf{x}P_\mathbf{y}$ is of the form $P_{\lambda\mathbf{x}+\mu\mathbf{y}}$.

From this last point, one can see the following: if $\mathbf{x}$, $\mathbf{y}$, $\mathbf{z}$ are three triples, none a linear combination of the others, and $\mathbf{w}$ is a triple not equal to zero ($\mathbf{w} \neq \mathbf{0}$), then $\mathbf{w} = \lambda\mathbf{x} + \mu\mathbf{y} + \nu\mathbf{z}$ for some λ, μ, ν. In fact, $\mathbf{x} \neq \mathbf{0}$, $\mathbf{y} \neq \mathbf{0}$, $\mathbf{z} \neq \mathbf{0}$ as one easily sees. Consider the points $P_\mathbf{x}$, $P_\mathbf{y}$, $P_\mathbf{z}$. Let $P_\mathbf{w} \neq P_\mathbf{x}$, say. The line $P_\mathbf{w}P_\mathbf{x}$ meets $P_\mathbf{y}P_\mathbf{z}$ in a point $\mathbf{v}$ (Fig. 8.1). Then $\mathbf{v}$ is a linear combination of $\mathbf{y}$ and $\mathbf{z}$, say $\mathbf{v} = \rho\mathbf{y} + \sigma\mathbf{z}$, and $\mathbf{w}$ is a linear combination of $\mathbf{x}$ and $\mathbf{v}$, say $\mathbf{w} = \alpha\mathbf{x} + \beta\mathbf{v}$. Then $\mathbf{w} = \alpha\mathbf{x} + \beta\rho\mathbf{y} + \beta\sigma\mathbf{z}$; i.e., $\mathbf{w}$ is a linear combination of $\mathbf{x}$, $\mathbf{y}$, $\mathbf{z}$, Q.E.D. We have then, $\mathbf{w} = \lambda\mathbf{x} + \mu\mathbf{y} + \nu\mathbf{z}$; and we say that λ, μ, ν are uniquely determined. Writing $\lambda\mathbf{x} + \mu\mathbf{y} + \nu\mathbf{z} = \lambda'\mathbf{x} + \mu'\mathbf{y} + \nu'\mathbf{z}$, we have to see that $\lambda = \lambda'$, $\mu = \mu'$, $\nu = \nu'$. We have $(\lambda - \lambda')\mathbf{x} + (\mu - \mu')\mathbf{y} + (\nu - \nu')\mathbf{z} = \mathbf{0}$; if, say, $\lambda - \lambda' \neq 0$, we conclude that $\mathbf{x} = (-1/(\lambda - \lambda'))((\mu - \mu')\mathbf{y} + (\nu - \nu')\mathbf{z})$, a contradiction. Thus every triple $\mathbf{w}$ can be written in one and only one way as a linear combination of $\mathbf{x}$, $\mathbf{y}$, $\mathbf{z}$. (This holds also for $\mathbf{w} = \mathbf{0}$, which was excluded above, namely, $\mathbf{0} = 0\mathbf{x} + 0\mathbf{y} + 0\mathbf{z}$).

DEFINITION. We say that a triple $\mathbf{w}$ *depends linearly* on triples $\mathbf{x}_1$, $\mathbf{x}_2$, $\mathbf{x}_3$, $\mathbf{x}_4, \cdots$ if there are numbers c_1, c_2, c_3, $c_4, \cdots$ such that $\mathbf{w} = c_1\mathbf{x}_1 + c_2\mathbf{x}_2 + \cdots$. We say that triples $\mathbf{x}_1$, $\mathbf{x}_2, \cdots$ are linearly *independent* if none of them is dependent on the others.

Exercise. Prove that $\mathbf{x}_1, \mathbf{x}_2, \cdots$ are independent if and only if $c_1\mathbf{x}_1 + c_2\mathbf{x}_2 + \cdots = 0$ implies $c_1 = c_2 = \cdots = 0$.

We can sum up the last paragraph as follows.

THEOREM. *Let* $\mathbf{x}, \mathbf{y}, \mathbf{z}$ *be linearly independent triples. Then any triple* $\mathbf{w}$ *can be written uniquely in the form* $\mathbf{w} = \lambda\mathbf{x} + \mu\mathbf{y} + \nu\mathbf{z}$.

Let $P_\mathbf{x}, P_\mathbf{y}, P_\mathbf{z}$ be three noncollinear points, $P_\mathbf{w}$ a fourth point. One could try to introduce *new* coordinates for $P_\mathbf{w}$ relative to $P_\mathbf{x}, P_\mathbf{y}, P_\mathbf{z}$ as follows: write $\mathbf{w} = \lambda\mathbf{x} + \mu\mathbf{y} + \nu\mathbf{z}$ and take (λ, μ, ν) as the new coordinates. This fails, though, because by varying the representative $\mathbf{x}$ of $P_\mathbf{x}$ one can vary λ at will; similarly for μ, ν. Clearly, if we can vary (λ, μ, ν) at will, they cannot serve as coordinates for $P_\mathbf{w}$.

One can remedy the defect as follows. Let $P_\mathbf{g}$ be a fourth point not collinear with any two of $P_\mathbf{x}, P_\mathbf{y}, P_\mathbf{z}$. Let $\mathbf{g}$ be a representative for $P_\mathbf{g}$ and let us impose the condition that representatives $\mathbf{x}, \mathbf{y}, \mathbf{z}$ be taken so that $\mathbf{g} = 1\cdot\mathbf{x} + 1\cdot\mathbf{y} + 1\cdot\mathbf{z}$, i.e., the multipliers should be 1. We can do this, because if $\mathbf{x}, \mathbf{y}, \mathbf{z}$ are any representatives of $P_\mathbf{x}, P_\mathbf{y}, P_\mathbf{z}$, we have $\mathbf{g} = \rho\mathbf{x} + \sigma\mathbf{y} + \tau\mathbf{z}$ and it is only necessary to take $\rho\mathbf{x}, \sigma\mathbf{y}, \tau\mathbf{z}$ instead of $\mathbf{x}, \mathbf{y}, \mathbf{z}$ as representatives to get the multipliers to be equal to 1. Moreover, since ρ, σ, τ are uniquely determined, we see that there is (one and) only one way to select the representatives of $P_\mathbf{x}, P_\mathbf{y}, P_\mathbf{z}$, so that $\mathbf{g}$ will be the sum of them. For any point $P_\mathbf{w}$, we write (for the $\mathbf{x}, \mathbf{y}, \mathbf{z}$ selected as stated) $\mathbf{w} = \lambda\mathbf{x} + \mu\mathbf{y} + \nu\mathbf{z}$ and call $\begin{pmatrix}\lambda\\ \mu\\ \nu\end{pmatrix}$ *the coordinates of* $P_\mathbf{w}$ *relative to* $P_\mathbf{x}, P_\mathbf{y}, P_\mathbf{z}, P_\mathbf{g}$. We also note that if $\kappa\mathbf{w}$ instead of $\mathbf{w}$ were taken as representative of $P_\mathbf{w}$, then we come out with coordinates $(\kappa\lambda, \kappa\mu, \kappa\nu)$ instead of (λ, μ, ν). But this does not disturb us, since even for the *old*, or original coordinates, there was an arbitrary factor coming in. Also observe that if $\kappa\mathbf{g}$ instead of $\mathbf{g}$ is taken as representative of $P_\mathbf{g}$, then this changes the $\mathbf{x}, \mathbf{y}, \mathbf{z}$ to be selected to $\kappa\mathbf{x}, \kappa\mathbf{y}, \kappa\mathbf{z}$, where $\mathbf{x}, \mathbf{y}, \mathbf{z}$ are the representatives first selected. Writing $\mathbf{w} = \left(\frac{\lambda}{\kappa}\right)(\kappa\mathbf{x}) + \left(\frac{\mu}{\kappa}\right)(\kappa\mathbf{y}) + \left(\frac{\nu}{\kappa}\right)(\kappa\mathbf{z})$ we get $(\lambda/\kappa, \mu/\kappa, \nu/\kappa)$ instead of (λ, μ, ν), again a change by a factor. (The last three sentences show that the variable elements entering the definition influence the result (λ, μ, ν) only by a factor.) To sum up:

DEFINITION. Let $P_\mathbf{x}, P_\mathbf{y}, P_\mathbf{z}, P_\mathbf{g}$ be four points, no three collinear. Let $\mathbf{x}, \mathbf{y}, \mathbf{z}, \mathbf{g}$ be representatives for them such that $\mathbf{g} = \mathbf{x} + \mathbf{y} + \mathbf{z}$. Let $P_\mathbf{w}$ be an arbitrary point and write $\mathbf{w} = \lambda\mathbf{x} + \mu\mathbf{y} + \nu\mathbf{z}$. Then $\begin{pmatrix}\lambda\\ \mu\\ \nu\end{pmatrix}$ are called *coordinates of* P_w *relative to the coordinate system* P_x, P_y, P_z, P_g. The points P_x, P_y, P_z are called the *vertices*, P_g the *unit point*, of the system.

Note that any triple of numbers $\begin{pmatrix}\lambda\\ \mu\\ \nu\end{pmatrix}$, not all zero, are (new) coordinates of some point, and that two such triples are coordinates of the same point if and only if each is a multiple of the other.

Given a point P, we speak of its *absolute coordinates* and its *relative coordinates* (coordinates relative to a system P_x, P_y, P_z, P_g). A point (of the analytic model) is by definition a collection of triples of numbers, and the entries of any one of these triples are its absolute coordinates: they are the numbers which specify it in the first place. Note, however, that the absolute coordinates are also relative coordinates relative to E_0, E_1, E_2, E_3, where these points have coordinates

$$\begin{pmatrix}1\\0\\0\end{pmatrix},\quad \begin{pmatrix}0\\1\\0\end{pmatrix},\quad \begin{pmatrix}0\\0\\1\end{pmatrix},\quad \begin{pmatrix}1\\1\\1\end{pmatrix},$$

respectively.

Instead of using $\mathbf{x}$, $\mathbf{y}$, $\mathbf{z}$, $\mathbf{g}$ it will be more convenient now to use $\mathbf{p}_0$, $\mathbf{p}_1$, $\mathbf{p}_2$, $\mathbf{p}_3$. Here

$$\mathbf{p}_0 = \begin{pmatrix}p_{00}\\p_{10}\\p_{20}\end{pmatrix},\qquad \mathbf{p}_1 = \begin{pmatrix}p_{01}\\p_{11}\\p_{21}\end{pmatrix},\qquad \mathbf{p}_2 = \begin{pmatrix}p_{02}\\p_{12}\\p_{22}\end{pmatrix}:$$

the second subscript gives the point, the first subscript gives the coordinates to which one is referring.

Consider Table 8.1. Here $\mathbf{p}_0, \mathbf{p}_1, \mathbf{p}_2, \mathbf{p}_3, \mathbf{x}, \mathbf{e}_0, \mathbf{e}_1, \mathbf{e}_2, \mathbf{e}_3$ are (chosen) abso-

TABLE 8.1

	P_0	P_1	P_2	P_3	P	E_0	E_1	E_2	E_3
	$\mathbf{p}_0$	$\mathbf{p}_1$	$\mathbf{p}_2$	$\mathbf{p}_3$	$\mathbf{x}$	$\mathbf{e}_0$	$\mathbf{e}_1$	$\mathbf{e}_2$	$\mathbf{e}_3$
Absolute Coordinates	p_{00}	p_{01}	p_{02}		x_0	1	0	0	1
	p_{10}	p_{11}	p_{12}		x_1	0	1	0	1
	p_{20}	p_{21}	p_{22}		x_2	0	0	1	1
Relative Coordinates	1	0	0	1	x_0'	c_{00}	c_{01}	c_{02}	
	0	1	0	1	x_1'	c_{10}	c_{11}	c_{12}	
	0	0	1	1	x_2'	c_{20}	c_{21}	c_{22}	
	$\mathbf{e}_0$	$\mathbf{e}_1$	$\mathbf{e}_2$	$\mathbf{e}_3$	$\mathbf{x}'$	$\mathbf{c}_0$	$\mathbf{c}_1$	$\mathbf{c}_2$	$\mathbf{c}_3$

lute coordinates of P_0, P_1, P_2, P_3, P, E_0, E_1, E_2, E_3, respectively; and $\mathbf{e}_0$, $\mathbf{e}_1$, $\mathbf{e}_2$, $\mathbf{e}_3$, $\mathbf{x}'$, $\mathbf{c}_0$, $\mathbf{c}_1$, $\mathbf{c}_2$, $\mathbf{c}_3$ are the respective relative coordinates (uniquely determined through $\mathbf{p}_0$, $\mathbf{p}_1$, $\mathbf{p}_2$). We set the following two problems:

(a) Given the absolute coordinates $\mathbf{x}$ of an arbitrary point P, how does one compute its relative coordinates $\mathbf{x}'$?

(b) Given the relative coordinates $\mathbf{x}'$ of an arbitrary point P, how does one compute its absolute coordinates $\mathbf{x}$?

Before entering into the solution proper, we consider some computational rules that ease what has to be written. These rules concern *matrices*: a matrix is simply a rectangular array of numbers (entries

$$\begin{pmatrix} a_{11} & a_{12} & \cdots & a_{1n} \\ a_{21} & a_{22} & \cdots & a_{2n} \\ & \vdots & & \\ a_{m1} & a_{m2} & \cdots & a_{mn} \end{pmatrix}$$

from K). Note that the first subscript gives the row, the second subscript the column. Let $(a_1, \cdots, a_n)$ be a $1 \times n$ matrix and $\begin{pmatrix} b_1 \\ b_2 \\ \vdots \\ b_n \end{pmatrix}$ an $n \times 1$ matrix (number of rows is given first). We define the product of these two matrices as follows:

$$(a_1, \cdots, a_n) \cdot \begin{pmatrix} b_1 \\ \vdots \\ b_n \end{pmatrix} = (a_1b_1 + a_2b_2 + \cdots + a_nb_n),$$

in other words, as the number (or more strictly speaking, the 1×1 matrix) $a_1b_1 + \cdots + a_nb_n$. This way of combining two n-tuples is very frequently encountered and is called the *inner product* of $\mathbf{a}$ and $\mathbf{b}$; notation $\mathbf{a} \cdot \mathbf{b}$; it is also called the *scalar product*. More generally we will define the product $A \cdot B$ of an $r \times s$ matrix A by an $s \times t$ matrix B. Note that we require the number of columns of A to be equal to the number of rows of B. This will be necessary because the product $A \cdot B$ is defined in terms of the inner products of the rows of A by the columns of B. In fact, by definition *$A \cdot B$ is an $r \times t$ matrix whose entry c_{ij} in the* ith *row and* jth *column is obtained as follows: c_{ij} is the inner product of the* ith *row of A and the* jth *column of B.*

Exercises

1. Compute $\begin{pmatrix} 2 & 3 & 4 \\ -1 & 2 & 3 \end{pmatrix} \begin{pmatrix} 5 & 1 & 0 & 2 \\ 6 & 1 & 0 & -1 \\ 1 & 1 & 0 & 3 \end{pmatrix}$

2. Find two 2×2 matrices A, B with numerical (not literal) entries such that $AB \neq BA$.

Let A, B, C be matrices, $r \times s$, $s \times t$, and $t \times u$, respectively. One can form $A \cdot B$; the result is $r \times t$; and this can be combined with C to give $(A \cdot B) \cdot C$. Similarly we can form $B \cdot C$, then combine it with A to get $A \cdot (B \cdot C)$. The following theorem is a basic one, called the associativity of matrix multiplication.

THEOREM. $(A \cdot B) \cdot C = A \cdot (B \cdot C)$.

Exercise. Verify the theorem for

$$A = \begin{pmatrix} a_{11} & a_{12} \\ a_{21} & a_{22} \end{pmatrix}, \qquad B = \begin{pmatrix} b_{11} & b_{12} \\ b_{21} & b_{22} \end{pmatrix}, \qquad C = \begin{pmatrix} c_{11} & c_{12} \\ c_{21} & c_{22} \end{pmatrix}.$$

PROOF OF THE THEOREM. We compute the element in the ith row and the lth column of $(A \cdot B) \cdot C$; and similarly compute the element in the ith row and the lth column of $A \cdot (B \cdot C)$. Then we compare the two, and see that they are equal: this will give the proof.

To get the element in the ith row and lth column of $(A \cdot B) \cdot C$, we need to know the whole ith row of $A \cdot B$. This is obtained by multiplying the ith row of A by the various columns of B; we get

$$\sum_{j=1}^{j=s} a_{ij} b_{j1}, \quad \sum_{j=1}^{j=s} a_{ij} b_{j2}, \cdots, \quad \sum_{j=1}^{j=s} a_{ij} b_{jt},$$

where $\sum_{j=1}^{j=s} a_{ij} b_{j1}$ abbreviates $a_{i1} b_{11} + a_{i2} b_{21} + \cdots + a_{is} b_{s1}$. Now we have to take the inner product of this row with the lth column of C. This gives $\sum_{k=1}^{k=t} \left[\left(\sum_{j=1}^{s} a_{ij} b_{jk} \right) c_{kl} \right]$. Similarly we compute the element in the ith row and lth column of $A \cdot (B \cdot C)$, and find it to be

$$\sum_{j=1}^{s} \left[a_{ij} \left(\sum_{k=1}^{t} b_{jk} c_{kl} \right) \right].$$

Now the question is: Why are these two expressions equal. Applying familiar rules on the first expression, i.e., expanding it and eliminating parentheses, one finds that it is the sum of the terms $a_{ij} b_{jk} c_{kl}$ for all possible values of $j (j = 1, \cdots, s)$ and $k (k = 1, \cdots, t)$. Similarly for the other expression. The two expressions are the sums of the same terms, hence equal. Q.E.D.

We go back to the questions raised on p. 145. Let $P = (\mathbf{p}_0, \mathbf{p}_1, \mathbf{p}_2)$; P is a 3×3 matrix. Consider the point P, which has relative coordinates $\mathbf{x}'$. By definition, then, $\mathbf{p}_0 x_0' + \mathbf{p}_1 x_1' + \mathbf{p}_2 x_2'$ are absolute coordinates for P. This last column can be written $P\mathbf{x}'$, i.e., $(\mathbf{p}_0, \mathbf{p}_1, \mathbf{p}_2)\mathbf{x}'$. The result can be put as follows.

THEOREM. *To get the absolute coordinates of a point, multiply its (column of) relative coordinates (on the left) by the matrix* $P = (\mathbf{p}_0, \mathbf{p}_1, \mathbf{p}_2)$. *Here* $\mathbf{p}_0$, $\mathbf{p}_1$, $\mathbf{p}_2$ *are the absolute coordinates selected for the vertices of the coordinate system* $P_0P_1P_2P_3$.

This theorem can be easily remembered. One cannot forget that the change in coordinates takes place by multiplying by a 3×3 matrix; the only question is: Which matrix? Now note that any 3×3 matrix times $\begin{pmatrix}1\\0\\0\end{pmatrix}$ gives its first column, times $\begin{pmatrix}0\\1\\0\end{pmatrix}$ its second, times $\begin{pmatrix}0\\0\\1\end{pmatrix}$ its third. On the other hand, the change transforms

$$\begin{pmatrix}1\\0\\0\end{pmatrix} \text{ to } \begin{pmatrix}p_{00}\\p_{10}\\p_{20}\end{pmatrix}, \quad \begin{pmatrix}0\\1\\0\end{pmatrix} \text{ to } \begin{pmatrix}p_{01}\\p_{11}\\p_{21}\end{pmatrix}, \quad \begin{pmatrix}0\\0\\1\end{pmatrix} \text{ to } \begin{pmatrix}p_{02}\\p_{12}\\p_{22}\end{pmatrix},$$

so that the matrix must be $\begin{pmatrix}p_{00} & p_{01} & p_{02}\\ p_{10} & p_{11} & p_{12}\\ p_{20} & p_{21} & p_{22}\end{pmatrix}$.

Now to go from the absolute to the relative. We have:

$$\begin{pmatrix}x_0\\x_1\\x_2\end{pmatrix} = \begin{pmatrix}1 & 0 & 0\\0 & 1 & 0\\0 & 0 & 1\end{pmatrix}\begin{pmatrix}x_0\\x_1\\x_2\end{pmatrix} = (\mathbf{e}_0, \mathbf{e}_1, \mathbf{e}_2)\mathbf{x}$$

$$= (\mathbf{p}_0c_{00} + \mathbf{p}_1c_{10} + \mathbf{p}_2c_{20},\ \mathbf{p}_0c_{01} + \mathbf{p}_1c_{11} + \mathbf{p}_2c_{21},\ \mathbf{p}_0c_{02} + \mathbf{p}_1c_{12} + \mathbf{p}_2c_{22})\mathbf{x}$$

$$= \left[\begin{pmatrix}p_{00} & p_{01} & p_{02}\\ p_{10} & p_{11} & p_{12}\\ p_{20} & p_{21} & p_{22}\end{pmatrix}\begin{pmatrix}c_{00} & c_{01} & c_{02}\\ c_{10} & c_{11} & c_{12}\\ c_{20} & c_{21} & c_{22}\end{pmatrix}\right]\begin{pmatrix}x_0\\x_1\\x_2\end{pmatrix} = (PC)\mathbf{x} = P(C\mathbf{x})$$

(by associativity)

$$= (\mathbf{p}_0, \mathbf{p}_1, \mathbf{p}_2)\begin{pmatrix}z_0\\z_1\\z_2\end{pmatrix},$$

where $\begin{pmatrix}z_0\\z_1\\z_2\end{pmatrix} = C\begin{pmatrix}x_0\\x_1\\x_2\end{pmatrix}$; but then, by definition, $\begin{pmatrix}z_0\\z_1\\z_2\end{pmatrix}$ are relative coordinates of P, and these, in Table 8.1, we called $\begin{pmatrix}x_0'\\x_1'\\x_2'\end{pmatrix}$. Therefore $\mathbf{x}' = C\mathbf{x}$.

THEOREM. *To get the relative coordinates of a point, multiply its (column of) absolute coordinates (on the left) by the matrix* $C = (\mathbf{c}_0, \mathbf{c}_1, \mathbf{c}_2)$. *Here* $(\mathbf{c}_0, \mathbf{c}_1, \mathbf{c}_2)$ *are the relative coordinates for the points* E_0, E_1, E_2.

Of course, for this theorem really to answer the question of how to pass from absolute to relative coordinates, we should know how to find C. We shall take this up systematically; meanwhile, here is an exercise.

Exercise. Let $P_0\colon \begin{pmatrix}1\\2\\3\end{pmatrix}$, $P_1\colon \begin{pmatrix}0\\0\\1\end{pmatrix}$, $P_2\colon \begin{pmatrix}2\\3\\4\end{pmatrix}$, $P_3\colon \begin{pmatrix}6\\6\\7\end{pmatrix}$. Find the coordinates of E_0, E_1, E_2, E_3 relative to the coordinate system $P_0P_1P_2P_3$.

Let A be a matrix with 3 columns. Note that $A\begin{pmatrix}1\\0\\0\end{pmatrix}$ gives the first column of A, and that $A\mathbf{x}$ will yield the first column of A for all A only if $\mathbf{x} = \begin{pmatrix}1\\0\\0\end{pmatrix}$. Similarly for the second and third columns. We designate the matrix $\begin{pmatrix}1&0&0\\0&1&0\\0&0&1\end{pmatrix}$ by E. It is neutral on the right: $AE = A$; and it is the only matrix with this property for all A. Similarly $EB = B$ for any 3-rowed matrix B.

One further remark on matrix notation. Just as for vectors we write $c\begin{pmatrix}x_0\\x_1\\x_2\end{pmatrix} = \begin{pmatrix}cx_0\\cx_1\\cx_2\end{pmatrix}$, so for matrices we write

$$c\begin{pmatrix}a_{00}&a_{01}&a_{20}\\a_{10}&a_{11}&a_{21}\\a_{20}&a_{21}&a_{22}\end{pmatrix} = \begin{pmatrix}ca_{00}&ca_{10}&ca_{20}\\ca_{10}&ca_{11}&ca_{21}\\ca_{20}&ca_{21}&ca_{22}\end{pmatrix}.$$

A similar notation holds for arbitrary matrices. Note that $cAB = AcB$ for any matrices A, B that can be multiplied.

THEOREM. $PC = E$. $CP = E$.

PROOF. Let P be a point having absolute coordinates $\mathbf{x}$. Then $C\mathbf{x}$ are relative and therefore $P(C\mathbf{x})$ are absolute coordinates of P. This does not quite give $PC\mathbf{x} = \mathbf{x}$ but only $PC\mathbf{x} = \kappa\mathbf{x}$. This is enough to show that PC is of the form $\begin{pmatrix}\kappa&0&0\\0&\kappa&0\\0&0&\kappa\end{pmatrix}$; moreover κ does not depend on $\mathbf{x}$, since forming PC does not depend on $\mathbf{x}$. It remains to be seen that

$\kappa = 1$. This will be seen if we see that for some point with coordinates $\mathbf{x}$ we get $PC\mathbf{x} = \mathbf{x}$. Consider the point with absolute coordinates $\begin{pmatrix}1\\0\\0\end{pmatrix}$. Then $C\begin{pmatrix}1\\0\\0\end{pmatrix} = \begin{pmatrix}c_{00}\\c_{10}\\c_{20}\end{pmatrix}$, and these numbers were taken so that $\mathbf{p}_0c_{00} + \mathbf{p}_1c_{10} + \mathbf{p}_2c_{20} = \begin{pmatrix}1\\0\\0\end{pmatrix}$. Hence for $\mathbf{x} = \mathbf{e}_0$, $PC\mathbf{e}_0 = \mathbf{e}_0$, so $\kappa = 1$. (Computations on p. 147 also show that $PC = E$.) Similarly we find

$$CP = \begin{pmatrix}\lambda & 0 & 0\\0 & \lambda & 0\\0 & 0 & \lambda\end{pmatrix},$$

and want still to prove that $\lambda = 1$. We get $PCP = P\lambda$, whence $P = P\lambda$ (since $PC = E$). This gives $\lambda = 1$.

THEOREM. *$PD = E$ implies $D = C$. $DP = E$ implies $D = C$.*

PROOF. $PD = E$ implies $CPD = C$; $CP = E$, whence $D = C$; similarly for the second statement.

The matrix P arose from a geometrical situation, but obviously any matrix P with its 3 columns linearly independent can occur in such a situation; we simply take for P_0, P_1, P_2 the points whose coordinates are given by the columns of P. Hence:

THEOREM. *For any matrix P with linearly independent columns, there is a matrix C such that $PC = E$, $CP = E$.*

THEOREM. *For P as in the last theorem, there is only one C such that $PC = E$; and there is only one C such that $CP = E$.*

PROOF. From $PC = E$ and $PD = E$ we are to prove $C = D$; here we may assume C to be the C of the last theorem. Then $CPD = CE = C$, so $ED = C$ or $D = C$. Similarly suppose $DP = E$. Then $DPC = C$ and $D = C$.

THEOREM. *The columns of C (where $PC = E$) are linearly independent.*

PROOF. We have to see that $\mathbf{c}_0x_0 + \mathbf{c}_1x_1 + \mathbf{c}_2x_2 = 0$ implies $x_0 = x_1 = x_2 = 0$. The given equality can be written $C\mathbf{x} = 0$. Multiplying by P, we get $PC\mathbf{x} = 0$, whence $\mathbf{x} = 0$. Q.E.D.

The relations between P and C are symmetric: P and C both have their columns linearly independent and each times the other, in either order, is E. Each is called the inverse of the other, and one writes $C=P^{-1}$, $P=C^{-1}$.

2. Determinants. We suppose that determinants and their basic properties are known from other courses, but let us recall some of them. First a determinant is a number associated in a certain way with a square matrix. For the matrix

$$\begin{pmatrix} a_{11} & \cdots & a_{1n} \\ & \vdots & \\ a_{n1} & \cdots & a_{nn} \end{pmatrix}, \quad \text{the notation is} \quad \begin{vmatrix} a_{11} & \cdots & a_{1n} \\ & \vdots & \\ a_{n1} & \cdots & a_{nn} \end{vmatrix}.$$

For a 2×2 matrix $\begin{pmatrix} a & b \\ c & d \end{pmatrix}$ the determinant has the value $ad-bc$. Notationally $\begin{vmatrix} a & b \\ c & d \end{vmatrix}=ad-bc$. We may regard this as a definition. To give the value of a 3×3 determinant $\begin{vmatrix} a_{11} & a_{12} & a_{13} \\ a_{21} & a_{22} & a_{23} \\ a_{31} & a_{32} & a_{33} \end{vmatrix}$, it is convenient to define the *cofactors*: the cofactor of any element a_{ij} is $\pm$ the 2×2 determinant obtained upon deleting the row and column containing a_{ij}; the $+$ or $-$ is taken according to the pattern $\begin{vmatrix} + & - & + \\ - & + & - \\ + & - & + \end{vmatrix}$. Thus the cofactor of a_{13} is $+\begin{vmatrix} a_{21} & a_{22} \\ a_{31} & a_{32} \end{vmatrix}$; the cofactor of a_{23} is $-\begin{vmatrix} a_{11} & a_{12} \\ a_{31} & a_{32} \end{vmatrix}$. The value of the determinant is the inner product of any row or column by the corresponding row or column of cofactors. For example, taking the second column:

$$\Delta = \begin{vmatrix} a_{11} & a_{12} & a_{13} \\ a_{21} & a_{22} & a_{23} \\ a_{31} & a_{32} & a_{33} \end{vmatrix} = -a_{12}\begin{vmatrix} a_{21} & a_{23} \\ a_{31} & a_{33} \end{vmatrix} + a_{22}\begin{vmatrix} a_{11} & a_{13} \\ a_{31} & a_{33} \end{vmatrix} - a_{32}\begin{vmatrix} a_{11} & a_{13} \\ a_{21} & a_{23} \end{vmatrix};$$

or taking the third row:

$$\Delta = a_{31}\begin{vmatrix} a_{12} & a_{13} \\ a_{22} & a_{23} \end{vmatrix} - a_{32}\begin{vmatrix} a_{11} & a_{13} \\ a_{21} & a_{23} \end{vmatrix} + a_{33}\begin{vmatrix} a_{11} & a_{12} \\ a_{21} & a_{22} \end{vmatrix}.$$

Letting A_{ij} represent the cofactor of a_{ij}, the first equality can be written $\Delta=a_{12}A_{12}+a_{22}A_{22}+a_{32}A_{32}$; the second $\Delta=a_{31}A_{31}+a_{32}A_{32}+a_{33}A_{33}$. In this way we get six expressions for Δ. We could take any one of them as the definition, but then it would be a theorem that the other five give the same result. For the time being, however, we are not concerned with an elegant treatment of determinants, but only a rapid recollection of some of the basic facts.

We recall:

(I) $$\begin{vmatrix} a_{11} & ca_{12} & a_{13} \\ a_{21} & ca_{22} & a_{23} \\ a_{31} & ca_{32} & a_{33} \end{vmatrix} = c\begin{vmatrix} a_{11} & a_{12} & a_{13} \\ a_{21} & a_{22} & a_{23} \\ a_{31} & a_{32} & a_{33} \end{vmatrix};$$

and similarly, a factor can be removed from any row or column.

(II) $$\begin{vmatrix} a_{11} & a_{12}+b_{12} & a_{13} \\ a_{21} & a_{22}+b_{22} & a_{23} \\ a_{31} & a_{32}+b_{32} & a_{33} \end{vmatrix} = \begin{vmatrix} a_{11} & a_{12} & a_{13} \\ a_{21} & a_{22} & a_{23} \\ a_{31} & a_{32} & a_{33} \end{vmatrix} + \begin{vmatrix} a_{11} & b_{12} & a_{13} \\ a_{21} & b_{22} & a_{23} \\ a_{31} & b_{32} & a_{33} \end{vmatrix};$$

and similarly for any row or column.

(III) Interchange of two rows or two columns negates the determinant.

(IV) $$\begin{vmatrix} 1 & 0 & 0 \\ 0 & 1 & 0 \\ 0 & 0 & 1 \end{vmatrix} = 1.$$

(V) determinant (AB) = (determinant A) (determinant B).

$$|AB| = |A|\ |B|.$$

(VI) Inner product of any row of A by cofactors of the same row gives $|A|$; and inner product of any row by cofactors of a different row $=0$; similarly for columns.

(VII) With appropriate notational changes, properties I–VI hold also for $n \times n$ determinants.

THEOREM. *If P has linearly independent columns, then* $\det P \neq 0$.

PROOF. $PC = E$ for some C. $\det P \cdot \det C = \det E$ (by V) and $=1$ (by (IV). Hence, $\det P \neq 0$. Q.E.D.

The *transpose* of $\begin{pmatrix} a_{11} & a_{12} & a_{13} \\ a_{21} & a_{22} & a_{23} \\ a_{31} & a_{32} & a_{33} \end{pmatrix}$ is obtained by interchanging rows and columns; in other words, it is $\begin{pmatrix} a_{11} & a_{21} & a_{31} \\ a_{12} & a_{22} & a_{32} \\ a_{13} & a_{23} & a_{33} \end{pmatrix}$. A square matrix A and its transpose A^T have equal determinants: $|A| = |A^T|$.

THEOREM. $$\begin{pmatrix} a_{11} & a_{12} & a_{13} \\ a_{21} & a_{22} & a_{23} \\ a_{31} & a_{32} & a_{33} \end{pmatrix}\begin{pmatrix} A_{11} & A_{21} & A_{31} \\ A_{12} & A_{22} & A_{32} \\ A_{13} & A_{23} & A_{33} \end{pmatrix} = \begin{pmatrix} \Delta & 0 & 0 \\ 0 & \Delta & 0 \\ 0 & 0 & \Delta \end{pmatrix},$$

where $\Delta = \begin{vmatrix} a_{11} & a_{12} & a_{13} \\ a_{12} & a_{22} & a_{23} \\ a_{31} & a_{32} & a_{33} \end{vmatrix}$.

PROOF. The entries in the product are obtained by multiplying the rows of the first matrix by the columns of the second; notice that these columns are the cofactors of the *rows* of the first matrix. By properties recalled above, the result is as stated.

THEOREM. *Assuming* $\Delta \neq 0$,

$$\begin{pmatrix} a_{11} & a_{12} & a_{13} \\ a_{21} & a_{22} & a_{23} \\ a_{31} & a_{32} & a_{33} \end{pmatrix} \begin{pmatrix} A_{11}/\Delta & A_{21}/\Delta & A_{31}/\Delta \\ A_{12}/\Delta & A_{22}/\Delta & A_{32}/\Delta \\ A_{13}/\Delta & A_{23}/\Delta & A_{33}/\Delta \end{pmatrix} = \begin{pmatrix} 1 & & \\ & 1 & \\ & & 1 \end{pmatrix} = E.$$

PROOF. This is just the last theorem with one slight modification. *This theorem answers the question of how to find C when P is given.*

Exercises

1. Compute $\begin{vmatrix} 1 & 0 & 0 \\ 0 & 1 & 0 \\ 0 & 0 & 1 \end{vmatrix}$; $\begin{vmatrix} 1 & 2 & -3 \\ 4 & 5 & -7 \\ 8 & 2 & -1 \end{vmatrix}$; $\begin{vmatrix} 0 & 0 & 0 \\ 1 & 2 & 3 \\ 4 & 5 & 6 \end{vmatrix}$; $\begin{vmatrix} 1 & 2 & 3 \\ 2 & 4 & 6 \\ 1 & 3 & 2 \end{vmatrix}$.

2. Show that $\begin{vmatrix} 1 & x & x^2 \\ 1 & y & y^2 \\ 1 & z & z^2 \end{vmatrix} = (x-y)(y-z)(z-x)$.

3. Show that $\begin{vmatrix} 1 & 2 & 3 \\ 2 & 4 & 7 \\ 3 & 0 & 0 \end{vmatrix} \neq 0$ and taking $P = \begin{pmatrix} 1 & 2 & 3 \\ 2 & 4 & 7 \\ 3 & 0 & 0 \end{pmatrix}$,

compute C such that $PC = E$.

3. Coordinate systems resumed. Let $P_0P_1P_2P_3$, $Q_0Q_1Q_2Q_3$ be two coordinate systems, systems I, II. How does one pass from the coordinates relative to system I to coordinates relative to system II? Let III represent the coordinate system $E_0E_1E_2E_3$, i.e., the system relative to which we get the absolute coordinates. To get from the coordinates relative to I to those relative to III, we multiply by a matrix (see p. 147); to get from III to II we multiply by a matrix again; hence to get from coordinates relative to I to those relative to II, we multiply by a matrix. Now which matrix? Consider Table 8.2.

TABLE 8.2

	P_0	P_1	P_2	P_3	Q_0	Q_1	Q_2	Q_3
Coordinates relative to system II	p_{00}	p_{01}	p_{02}		1	0	0	1
	p_{10}	p_{11}	p_{12}		0	1	0	1
	p_{20}	p_{21}	p_{22}		0	0	1	1
Coordinates relative to system I	1	0	0	1	q_{00}	q_{01}	q_{02}	
	0	1	0	1	q_{10}	q_{11}	q_{12}	
	0	0	1	1	q_{20}	q_{21}	q_{22}	

THEOREM. (Compare pp. 147–148). *To get the coordinates of a point relative to system* I, *multiply its column of coordinates relative to system* II (*on the left*) *by the matrix whose columns are the coordinates in system* I *of the vertices of the coordinate system* II.

PROOF. We need the matrix which times $\mathbf{e}_0$ gives $\mathbf{q}_0$, times $\mathbf{e}_1$ gives $\mathbf{q}_1$, times $\mathbf{e}_2$ gives $\mathbf{q}_2$. This is the matrix $(\mathbf{q}_0, \mathbf{q}_1, \mathbf{q}_2)$.

4. Coordinate changes, alias linear transformations. Let $\mathbf{x}$, $\mathbf{x}'$ be the coordinates of a point P in two coordinate systems. Then we have seen that $\mathbf{x}$, $\mathbf{x}'$ are related by equations: $\mathbf{x}' = A\mathbf{x}$. Here the matrix A has linearly independent columns.

The equations $\mathbf{x}' = A\mathbf{x}$ tell us how to pass from the old name of a point to a new name of the point. There is another way, however, of looking at these very same equations, namely, as a way of passing from the point $\mathbf{x}$ to another point $\mathbf{x}'$ whose coordinates (in the original coordinate system) are given by $A\mathbf{x}$. In the first way of looking at $\mathbf{x}' = A\mathbf{x}$ we have two coordinate systems and no transformation of points; in the second way, we have only one coordinate system, but the points are moved about according to the rule: $\mathbf{x} \to A\mathbf{x}$. In the first way, points are assigned new *names*, and one speaks of an *alias* transformation; in the second way, points are assigned new *locations*, and one speaks of an *alibi* transformation. We thus have two distinct interpretations of the formula $\mathbf{x}' = A\mathbf{x}$. It is clear that results for one interpretation can be translated into results for the other; because of this, in considering the formula $\mathbf{x}' = A\mathbf{x}$ one may shift back and forth between the two interpretations, and sometimes it is convenient to do this.

The transformation $\mathbf{x}' = A\mathbf{x}$, regarded as an alibi transformation, is called a *linear transformation*. In the above discussion of coordinate changes, the matrix A had, and necessarily had, linearly independent columns; but one can also consider a transformation of points given by $\mathbf{x}' = A\mathbf{x}$, where A is arbitrary. It is true, here, that if A has linearly dependent columns, then for some point (or points) $\mathbf{x}$ we shall have $A\mathbf{x} = \begin{pmatrix} 0 \\ 0 \\ 0 \end{pmatrix}$, so that $\mathbf{x}$ is not assigned to a point: all the same one can study such transformations—they are called *singular*. However, we will confine ourselves to the so-called *nonsingular linear transformations*.

Of course, distinct points in a coordinate change get distinct names. Therefore in terms of alibi transformations we have:

THEOREM. *A* (*nonsingular*) *linear transformation sends distinct points into distinct points.* (*Transformations that map distinct points into distinct points are called univalent or one to one.*)

Exercise. Prove this anew and directly.

Having chosen $P_{\mathbf{x}}$, $P_{\mathbf{y}}$, $P_{\mathbf{z}}$, $P_{\mathbf{g}}$ as the vertices and unit point of a new coordinate system, one obtains the new coordinates of a point $P_{\mathbf{w}}$ by writing $\mathbf{w}$ in the form $\mathbf{w}=\lambda\mathbf{x}+\mu\mathbf{y}+\nu\mathbf{z}$, and then λ, μ, ν are the new coordinates of $\mathbf{w}$. But conversely, any three numbers λ, μ, ν not all $=0$ are the new coordinates of some point, namely, of $\mathbf{w}=\lambda\mathbf{x}+\mu\mathbf{y}+\nu\mathbf{z}$ ($\mathbf{w}\neq\mathbf{0}$, since $\mathbf{x}$, $\mathbf{y}$, $\mathbf{z}$ are linearly independent). Thus every name is the alias of some point. In terms of alibi transformations we have:

THEOREM. *A linear transformation maps the plane onto itself, i.e., every point arises as an image. In terms of a formula, this says that if* $\mathbf{x}'\neq\mathbf{0}$, *then* $\mathbf{x}'=A\mathbf{x}$ *has a nontrivial solution (where, since we have agreed to confine ourselves to the so-called nonsingular linear transformations, it is understood that the columns of A are linearly independent).*

Exercises

1. Prove the last theorem anew and directly.

2. Translations, i.e., the transformations given in the affine plane by equations of the form $x'=x+a$, $y'=y+b$ are linear transformations. This is true also for rotations of the (real) Euclidean plane.

THEOREM. *A linear transformation sends collinear points into collinear points.*

PROOF. We may as well assume that the three points $\mathbf{u}$, $\mathbf{v}$, $\mathbf{w}$ are distinct. Then $\mathbf{w}=\lambda\mathbf{u}+\mu\mathbf{v}$. The transformed points are $A\mathbf{u}$, $A\mathbf{v}$, and $A\mathbf{w}=A(\lambda\mathbf{u}+\mu\mathbf{v})=A(\lambda\mathbf{u})+A(\mu\mathbf{v})=\lambda(A\mathbf{u})+\mu(A\mathbf{v})$, whence the transformed points are collinear. Q.E.D.

Thus a linear transformation induces a transformation of the lines, namely, we send a line l into the line l' that carries the images of the points of l. How does one obtain the equation of l'?

Let (ξ_0, ξ_1, ξ_2) be the coordinates of l, so that its equation is $\xi_0x_1+\xi_1x_1+\xi_2x_2=0$, or more compactly, $\boldsymbol{\xi}\cdot\mathbf{x}=0$. Let the linear transformation be given by $\mathbf{x}\rightarrow\mathbf{x}'=A\mathbf{x}$. Then $\mathbf{x}=A^{-1}\mathbf{x}'$ and we obtain $\boldsymbol{\xi}\cdot A^{-1}\mathbf{x}'=0$; conversely, from $\boldsymbol{\xi}\cdot A^{-1}\mathbf{x}'=0$ one obtains $\boldsymbol{\xi}\cdot\mathbf{x}=0$. Thus the condition on $\mathbf{x}'$ is that $\boldsymbol{\xi}A^{-1}\cdot\mathbf{x}'=0$; in other words, $\boldsymbol{\xi}A^{-1}$ are the coordinates of l' (and $\boldsymbol{\xi}A^{-1}\mathbf{x}=0$ is its equation).

Exercise. Carry out all the above considerations for 3-dimensional projective space.

We have defined a linear transformation of a plane as a (certain kind of) transformation of its points; but in a plane, the points and the lines appear on the same level, dually, neither to be preferred over the other.

The linear transformation does, indeed, induce a transformation on the lines, but if we want to maintain our impartiality to points and lines, then a linear transformation should be defined as a transformation on the points and on the lines simultaneously. It is a matter of a slight change in point of view, but if we want to proceed that way, we come to the following definition:

DEFINITION. *A linear transformation* is a transformation given by equations of the form

$$\mathbf{x} \to \mathbf{x}' = A\mathbf{x}$$
$$\boldsymbol{\xi} \to \boldsymbol{\xi}' = \boldsymbol{\xi} A^{-1},$$

where the first line gives the effect on points, the second the effect on lines.

From $\boldsymbol{\xi}' = \boldsymbol{\xi} A^{-1}$ we see that lines are transformed by a matrix multiplication, just as points are, and hence the last three theorems apply to lines also: i.e., the set of lines is mapped univalently onto itself, concurrent lines being mapped into concurrent lines.

Exercise. Since the linear transformation $\mathbf{x} \to \mathbf{x}' = A\mathbf{x}$, $\boldsymbol{\xi} \to \boldsymbol{\xi}' = A^{-1}\boldsymbol{\xi}$ maps the set of points univalently onto itself, and likewise for lines, it has an inverse transformation, namely, the one that sends $\mathbf{x}' \to \mathbf{x}$ and $\boldsymbol{\xi}' \to \boldsymbol{\xi}$ if the give transformation sends $\mathbf{x} \to \mathbf{x}'$, $\boldsymbol{\xi} \to \boldsymbol{\xi}'$. Show that this inverse is also a linear transformation.

Linear transformations are *incidence preserving*, i.e., if $\mathbf{x} \to \mathbf{x}'$ and $\boldsymbol{\xi} \to \boldsymbol{\xi}'$, then $\mathbf{x}$ incident with $\boldsymbol{\xi}$ implies $\mathbf{x}'$ is incident with $\boldsymbol{\xi}'$. This is practically obvious, since the definition was put down in accordance with the requirement of such an incidence preservation. Still we may note that $\boldsymbol{\xi}' \cdot \mathbf{x}' = \boldsymbol{\xi} A^{-1} A \mathbf{x} = \boldsymbol{\xi} \cdot \mathbf{x}$, in particular that $\boldsymbol{\xi} \cdot \mathbf{x} = 0$ implies $\boldsymbol{\xi}' \cdot \mathbf{x}' = 0$, which proves that incidence is preserved. Also note that if $\boldsymbol{\xi}$ is not incident with $\mathbf{x}$, that $\boldsymbol{\xi}'$ is not incident with $\mathbf{x}'$.

Summing up these remarks we see that a linear transformation is a univalent, onto, incidence preserving mapping of the set of points and set of lines of a plane, points going into points, lines into lines. Thus a linear transformation is an isomorphism of the plane onto itself. An "isomorphism onto self" is, however, an automorphism, so we can say:

THEOREM. *A linear transformation is an automorphism.*

The converse question of whether an automorphism is necessarily a linear transformation will be taken up in the next chapter.

Exercise: There is one and only one linear transformation taking the vertices and unit point of one coordinate system respectively into those of another.

5. A generalization from $n = 2$ to $n = 1$. For the most part the above considerations can be extended without difficulty to higher-dimensional spaces: instead of speaking of triples, one speaks of $(n+1)$-tuples. Except for some basic facts on the linear dependence and independence of such $(n+1)$-tuples, the considerations needed for the extension involve little more than changes in notation. Moreover, whatever the difficulties are, if we are content to stick to the plane, they need not be met. On the other hand, even confining ourselves to the plane, a 2-space, we cannot avoid generalization entirely, because within the 2-space, there are the 1-spaces, the lines; and as we shall see, one can speak of coordinates on a line. Here there already arises a basically new phenomenon, namely, one will deal not only with a space itself, but also with the space relative to its subspaces: in the case of the plane, with the plane in relation to its lines; in the case of 3-space, with the 3-space in relation to its planes and its lines.

Let then l be a line in a plane, and let $P_{\mathbf{c}_0}$, $P_{\mathbf{c}_1}$ be two points on l. For any third point $P_{\mathbf{c}_2}$, we can write $\mathbf{c}_2 = x_0\mathbf{c}_0 + x_1\mathbf{c}_1$, and try to take $\begin{pmatrix} x_0 \\ x_1 \end{pmatrix}$ as coordinates on l of $P_{\mathbf{c}_2}$; but since $\mathbf{c}_0$, $\mathbf{c}_1$ can be varied, still remaining representatives of $P_{\mathbf{c}_0}$, $P_{\mathbf{c}_1}$, we see, as similarly once before, that x_0 and x_1 can be varied arbitrarily. As before, we remedy the difficulty by taking three points $P_{\mathbf{c}_0}$, $P_{\mathbf{c}_1}$, $P_{\mathbf{c}_2}$ on l and limiting ourselves to representatives $\mathbf{c}_0$, $\mathbf{c}_1$, $\mathbf{c}_2$ such that $\mathbf{c}_2 = \mathbf{c}_0 + \mathbf{c}_1$. For any point $P_{\mathbf{c}}$ on l, if we write $\mathbf{c} = x_0\mathbf{c}_0 + x_1\mathbf{c}_1$, then we take x_0, x_1 as the coordinates of $P_{\mathbf{c}}$ on l; and the points $P_{\mathbf{c}_0}$, $P_{\mathbf{c}_1}$ are called the *vertices* of the coordinate system, $P_{\mathbf{c}_2}$ the *unit point*.

Let now $\mathbf{c}_0$, $\mathbf{c}_1$, $\mathbf{c}_2 = \mathbf{c}_0 + \mathbf{c}_1$ be the vertices and unit point of one system on l, call it system I; and let $\mathbf{p}_0$, $\mathbf{p}_1$, $\mathbf{p}_2 = \mathbf{p}_0 + \mathbf{p}_1$ be the vertices and unit point of another system, system II, on the same line. Writing any point $\mathbf{x}$ of l in the form $x_0\mathbf{c}_0 + x_1\mathbf{c}_1$ and in the form $x_0'\mathbf{p}_0 + x_1'\mathbf{p}_1$, $\begin{pmatrix} x_0 \\ x_1 \end{pmatrix}$ will be coordinates of $\mathbf{x}$ in system I and $\begin{pmatrix} x_0' \\ x_1' \end{pmatrix}$ will be coordinates of $\mathbf{x}$ in system II. Let $\mathbf{c}_0 = \lambda_{00}\mathbf{p}_0 + \lambda_{01}\mathbf{p}_1$, $\mathbf{c}_1 = \lambda_{10}\mathbf{p}_0 + \lambda_{11}\mathbf{p}_1$, so that

$$x_0\mathbf{c}_0 + x_1\mathbf{c}_1 = (x_0\lambda_{00} + x_1\lambda_{10})\mathbf{p}_0 + (x_0\lambda_{01} + x_1\lambda_{11})\mathbf{p}_1;$$

whence one finds that $\begin{pmatrix} x_0' \\ x_1' \end{pmatrix} = \begin{pmatrix} \lambda_{00} & \lambda_{10} \\ \lambda_{01} & \lambda_{11} \end{pmatrix} \begin{pmatrix} x_0 \\ x_1 \end{pmatrix}$. Thus, just as for coordinate systems of the plane, one passes from one system to another by a matrix multiplication; and one finds that, having fixed on the representatives $\mathbf{c}_0$, $\mathbf{c}_1$, $\mathbf{p}_0$, $\mathbf{p}_1$, the matrices for passing from system I to system II and from system II to system I are inverses of each other.

Exercises

1. Let $\mathbf{c}_0=\begin{pmatrix}1\\0\\0\end{pmatrix}$, $\mathbf{c}_1=\begin{pmatrix}0\\1\\0\end{pmatrix}$, $\mathbf{c}_2=\begin{pmatrix}1\\1\\0\end{pmatrix}$. What are the coordinates, with respect to this choice of vertices and unit point, of the point $\begin{pmatrix}x_0\\x_1\\0\end{pmatrix}$? Let $\mathbf{p}_0=\begin{pmatrix}0\\1\\0\end{pmatrix}$, $\mathbf{p}_1=\begin{pmatrix}1\\0\\0\end{pmatrix}$, $\mathbf{p}_2=\begin{pmatrix}1\\1\\0\end{pmatrix}$. What are the matrices connecting the coordinates of a point in these two systems?

2. Given the vertices and unit point $P_{\mathbf{c}_0}$, $P_{\mathbf{c}_1}$, $P_{\mathbf{c}_0+\mathbf{c}_1}$ of a coordinate system on l, show that one can take a coordinate system $P_{\mathbf{c}_0}, P_{\mathbf{c}_1}, P_{\mathbf{c}_2}, P_{\mathbf{c}_3}$ in the plane so that the coordinates of a point $P_{\mathbf{x}}$ on l relative to $P_{\mathbf{c}_0}$, $P_{\mathbf{c}_1}$, $P_{\mathbf{c}_0+\mathbf{c}_1}$ are the first two of the three coordinates of $P_{\mathbf{x}}$ relative to $P_{\mathbf{c}_0}, P_{\mathbf{c}_1}, P_{\mathbf{c}_2}, P_{\mathbf{c}_3}$.

6. Linear transformations on a line and from one line to another. Just as in the plane, the matrix multiplication applied to coordinates $\begin{pmatrix}x_0\\x_1\end{pmatrix}$ on a line may be regarded in the *alias* or *alibi* sense. In the former case we have a change of coordinates, of names; in the latter, a change of place, a transformation, and such transformation is called *linear.*

Let l, l' be two lines and fix coordinate systems on each. (l may be equal to l' but the coordinate systems may even then be distinct.) Let $A=\begin{pmatrix}a_{00} & a_{01}\\a_{10} & a_{11}\end{pmatrix}$ be a 2×2 matrix having an inverse, and consider the transformation from l to l' defined by the rule

$$\begin{pmatrix}x_0\\x_1\end{pmatrix}\to\begin{pmatrix}x_0'\\x_1'\end{pmatrix} = A\begin{pmatrix}x_0\\x_1\end{pmatrix}.$$

Such a transformation is called *linear,* a linear transformation from l to l'. It has an inverse, and the inverse is also linear.

Exercises

1. Show that coordinate systems on l and l' may be taken so that a given linear transformation takes the form $x_0'=x_0$, $x_1'=x_1$.

2. Every linear transformation from l to l' is induced by a linear transformation of the plane.

THEOREM. *Every perspectivity and every projectivity from one line to another (or the same) line is a linear transformation.*

PROOF. Let $\mathbf{a}$ be the center of a perspectivity from l to l'. Let $\mathbf{x}_0$, $\mathbf{x}_1$, $\mathbf{x}_0+\mathbf{x}_1$ be the vertices and unit point of a given coordinate system on l. Then the line joining $\mathbf{a}$ and $\mathbf{x}_0$ meets l' in a point $\rho\mathbf{a}+\sigma\mathbf{x}_0$. Since

$\sigma \neq \mathbf{0}$, one may divide by σ and take as representative for this point a triple of the form $c_0\mathbf{a}+\mathbf{x}_0$. Similarly we may say that $\mathbf{a}$ joined to $\mathbf{x}_1$ meets l' in $c_1\mathbf{a}+\mathbf{x}_1$. Now take any third point $\lambda\mathbf{x}_0+\mu\mathbf{x}_1$ on l. Where does it meet l'? Clearly $\lambda(c_0\mathbf{a}+\mathbf{x}_0)+\mu(c_1\mathbf{a}+\mathbf{x}_1)$ is on l'; but since this triple equals $(\lambda c_0+\mu c_1)\mathbf{a}+(\lambda\mathbf{x}_0+\mu\mathbf{x}_1)$, it represents a point on the join of $\mathbf{a}$ to $\lambda\mathbf{x}_0+\mu\mathbf{x}_1$. Thus the sought point is $\lambda(c_0\mathbf{a}+\mathbf{x}_0)+\mu(c_1\mathbf{a}+\mathbf{x}_1)$. If we take $c_0\mathbf{a}+\mathbf{x}_0$, $c_1\mathbf{a}+\mathbf{x}_1$, and their sum $(c_0\mathbf{a}+\mathbf{x}_0)+(c_1\mathbf{a}+\mathbf{x}_1)$ as the vertices and unit point of a coordinate system on l', the coordinates of that point in this system are λ, μ. Thus if $\begin{pmatrix}\lambda\\ \mu\end{pmatrix}$ are the coordinates of a point in the first system, and $\begin{pmatrix}\lambda'\\ \mu'\end{pmatrix}$ coordinates in the second, the transformation is given by $\lambda'=\lambda$, $\mu'=\mu$, or in matrix notation

$$\begin{pmatrix}\lambda'\\ \mu'\end{pmatrix} = \begin{pmatrix}1 & 0\\ 0 & 1\end{pmatrix}\begin{pmatrix}\lambda\\ \mu\end{pmatrix},$$

which proves that a perspectivity is a linear transformation. If we now pass from l' to l'' by a perspectivity, we may suppose the equations of this transformation are given by

$$\begin{pmatrix}\lambda''\\ \mu''\end{pmatrix} = \begin{pmatrix}1 & 0\\ 0 & 1\end{pmatrix}\begin{pmatrix}\lambda'\\ \mu'\end{pmatrix},$$

where $\begin{pmatrix}\lambda''\\ \mu''\end{pmatrix}$ are coordinates on l''. Thus, for an appropriate coordinate system on l'', the product of the two transformations is given by

$$\lambda'' = \lambda, \qquad \mu'' = \mu, \quad \text{or} \quad \begin{pmatrix}\lambda''\\ \mu''\end{pmatrix} = \begin{pmatrix}1 & 0\\ 0 & 1\end{pmatrix}\begin{pmatrix}\lambda\\ \mu\end{pmatrix};$$

therefore the transformation is linear. In this way, we see that any projectivity is linear. In fact, if the projectivity is from l to $\bar{l}$, then by an appropriate choice of a coordinate system on $\bar{l}$, the multiplying matrix is simply $\begin{pmatrix}1 & 0\\ 0 & 1\end{pmatrix}$. But note, further, that whatever the coordinate system on $\bar{l}$, the transformation from l to $\bar{l}$ is given by a matrix multiplication, since a shift of coordinates on $\bar{l}$ merely involves multiplication by a 2×2 matrix; a product of 2×2 matrices is a 2×2 matrix.

Exercise. In 3-space, every projectivity from one plane to another (or itself) is a linear transformation.

Converse to the last theorem, we have the following.

THEOREM. *Every linear transformation between two lines of a plane is a projectivity.*

PROOF. Let A, B, C be three points of the first line, A', B', C' the points on the second line into which they are sent. The main point is that *there is only one linear transformation taking* A, B, C, *respectively, into* A', B', C': since there is a projectivity which takes A, B, C into A', B', C', and the projectivity is linear, this will prove that every linear transformation is a projectivity. To see that there is only one linear transformation taking A, B, C into A', B', C', let us take A, B, C as vertices and unit point of a coordinate system on first line; and similarly with A', B', C' for the second line. Then we are looking for the 2×2 matrices that will take $\begin{pmatrix}1\\0\end{pmatrix}$, $\begin{pmatrix}0\\1\end{pmatrix}$, $\begin{pmatrix}1\\1\end{pmatrix}$, respectively, into $\begin{pmatrix}1\\0\end{pmatrix}$, $\begin{pmatrix}0\\1\end{pmatrix}$, $\begin{pmatrix}1\\1\end{pmatrix}$. The matrices taking $\begin{pmatrix}1\\0\end{pmatrix}$, $\begin{pmatrix}0\\1\end{pmatrix}$, respectively, into $\begin{pmatrix}1\\0\end{pmatrix}$, $\begin{pmatrix}0\\1\end{pmatrix}$ are of the form $\begin{pmatrix}\rho & 0\\0 & \sigma\end{pmatrix}$; and if in addition $\begin{pmatrix}1\\1\end{pmatrix}$ goes into $\begin{pmatrix}1\\1\end{pmatrix}$, we must have $\rho=\sigma$. Thus the desired matrices are of the form $\begin{pmatrix}\rho & 0\\0 & \rho\end{pmatrix}$. All of these represent the same linear transformation. This completes the proof.

We shall resume the present topic in higher-dimensional space after the next section.

REMARK. Even in the case of a noncommutative field K, a linear transformation sending $\begin{pmatrix}1\\0\end{pmatrix}$, $\begin{pmatrix}0\\1\end{pmatrix}$, $\begin{pmatrix}1\\1\end{pmatrix}$ into themselves is given by a matrix of the form $\begin{pmatrix}\rho & 0\\0 & \rho\end{pmatrix}$, but a matrix of this form need not give the identity transformation.

7. Cross-ratio. Let $P_{\mathbf{a}}$, $P_{\mathbf{b}}$, $P_{\mathbf{c}}$, $P_{\mathbf{d}}$ be four collinear points in PNK^n. Write $\mathbf{c}=\lambda\mathbf{a}+\mu\mathbf{b}$, $\mathbf{d}=\rho\mathbf{a}+\sigma\mathbf{b}$. For the present we define the cross-ratio $R(P_{\mathbf{a}}, P_{\mathbf{b}}; P_{\mathbf{c}}, P_{\mathbf{d}})$ to be $(\mu/\lambda)/(\sigma/\rho)$, leaving aside for the moment whether this agrees with our previous definition.

The first point to be observed is that the cross-ratio (though defined in terms of representatives) does not depend on the representatives chosen. Thus, if we take $\alpha\mathbf{a}$, $\beta\mathbf{b}$, $\gamma\mathbf{c}$, $\delta\mathbf{d}$ as representatives of $P_{\mathbf{a}}$, $P_{\mathbf{b}}$, $P_{\mathbf{c}}$, $P_{\mathbf{d}}$, and write

$$\gamma\mathbf{c} = (\gamma\lambda/\alpha)\alpha\mathbf{a}+(\gamma\mu/\beta)(\beta\mathbf{b}), \qquad \delta\mathbf{d} = (\delta\rho/\alpha)\alpha\mathbf{a}+(\delta\sigma/\beta)(\beta\mathbf{b}),$$

we find

$$[(\gamma\mu/\beta)/(\gamma\lambda/\alpha)]/[(\delta\sigma/\beta)/(\delta\rho/\alpha)] = (\mu/\lambda)/(\sigma/\rho).$$

The next thing to observe is that cross-ratio is invariant under a (nonsingular) linear transformation. If the transformation is effected by the matrix A, then $P_{\mathbf{a}}$, $P_{\mathbf{b}}$, $P_{\mathbf{c}}$, $P_{\mathbf{d}}$ go over into $P_{A\mathbf{a}}$, $P_{A\mathbf{b}}$, $P_{A\mathbf{c}}$, $P_{A\mathbf{d}}$, and since $A\mathbf{c}=\lambda A\mathbf{a}+\mu A\mathbf{b}$, $A\mathbf{d}=\rho A\mathbf{a}+\sigma A\mathbf{b}$, we see that the cross-ratio of the second four points is the same as the cross-ratio of the first four points.

Another way of expressing the content of the last paragraph is to say that cross-ratio is independent of the coordinate system. Actually, in the next chapter we shall define coordinate system abstractly and shall see that the coordinate systems of this chapter form a proper subset of all coordinate systems. However, the coordinate systems of this chapter are sufficient for many purposes, and cross-ratio is invariant under shift of coordinate system in the sense of the present chapter.

Exercise. Show that the cross-ratio, as defined above, agrees (when applicable) with our previous definition (in terms of distances).

8. Coordinate systems and linear transformations in higher-dimensional spaces. The facts considered above for planes have easy generalizations to PNK^n. First, a coordinate system is determined by $n+2$ points, any $n+1$ of which are independent. Let $P_{\mathbf{y}^0}$, $P_{\mathbf{y}^1}, \cdots$, $P_{\mathbf{y}^n}$, $P_{\mathbf{g}}$ be such points and choose representatives for them such that $\mathbf{g}=\mathbf{y}^0+\cdots+\mathbf{y}^n$. If $\mathbf{y}=\lambda_0\mathbf{y}^0+\cdots+\lambda_n\mathbf{y}^n$, then $\lambda_0,\cdots,\lambda_n$ are called the coordinates of $P_{\mathbf{y}}$ with respect to the coordinate system having $P_{\mathbf{y}^0},\cdots,P_{\mathbf{y}^n}$ as vertices and $P_{\mathbf{g}}$ as unit point. Passage from one coordinate system to another is given by matrix multiplication. Making the usual distinction between alibi and alias transformations, the alibi transformations effected by nonsingular matrix multiplication define linear transformations. As in the plane, there is one and only one linear transformation taking the vertices and unit point of one coordinate system respectively into those of another.

Let S^m be an m-dimensional subspace of PNK^n. Let $P_{\mathbf{y}^0},\cdots$, $P_{\mathbf{y}^m}$ be $m+1$ independent points of S^m and let an $(m+2)$-th point $P_{\mathbf{g}'}$ of S^m be chosen. Let representatives of $P_{\mathbf{y}^0},\cdots$, $P_{\mathbf{y}^m}$ be chosen so that $\mathbf{g}'=\mathbf{y}^0+\cdots+\mathbf{y}^m$. The points $P_{\mathbf{y}^0},\cdots$, $P_{\mathbf{y}^m}$ $P_{\mathbf{g}}$ are said to be the vertices and unit point of a coordinate system of S^m (immersed in PNK^n). If $P_{\mathbf{y}}$ is a point of S^m and $\mathbf{y}=\lambda_0\mathbf{y}^0+\cdots+\lambda_m\mathbf{y}^m$, then $\lambda_0,\cdots,\lambda_m$ are the coordinates of $P_{\mathbf{y}}$ relative to the coordinate system $P_{\mathbf{y}^0},\cdots$, $P_{\mathbf{y}^m}$, $P_{\mathbf{g}'}$. If $\bar{S}^m$ is another (or the same) m-dimensional subspace of PNK^n and $P_{\bar{\mathbf{y}}^0},\cdots$, $P_{\bar{\mathbf{y}}^m}$, $P_{\bar{\mathbf{g}}'}$ is a coordinate system of $\bar{S}^m$, then the formula $\boldsymbol{\lambda}=(\lambda_0,\cdots,\lambda_m)\to A\boldsymbol{\lambda}=\bar{\boldsymbol{\lambda}}=(\bar{\lambda}_0,\cdots,\bar{\lambda}_m)$, i.e.,

$$P_{\lambda_0\mathbf{y}^0+\cdots+\lambda_m\mathbf{y}^m}\to P_{\bar{\lambda}_0\bar{\mathbf{y}}^0+\cdots+\bar{\lambda}_m\bar{\mathbf{y}}^m},$$

where A is a nonsingular $(m+1)\times(m+1)$-matrix, defines a linear transformation from S^m onto $\bar{S}^m$. There is one and only one linear transformation taking the vertices of a given coordinate system of S^m respectively into the vertices of a given coordinate system of $\bar{S}^m$.

Exercises

1. The coordinates in S^m relative to $P_{y^0}, \cdots, P_{y^m}, P_{g'}$ are the first $m+1$ coordinates in S^n relative to some coordinate system.

2. Let P, Q, R, S be four collinear points in S^m. The cross-ratio $R(P, Q; R, S)$ is defined relative to S^m and to S^n $(=PNK^n)$. Show that the definitions coincide.

Let S^m, $\bar{S}^m$ be two m-dimensional spaces in PNK^n and let S^{n-m-1} be a space skew to both of them. By the formula $\dim L + \dim M = \dim [L, M] + \dim (L \cap M)$, one sees that the join of S^m and S^{n-m-1} is the whole space; this is also true for the join of $\bar{S}^m$ and S^{n-m-1}. Now let y be a point of S^m. The join of $\mathbf{y}$ and S^{n-m-1} is an S^{n-m}; and the formula just referred to shows that the intersection of S^{n-m} with $\bar{S}^m$ is a single point $\bar{y}$. *The transformation* $\mathbf{y} \rightarrow \bar{\mathbf{y}}$ *is called a perspectivity from* S^m *to* $\bar{S}^m$ *with axis* S^{n-m-1}. *A projectivity is by definition a product of perspectivities.*

THEOREM. *A perspectivity is a (nonsingular) linear transformation.*

PROOF. S^{n-m-1} is the intersection of $n-(n-m-1)=m+1$ hyperplanes. Let these have the equations

$$\mathbf{u}^0\mathbf{x} = 0, \quad \cdots, \quad \mathbf{u}^m\mathbf{x} = 0,$$

where $\mathbf{u}^i\mathbf{x}=0$ abbreviates $u_0{}^ix_0+u_1{}^ix_1+\cdots+u_n{}^ix_n=0$. Let $\mathbf{x}^1, \cdots, \mathbf{x}^{n-m}$ be a basis of S^{n-m-1}. Then $\mathbf{y}, \mathbf{x}^1, \cdots, \mathbf{x}^{n-m}$ form a basis of the join of $\mathbf{y}$ and S^{n-m-1} and hence

$$\bar{\mathbf{y}} = \lambda\mathbf{y} + \lambda_1\mathbf{x}^1 + \cdots + \lambda_{n-m}\mathbf{x}^{n-m}.$$

Moreover we may suppose $\lambda = 1$ ($\lambda \neq 0$, since $\bar{\mathbf{y}}$ is not in S^{n-m-1}). Since $\mathbf{x}^1, \cdots, \mathbf{x}^{n-m}$ are in S^{n-m-1}, taking the inner product of both sides of the last equality successively by $\mathbf{u}^0, \cdots, \mathbf{u}^m$, we find

$$\begin{aligned} \mathbf{u}^0\bar{\mathbf{y}} &= \mathbf{u}^0\mathbf{y} = \beta_0 \\ \mathbf{u}^1\bar{\mathbf{y}} &= \mathbf{u}^1\mathbf{y} = \beta_1 \\ &\cdots \\ \mathbf{u}^m\bar{\mathbf{y}} &= \mathbf{u}^m\mathbf{y} = \beta_m \end{aligned}$$

Having chosen a coordinate system in S^m, let $\gamma_0, \cdots, \gamma_m$ be the coordinates of $\mathbf{y}=(y_0, \cdots, y_n)$. Then the y_i are certain linear combinations of the γ_j and hence the β_i are linear combinations of the γ_j:

$$\beta_i = \delta_{i0}\gamma_0 + \cdots + \delta_{im}\gamma_m.$$

By a similar argument we find

$$\beta_i = \epsilon_{i0}\bar{\gamma}_0 + \cdots + \epsilon_{im}\bar{\gamma}_m,$$

where $(\bar{\gamma}_0, \cdots, \bar{\gamma}_m)$ are the coordinates of $\bar{y}$ with respect to some coordinate system in $\bar{S}^m$. In matrix notation, we have $\boldsymbol{\beta} = (\delta)\boldsymbol{\gamma}$ and $\boldsymbol{\beta} = (\epsilon)\bar{\boldsymbol{\gamma}}$, where (δ) and (ϵ) are $(m+1)\times(m+1)$ matrices. Moreover the columns of (δ) are linearly independent; for, if there were a nontrivial set of values $\gamma_0,\cdots,\gamma_m$ such that $(\delta)\boldsymbol{\gamma}=\mathbf{0}$, then we would find that $\mathbf{u}^0\mathbf{y}=0,\cdots,\mathbf{u}^m\mathbf{y}=0$, i.e., that $\mathbf{y}$ lies in S^{n-m-1}, which is not so. Hence the γ's can be solved for in terms of the β's: $\boldsymbol{\gamma}=(\delta^{-1})\boldsymbol{\beta}$. Similarly $\bar{\boldsymbol{\gamma}}=(\epsilon^{-1})\boldsymbol{\beta}$. We have $\boldsymbol{\gamma}=(\delta^{-1}\epsilon)\bar{\boldsymbol{\gamma}}$ and $\bar{\boldsymbol{\gamma}}=(\epsilon^{-1}\delta)\boldsymbol{\gamma}$. From our discussion on determinants (extended to $n\times n$ matrices) one sees that $\boldsymbol{\delta}^{-1}\epsilon$ is nonsingular. Hence $\boldsymbol{\gamma}$ and $\bar{\boldsymbol{\gamma}}$ are connected by a nonsingular linear transformation. Q.E.D.

COROLLARY. *A projectivity between S^m and $S^{m'}$ is a (nonsingular) linear transformation.*

Let $P_{\mathbf{y}^0},\cdots,P_{\mathbf{y}^m},P_{\mathbf{g}}$ and $P_{\bar{\mathbf{y}}^0},\cdots,P_{\bar{\mathbf{y}}^m},P_{\bar{\mathbf{g}}}$ be the vertices of coordinate systems in S^m, $\bar{S}^m$, respectively, $m<n$. We leave it as an exercise to show that there is a projectivity sending $P_{\mathbf{y}^0}\to P_{\bar{\mathbf{y}}^0},\cdots,P_{\mathbf{g}}\to P_{\bar{\mathbf{g}}}$. This transformation is linear, and since there is only one linear transformation sending $P_{\mathbf{y}^0}\to P_{\bar{\mathbf{y}}^0},\cdots,P_{\mathbf{g}}\to P_{\bar{\mathbf{g}}}$, we see that every linear transformation between an S^m and an $\bar{S}^m$, $m<n$, is a projectivity. Thus:

THEOREM. *The notions of linear transformation and projectivity from an S^m to an $\bar{S}^m$, $m<n$, coincide.*

REMARK. To see what a linear transformation of S^n onto itself is in synthetic terms, embed the S^n in an S^{n+1}.

9. Coordinates in affine space. By definition, an affine n-space arises from a projective n-space by deletion of a hyperplane. For simplicity, let us restrict ourselves to projective spaces isomorphic to a PNK^n with K commutative. By appropriate choice of a coordinate system, the deleted hyperplane can be made to have $x_0=0$ as equation. Thus every point $(x_0,\cdots,x_n)$ in the affine space is such that $x_0\neq 0$. The numbers $\frac{x_1}{x_0},\cdots,\frac{x_n}{x_0}$ are called the nonhomogeneous coordinates of the point. Three points P: $(y_1,\cdots,y_n)$, P': $(y_1',\cdots,y_n')$, P'': $(y_1'',\cdots,y_n'')$ are collinear if and only if

$$(1,y_1,\cdots,y_n),\quad (1,y_1',\cdots,y_n'),\quad (1,y_1'',\cdots,y_n'')$$

are linearly dependent; in the case $P\neq P'$, if and only if, for some λ, μ,

$$(1,y_1'',\cdots,y_n'') = \lambda(1,y_1,\cdots,y_n)+\mu(y_1',\cdots,y_n').$$

Thus P, P', P'' are collinear if and only if either $P=P'$ or

$$(y_1'',\cdots,y_n'') = \lambda(y_1,\cdots,y_n)+\mu(y_1',\cdots,y_n')$$

with $\lambda+\mu=1$.

DEFINITION. By ANK^n we mean the set of ordered n-tuples $(y_1, \cdots, y_n)$ with $y_1, \cdots, y_n$ in K together with the specification that $P: (y_1, \cdots, y_n)$, $P': (y_1', \cdots, y_n')$, $P'': (y_1'', \cdots, y_n'')$ are collinear if $P = P'$ or

$$(y_1'', \cdots, y_n'') = \lambda(y_1, \cdots, y_n) + \mu(y_1', \cdots, y_n')$$

with $\lambda + \mu = 1$.

The previous paragraph can be summed by saying that every affine n-space (at least the ones to which we are restricting ourselves) is isomorphic to an ANK^n. Moreover, going over the steps of that paragraph in slightly different order, one sees that ANK^n is an affine space for every K. In short, the concept of an affine space coincides with that of an ANK^n.

By a *linear transformation of an affine space* we mean a transformation induced by a linear transformation in the associated projective space that leaves the deleted hyperplane invariant. Let

$$x_0' = a_{00}x_0 + \cdots + a_{0n}x_n, \quad \cdots, \quad x_n' = a_{n0}x_0 + \cdots + a_{nn}x_n$$

be equations of a linear transformation in PNK^n. If $x_0 = 0$ is invariant, the point $(0, 1, 0, \cdots, 0)$ goes into a point on $x_0 = 0$; since it goes into $(a_{01}, a_{11}, \cdots, a_{n1})$, we conclude that $a_{01} = 0$; similarly that $a_{02} = \cdots = a_{0n} = 0$. Then $a_{00} \neq 0$, since $\det(a_{ij}) \neq 0$. The transformation effected by matrix A is the same as that determined by ρA, $\rho \neq 0$; hence we may suppose $a_{00} = 1$. In terms of the nonhomogeneous coordinates, the linear transformation is given by the formulas:

$$\begin{aligned} y_1' &= a_{10} + a_{11}y_1 + \cdots + a_{1n}y_n \\ y_2' &= a_{20} + a_{21}y_1 + \cdots + a_{2n}y_n \\ &\vdots \\ y_n' &= a_{n0} + a_{n1}y_1 + \cdots + a_{nn}y_n, \end{aligned}$$

with $\begin{vmatrix} a_{11} & \cdots & a_{1n} \\ & \vdots & \\ a_{n1} & \cdots & a_{nn} \end{vmatrix} \neq 0.$

A *vector space of dimension* n is the notion made up of an affine n-space together with the specification of one of its points O. It is convenient, and possible, to take coordinates so that $O = (0, \cdots, 0)$: we suppose this done. By a linear transformation of the vector space we mean a linear transformation of the associated affine space which leaves O fixed: in the above formulas, then, $a_{10} = \cdots = a_{n0} = 0$. Thus the linear transformations of a vector space are given by: $\mathbf{y}' = A\mathbf{y}$, where A is a nonsingular $n \times n$ matrix. Alongside these linear transformations one considers coordinate system changes of the form $\mathbf{y}' = A\mathbf{y}$.

Exercise. Let A be an n-dimensional vector space. A projective $(n-1)$-space S can be defined as follows. The points of S are the lines of A through O; and the lines of S are the planes of A through O. Show that the linear transformations of A induce linear transformations of S.

CHAPTER IX

COORDINATE SYSTEMS ABSTRACTLY CONSIDERED

1. Definition of coordinate system. So far we have considered coordinate systems only in the analytic model, in the projective model built up from a given commutative field K:† we will refer to such a model as a PNK (=projective number-plane over the field K). Here the points are by definition triples of numbers, or rather classes of such, and the numbers are then referred to as coordinates; in addition to these absolute coordinates, we have also explained how to specify the points in other ways, by the relative coordinates. Each of these ways of assigning numbers to points so that the points can be specified by means of the numbers we can call a coordinate system.

Now we want to consider the matter somewhat more abstractly, i.e., directly in terms of a given projective plane π (defined by means of the axioms of alignment, Desargues' and Pappus' Theorems). There are no coordinates given at the start; these will have to be introduced. First a definition:

> Definition. By a *coordinate system* of a plane π one means an isomorphic mapping of π onto some PNK, i.e., an assignment, one to one, of the points P of π to the points P' of PNK and of the lines l of π to the lines l' of PNK in such a way that if P' is assigned to P and l' to l, then P on l implies P' on l'. If P' is the point $\begin{pmatrix} x_0 \\ x_1 \\ x_2 \end{pmatrix}$, then P is said to have coordinates $\begin{pmatrix} x_0 \\ x_1 \\ x_2 \end{pmatrix}$ in the coordinate system in question.

† The assumption that K is commutative is made solely in order to simplify initial study. The results of the section, as well as the proofs—aside from slight modifications—also hold for noncommutative K.

Exercise. If the conditions of the definition are met, then P not on l implies P' not on l'. (Recall that by saying *onto*, we mean that each point P' of PNK and each line l' of PNK arise from a point of π and a line of π.)

The definition can be made vivid by imagining the given plane π spread out in front of one, like a blank sheet of paper (see Fig. 9.1); the PNK is pictured as a sheet of graph paper, with the points (and lines) marked with numbers. The graph paper is placed on top of the blank sheet (or, rather, the blank sheet is placed under the graph paper). A point P of π is then specified by the point P' of the graph paper which

FIG. 9.1

lies over P. By moving the graph sheet around, one gets different coordinate systems.

Earlier we showed how,, on the basis of the axioms, one can introduce coordinates into the projective plane π. Our work there can be phrased as in the following theorem.

THEOREM. *Every projective plane has at least one coordinate system.*

The problem we set ourselves is this: *What are all the coordinate systems of a given projective plane π?*

Let $\mathfrak{C}_1$ be a coordinate system of π; the definition of $\mathfrak{C}_1$ involves a field K_1. Let $\mathfrak{C}_2$ be another coordinate system of π, with associated field K_2. There is nothing in the definition of coordinate system that states we must have $K_1 = K_2$ or K_1 isomorphic to K_2.

THEOREM. *Let $\mathfrak{C}_1, \mathfrak{C}_2$ be coordinate systems of π relative to fields K_1, K_2. Then K_1 and K_2 are isomorphic.*

PROOF. Let $E_0^{(1)}, E_1^{(1)}, E_2^{(1)}, E_3^{(1)}$ be the points $\begin{pmatrix}1\\0\\0\end{pmatrix}, \begin{pmatrix}0\\1\\0\end{pmatrix}, \begin{pmatrix}0\\0\\1\end{pmatrix}, \begin{pmatrix}1\\1\\1\end{pmatrix}$ of the plane PNK_1; here the entries 0, 1 are from K_1. Let $E_0^{(2)}, E_1^{(2)}, E_2^{(2)}, E_3^{(2)}$ be the points $\begin{pmatrix}1\\0\\0\end{pmatrix}, \begin{pmatrix}0\\1\\0\end{pmatrix}, \begin{pmatrix}0\\0\\1\end{pmatrix}, \begin{pmatrix}1\\1\\1\end{pmatrix}$ of the plane PNK_2; here the entries 0, 1 are from K_2. Let F_0, F_1, F_2, F_3 be the points of π which map under $\mathfrak{C}_1$ into $E_0^{(1)}, E_1^{(1)}, E_2^{(1)}, E_3^{(1)}$, respectively. It is not necessarily the case that under $\mathfrak{C}_2$ the points F_0, F_1, F_2, F_3 go into $E_0^{(2)}, E_1^{(2)}, E_2^{(2)}, E_3^{(2)}$, respectively; but we will show why we may assume they do. Let, in fact, $\mathfrak{C}_2$ send F_0, F_1, F_2, F_3 into G_0, G_1, G_2, G_3. Since $\mathfrak{C}_2$ is an isomorphism, no three of the points G_0, G_1, G_2, G_3 are collinear. Hence there is a linear transformation L_A (given by matrix A) that sends G_0, G_1, G_2, G_3, respectively, into $E_0^{(2)}, E_1^{(2)}, E_2^{(2)}, E_3^{(2)}$. One checks that $\mathfrak{C}_2 L_A$ (=the mapping $\mathfrak{C}_2$ followed by L_A) is also an isomorphic mapping of π onto PNK_2. Replacing $\mathfrak{C}_2$ by the coordinate system $\mathfrak{C}_2' = \mathfrak{C}_2 L_A$, we may suppose (writing $\mathfrak{C}_2$ for $\mathfrak{C}_2'$) that $\mathfrak{C}_2$ sends F_0, F_1, F_2, F_3, respectively, into $E_0^{(2)}, E_1^{(2)}, E_2^{(2)}, E_3^{(2)}$.

The transformation $\mathfrak{C}_1^{-1}\mathfrak{C}_2$ maps PNK_1 onto PNK_2 and sends $E_0^{(1)}, E_1^{(1)}, E_2^{(1)}, E_3^{(1)}$ into $E_0^{(2)}, E_1^{(2)}, E_2^{(2)}, E_3^{(2)}$ (Fig. 9.2). Since $\mathfrak{C}_1^{-1}\mathfrak{C}_2$ sends $E_0^{(1)}$ into $E_0^{(2)}$ and $E_1^{(1)}$ into $E_1^{(2)}$, it sends every point

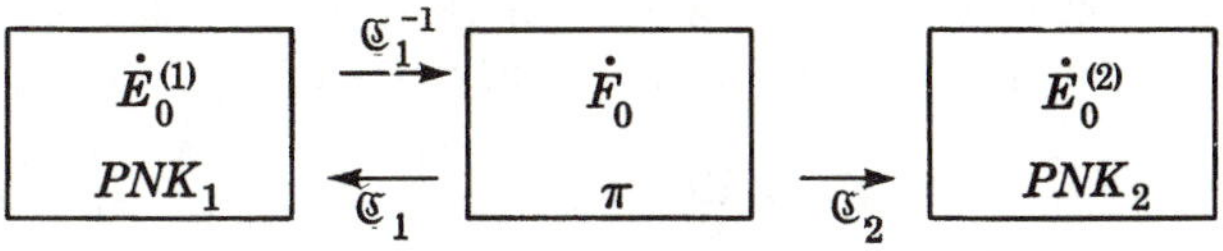

FIG. 9.2

$(1, x, 0)$ on line $E_0^{(1)}E_1^{(1)}$ into a point on $E_0^{(2)}E_1^{(2)}$, i.e., into a point $(1, x', 0)$. In this way we get a mapping $\alpha : x \to x'$ of the field K_1 onto K_2. We claim: α is an isomorphism. Proof: consider the construction by straight lines that leads from the points $(1, x, 0)$, $(1, y, 0)$ to the point $(1, x+y, 0)$; see p. 77. Let $\mathfrak{C}_1^{-1}\mathfrak{C}_2$ send these points, respectively, into $(1, x', 0)$, $(1, y', 0)$, $(1, u', 0)$. Then the configuration that connects $(1, x+y, 0)$ to $(1, x, 0)$ and $(1, y, 0)$ is mapped by $\mathfrak{C}_1^{-1}\mathfrak{C}_2$ into a configuration that connects $(1, u', 0)$ to $(1, x', 0)$, $(1, y', 0)$ in a similar way; this shows that $u' = x' + y'$, or what comes to the same thing $(x+y)\alpha = x\alpha + y\alpha$. Similarly one proves $(xy)(\alpha) = x(\alpha)\cdot y(\alpha)$. Thus α is an isomorphism, and maps K_1 isomorphically onto K_2. Q.E.D.

Let $\mathfrak{C}$ be a given coordinate system of π, say, into PNK. Then the above theorem tells us in effect that in looking for all the coordinate systems of π we may as well restrict ourselves to the one field K and the corresponding PNK. We can simply neglect the fields isomorphic to K, since whatever we can say for K we can say for the isomorphic fields, and vice versa. On the other hand, we cannot neglect the isomorphisms of K onto itself, the so-called *automorphisms* of K.

Exercise. Let K be the field of complex numbers. Each number in K can be written in one and only one way in the form $a+bi$, where a, b are real and i is a specified square root of -1. Show that the mapping $a+bi \to a-bi$ is an automorphism of K. (The number $a-bi$ is called the complex conjugate of $a+bi$.)

Starting with the given $\mathfrak{C}$, let $\mathfrak{C}'$ be another coordinate system into PNK. Then $\mathfrak{C}^{-1}\mathfrak{C}'$ is an isomorphic mapping $\mathfrak{D}$ of PNK onto itself. Since $\mathfrak{C}' = \mathfrak{C}\mathfrak{D}$ we would know all coordinate systems of π if we knew all isomorphisms of PNK onto itself, i.e., all automorphisms of PNK.

Let α be a given automorphism of the field K. We define a mapping of PNK as follows. Let $\begin{pmatrix} x_0 \\ x_1 \\ x_2 \end{pmatrix}$ be coordinates of a point. Then we send this point into $\begin{pmatrix} x_0^\alpha \\ x_1^\alpha \\ x_2^\alpha \end{pmatrix}$, where for notational purposes we have written the effect on a number x by α as x^α. Note that x_0^α, x_1^α, x_2^α are not all $= 0$, since not all of x_0, x_1, x_2 are $= 0$, and an isomorphism sends only 0 into 0. Thus x_0^α, x_1^α, x_2^α are the coordinates of some point. Note also that if we take other coordinates of $\begin{pmatrix} x_0 \\ x_1 \\ x_2 \end{pmatrix}$, say $\begin{pmatrix} \lambda x_0 \\ \lambda x_1 \\ \lambda x_2 \end{pmatrix}$, then the process described yields $\begin{pmatrix} \lambda^\alpha x_0^\alpha \\ \lambda^\alpha x_1^\alpha \\ \lambda^\alpha x_2^\alpha \end{pmatrix}$, i.e., the same point: so we really have a mapping of the points of PNK into points of PNK. One verifies that this is a univalent, onto mapping. Similarly we map the line (c_0, c_1, c_2) into the line $(c_0^\alpha, c_1^\alpha, c_2^\alpha)$.

This mapping preserves incidence: i.e., if $\begin{pmatrix} x_0 \\ x_1 \\ x_2 \end{pmatrix}$ is on (c_0, c_1, c_2), then $\begin{pmatrix} x_0^\alpha \\ x_1^\alpha \\ x_2^\alpha \end{pmatrix}$, the image of the point, is on $(c_0^\alpha, c_1^\alpha, c_2^\alpha)$, the image of the line. To see this we have to see that $c_0x_0 + c_1x_1 + c_2x_2 = 0$ implies $c_0^\alpha x_0^\alpha + c_1^\alpha x_1^\alpha + c_2^\alpha x_2^\alpha = 0$. Applying α to both sides of the equation $c_0x_0 + c_1x_1 + c_2x_2 = 0$, we get

$$(c_0x_0 + c_1x_1 + c_2x_2)^\alpha = 0^\alpha,$$

hence $(c_0x_0)^\alpha + (c_1x_1)^\alpha + (c_2x_2)^\alpha = 0^\alpha$, since an automorphism maps sums into sums; and $c_0^\alpha x_0^\alpha + c_1^\alpha x_1^\alpha + c_2^\alpha x_2^\alpha = 0$, since an automorphism maps products into products and 0 into 0.

Thus an automorphism α of K gives rise to an automorphism of PNK. We designate this automorphism P^α. Moreover we can recover α from P^α, because if P^α sends $(x, 1, 0)$ into $(x', 1, 0)$, then clearly $x' = x^\alpha$. Another way of stating this is that distinct automorphisms α_1, α_2 of K yield distinct automorphisms P^{α_1}, P^{α_2} of PNK.

The automorphism P^{α} leaves the points E_0, E_1, E_2, E_3 invariant. Now we prove the converse.

THEOREM. *If an automorphism T of PNK leaves the points E_0, E_1, E_2, E_3 invariant, then T is of the form P^{α} for some automorphism α of the field K.*

PROOF. Since T sends E_0 into E_0 and E_1 into E_1, it sends any point $(1, x, 0)$ of E_0E_1 into a point $(1, x', 0)$ of E_0E_1. The argument above that showed that any two coordinate systems of a plane π must be relative to isomorphic fields here shows that the mapping $x \to x'$ is an automorphism. Call this automorphism α, and consider the automorphism P^{α} of PNK. Then clearly T and P^{α} coincide on E_0E_1, i.e., have the same effects on the points of E_0E_1. One also sees that T and P^{α} coincide on the lines E_1E_2, E_2E_0. Since any line l cuts the sides of the triangle $E_0E_1E_2$ in at least two points, say C and D, the automorphisms T and P^{α} have the same effect on l: if T and P^{α} send C, D into C', D', then they must both send l into $C'D'$. Let now P be any point. On P take two lines m, n. If T and P^{α} take m, n into m', n', they must both take P into $m' \cdot n'$. Thus T and P^{α} have the same effect on all points and all lines, and hence $T = P^{\alpha}$. Q.E.D.

THEOREM. *Every automorphism of PNK is of the form L_AP^{α}, where L_A is the linear transformation of matrix A and α is an automorphism of K.*

PROOF. Let T be the given automorphism of PNK and let T^{-1} send E_0, E_1, E_2, E_3 into G_0, G_1, G_2, G_3, so that T sends G_0, G_1, G_2, G_3 into E_0, E_1, E_2, E_3. There also exists a linear transformation, L_A, taking G_0, G_1, G_2, G_3 into E_0, E_1, E_2, E_3. Hence, $L_A{}^{-1}T$ sends E_0, E_1, E_2, E_3 into E_0, E_1, E_2, E_3. By the last theorem, then, there exists an automorphism α such that $L_A{}^{-1}T = P^{\alpha}$, whence $T = L_AP^{\alpha}$. Q.E.D.

Exercise. Using the fact that the inverse of an automorphism is also an automorphism, show that every automorphism of PNK is also of the form $P^{\beta}L_B$.

The following theorem is a corollary of the last one, and answers the question we set ourselves above.

THEOREM. *Let $\mathfrak{C}$ be a given coordinate system of π, $\mathfrak{C}: \pi \to PNK$. Then every coordinate system of π is (except for isomorphisms of K) of the form $\mathfrak{C}L_AP^{\alpha}$, where L_A is a linear transformation of PNK and α is an automorphism of K.*

DEFINITION. Transformations of the form L_AP^{α} are called *semi-linear transformations.*

THEOREM. *Every semilinear transformation of PNK is an automorphism of PNK and, conversely, every automorphism is a semilinear transformation.*

PROOF. The first part is immediately verified. For the second, consider the identity coordinate system of *PNK*, i.e., the one which assigns to $\begin{pmatrix} x_0 \\ x_1 \\ x_2 \end{pmatrix}$ the coordinates $\begin{pmatrix} x_0 \\ x_1 \\ x_2 \end{pmatrix}$. Any automorphism of *PNK* is an isomorphism of *PNK* onto itself, i.e., it is a coordinate system, hence by the last theorem is of the form: identity $\times L_A P^\alpha$. Q.E.D.

2. Definition of a geometric object. Given a projective plane π, (plane) projective geometry is, of course, the study of the plane π and especially of the configurations in π. In studying π, however, we sometimes introduce a coordinate system, and the study becomes analytical. As long as the objects one is dealing with come directly from π, there is no question that these objects are of interest in a projective study. But having initiated an analytical study, one sometimes defines things analytically; for example, we defined a linear transformation as a transformation given in a *PNK* by matrix multiplication. Since any plane π can be viewed as a *PNK*, it would appear that linear transformations are of projective interest. Actually, however, "linear transformation" as thus defined is rather a property of π and a particular coordinate system than of π itself. To see that "linear transformation" is really of projective interest, we should see that a transformation given by matrix multiplication in one coordinate system is also given by matrix multiplication in any other. If that turns out to be so, then a linear transformation will be a property of π itself, and not of π and some special coordinate system. (We shall in a moment prove this to be so; meanwhile, we are making general remarks of an orientating character.) Thus we can say:

Analytic projective geometry is the study in PNK of properties that, though perhaps expressed in terms of a coordinate system, are independent of the coordinate system in which they are expressed.

Another way of stating the same thing is this:

Analytic projective geometry is the study in PNK of properties of analytic expressions that are invariant under a change of coordinate system.

Or still another way is:

Analytic projective geometry is the study in PNK of invariants of the semilinear transformations.

As an illustration, consider the line l whose equation in a given coordinate system is $x_0 = 0$. Usually a linear equation has three terms. Does the fact that the equation for l has only one term have any projective significance? No, since in other coordinate systems the equation of l may well have three terms. The fact that $x_0 = 0$ has only one term is a property of l relative to a certain coordinate system.

As another example, let us consider cross-ratio. In π take four collinear points P_1, P_2, P_3, P_4. Take a coordinate system in π such that these points lie on $x_2 = 0$ and let P_j have coordinates $(1, x_j, 0)$. Define $R(P_1, P_2; P_3, P_4)$ in the expected way as $\dfrac{x_3 - x_1}{x_4 - x_1} \Big/ \dfrac{x_3 - x_2}{x_4 - x_2}$. The question is: Does this concept have projective significance? Is it invariant under all automorphisms of π? The answer depends on the field K of π. Suppose K is the complex number field and

$$R(P_1, P_2; P_3, P_4) = i.$$

By an automorphism of π, the points P_j can be sent into points P_j' such that $R(P_1', P_2'; P_3', P_4') = -i$. Hence, $R(P_1, P_2; P_3, P_4)$ is not invariant under all semilinear transformations (though it is invariant under the linear transformations). For the real field, it does turn out that the cross-ratio is invariant under all semilinear transformations, as we shall see. (In fact, we shall also see that for the real field, any semilinear transformation is linear.)

On the other hand, let us take the property that $R(P_1, P_2; P_3, P_4) = -1$. Since $-1 \to -1$ under an automorphism, one finds that for any automorphism of π, which sends P_j into P_j', $R(P_1', P_2'; P_3', P_4') = -1$. Thus the property of four points expressed by saying

$$R(P_1, P_2; P_3, P_4) = -1$$

is definitely of projective interest.

Let us now check the assertion that *if a transformation T of π is given in one coordinate system by matrix multiplication, namely, $T: \mathbf{x} \to \mathbf{y} = B\mathbf{x}$, then it is given similarly in any other coordinate system.*

PROOF. Let $\mathfrak{C} = L_A P^\alpha$ be a coordinate system. If $\begin{pmatrix} x_0 \\ x_1 \\ x_2 \end{pmatrix}$ are the old coordinates, the new coordinates $\begin{pmatrix} x_0' \\ x_1' \\ x_2' \end{pmatrix}$ are given by $\mathbf{x}' = (A\mathbf{x})^\alpha$. Similarly, the new coordinates of $\mathbf{y}$ are $\mathbf{y}' = (A\mathbf{y})^\alpha$. From $\mathbf{x}' = (A\mathbf{x})^\alpha$, we find $(\mathbf{x}')^{\alpha^{-1}} = A\mathbf{x}$ and $\mathbf{x} = A^{-1}(\mathbf{x}')^{\alpha^{-1}}$. Similarly $\mathbf{y} = A^{-1}(\mathbf{y}')^{\alpha^{-1}}$. Hence, $A^{-1}(\mathbf{y}')^{\alpha^{-1}} = BA^{-1}(\mathbf{x}')^{\alpha^{-1}}$, so $(\mathbf{y}')^{\alpha^{-1}} = ABA^{-1}(\mathbf{x}')^{\alpha^{-1}}$ and $\mathbf{y}' = (ABA^{-1})^\alpha \mathbf{x}'$.

In other words, in the new coordinate system T is also given by a matrix multiplication, namely, $T: \mathbf{x}' \to \mathbf{y}' = (ABA^{-1})^{\alpha}\mathbf{x}'$. Q.E.D.

We sometimes express the fact just proved by saying that a linear transformation is a *geometric object*.

3. Algebraic curves. As another example, we will define the notion of an algebraic curve in the plane π. Here the field K is definitely assumed to be commutative. First we give a preliminary definition.

DEFINITION. By a *power-product* in three letters X_0, X_1, X_2 one means an expression of the form $X_0^{i_0}X_1^{i_1}X_2^{i_2}$, where i_0, i_1, i_2 are nonnegative integers; a *monomial* is an expression of the form $cX_0^{i_0}X_1^{i_1}X_2^{i_2}$, where the coefficient c is an element of a given field; a *polynomial* is a sum of monomials, therefore an expression of the form $\sum c_{i_0 i_1 i_2}X_0^{i_0}X_1^{i_1}X_2^{i_2}$. The *degree of* $cX_0^{i_0}X_1^{i_1}X^{i_2}$ is defined as $i_0+i_1+i_2$; thus every monomial except 0 has one and only one degree; 0 shall be assigned every nonnegative integer as a degree. Combining the monomials that are multiples of the same power-product, every polynomial ($\neq 0$) can be written uniquely as a linear combination of distinct power products with coefficients different from zero; and when this is done, the monomials involved are called the *terms* of the polynomial. The *degree of a polynomial* ($\neq 0$) is the maximum of the degrees of its terms; the *subdegree* is the minimum. A polynomial ($\neq 0$) is called *homogeneous* if all of its terms are of the same degree, or in other words, if the degree of the polynomial equals its subdegree; also 0 is called homogeneous.

We will not verify the following theorem.

THEOREM. *The product of two polynomials different from zero (the coefficients being in a given field), is different from zero; and the degree of the product is the sum of the degrees of the factors; likewise for the subdegrees—whence it follows that if a homogeneous polynomial is factored into two polynomials, the factors must also be homogeneous.*

By a *polynomial equation* one means the condition on numbers x_0, x_1, x_2 that they annihilate a given polynomial $F(X_0, X_1, X_2)$, i.e., that $F(x_0, x_1, x_2)=0$. The equation is usually written:

$$F(X_0, X_1, X_2) = 0$$

(which, of course, does not mean that $F(X_0, X_1, X_2)$ is the polynomial 0).

DEFINITION. Let π be a plane and $\mathfrak{C}: \pi \to PNK$, a coordinate system. By an *algebraic curve* one means the set of points $\begin{pmatrix} x_0 \\ x_1 \\ x_2 \end{pmatrix}$,

x_i in K, that satisfy a polynomial equation $F(X_0, X_1, X_2) = 0$, where F is a homogeneous polynomial different from zero with coefficients in K.

The homogeneity condition is put in in order to assure us that if $F(x_0, x_1, x_2) = 0$, then also $F(\lambda x_0, \lambda x_1, \lambda x_2) = 0$; in other words, the equation is a condition on *points* and not merely on triples.

The first question is whether an algebraic curve is a geometric object, or, more explicitly stated, whether an algebraic curve is given by a homogeneous polynomial equation in every coordinate system.

Let the change of coordinates be given by the semilinear transformation $L_A P^\alpha$, so that if $\mathbf{x}$ is the old and $\mathbf{x}'$ the new coordinates of a point, then $\mathbf{x}' = (A\mathbf{x})^\alpha$. It is convenient to break up this change into two changes: first from $\mathbf{x}$ to $A\mathbf{x}$; and then to $(A\mathbf{x})^\alpha$. Thus we consider changes of the form $\mathbf{x}' = A\mathbf{x}$ and of the form $\mathbf{x}' = \mathbf{x}^\alpha$; and this simplifies matters notationally a little.

Let $F(X_0, X_1, X_2) = \sum c_{i_0 i_1 i_2} X_0^{i_0} X_1^{i_1} X_2^{i_2}$ and let $F = 0$ be the given equation. We have $\mathbf{x}' = A\mathbf{x}$; whence also $\mathbf{x} = B\mathbf{x}'$, $B = A^{-1}$. Or in expanded notation:

$$\begin{aligned} x_0 &= b_{00}x_0' + b_{01}x_1' + b_{02}x_2' \\ x_1 &= b_{10}x_0' + b_{11}x_1' + b_{12}x_2' \\ x_2 &= b_{20}x_0' + b_{21}x_1' + b_{22}x_2'. \end{aligned}$$

Clearly $\sum c_{i_0 i_1 i_2} x_0^{i_0} x_1^{i_1} x_2^{i_2} = 0$ if and only if

$$\sum c_{i_0 i_1 i_2}(b_{00}x_0' + b_{01}x_1' + b_{02}x_2')^{i_0} (b_{10}x_0' + b_{11}x_1' + b_{12}x_2')^{i_1} (b_{20}x_0' + b_{21}x_1' + b_{22}x_2')^{i_2} = 0,$$

i.e., if and only if x_0', x_1', x_2' satisfy the polynomial condition

$$\sum c_{i_0 i_1 i_2}(b_{00}X_0' + \cdots)^{i_0}(b_{10}X_0' + \cdots)^{i_1}(b_{20}X_0' + \cdots)^{i_2} = 0;$$

and this is a homogeneous polynomial condition. Note that we can recover the polynomial $F(X_0, X_1, X_2)$ from the polynomial

$$F'(X_0', X_1', X_2') = \sum c_{i_0 i_1 i_2}(b_{00}X_0' + \cdots)^{i_0}(b_{10}X_0' + \cdots)^{i_1}(b_{20}X_0' + \cdots)^{i_2}$$

by replacing X_0', X_1', X_2' by

$$a_{00}X_0 + a_{01}X_1 + a_{02}X_2, \quad a_{10}X_0 + a_{11}X_1 + a_{12}X_2, \quad a_{20}X_0 + a_{21}X_1 + a_{22}X_2,$$

respectively; hence $F'(X_0', X_1', X_2')$ is not the polynomial zero, as this would imply that $F(X_0, X_1, X_2)$ is zero.

(If we look at $\mathbf{x}' = A\mathbf{x}$ not as a coordinate change but as a linear transformation, then the given curve goes over into the curve given by the equation $F'(X_0, X_1, X_2) = 0$.)

We have still to consider transformations of the form $\mathbf{x}' = \mathbf{x}^\alpha$. Here we see that

$$\sum c_{i_0 i_1 i_2} x_0^{i_0} x_1^{i_1} x_2^{i_2} = 0$$

if and only if

$$\sum c^\alpha{}_{i_0 i_1 i_2}\, x_0'^{i_0} x_1'^{i_1} x_2'^{i_2} = 0;$$

or in other words, if and only if x_0', x_1', x_2' satisfy the polynomial condition

$$\sum c^\alpha{}_{i_0 i_1 i_2} X_0'^{i_0} X_1'^{i_1} X_2'^{i_2} = 0.$$

In this way we see that an algebraic curve is a geometric object. One could also put the matter thus:

THEOREM. *Under a semilinear transformation, an algebraic curve goes into an algebraic curve.*

REMARK. In studying a problem analytically, the object of making a coordinate change is to simplify the analytical situation. It may be convenient, then, to consider a proper subset of all coordinate changes, rather than all of them. For example, we may consider all coordinate changes of the form $\mathbf{x}' = B\mathbf{x}$: this has the advantage that we need not then take into account the automorphisms of the base field. This set of coordinate changes is often sufficient, so much so that projective geometry has been defined in terms of this set of coordinate changes rather than in terms of the full set. For convenience of exposition we will also adopt this point of view in later chapters: in other words, we will not bother with the automorphisms of the base field.

4. A short cut to *PNK*. A system of type Σ satisfying certain axioms is a *PNK*—the axioms have been so selected that we come out with a *PNK*. The reason for the choice is that one knows that a *PNK* is an interesting object of study. Buf if what we want to study is a *PNK*, why not say so to start with, and thus make a short cut to the desired object of study? One objection is that a certain coordinate system is selected as absolute, and this is undesirable. To get around this difficulty, we can define a projective plane to be a system of type Σ that is *isomorphic* to some *PNK*. Then we define coordinate system, as above, as an isomorphic mapping of the plane into some *PNK*. In this way one comes immediately to the main object of study, the *PNK*, without being saddled with an absolute coordinate system.

5. A result for the field of real numbers. Having made the assumption that Desargues' Theorem holds, we know that the study of projective geometry comes to the study of space over a field K. One can therefore abandon, if one wishes, axioms of a synthetic type and instead make assumptions on the field K. Thus, instead of assuming that Pappus' Theorem holds, we may equivalently assume that K is commutative. The intuitive geometric notion of order on a line can be made explicit by means of synthetic axioms and corresponds to the idea that the field K is a so-called ordered field; further notions, of a geometric sort, lead to the study of space over the real field. A short cut to this stage of study is simply to assume that the field K is the real field. The main line of development of projective geometry, especially its continuation into so-called algebraic geometry, does not give to the real field any dominating role; but because of the general significance of the real field for mathematics, the study of this particular case—though aside from the main line—is undoubtedly significant. In this regard, we prove a result promised above, namely the following.

THEOREM. *The real field has no automorphisms but the identity.*

PROOF. We first observe that in any automorphism $0 \to 0$. The reason is that 0 may be characterized as the unique solution of the equation $a+x=a$ (for any a). Applying the automorphism $\tau : u \to u'$ to both sides of this equality, we get $a'+x'=a'$, whence the image x' of x is seen to be zero. In a similar way one proves that $1 \to 1$. Then $1+1 \to 1+1$, $1+1+1 \to 1+1+1$, etc.; therefore all the positive integers, as well as zero, are fixed under τ. From $a+x=0$ one gets $a'+x'=0$, whence the image of $-a$ is minus the image of a; hence also the negative integers, hence all integers, are fixed under τ.

Next we show that τ sends any rational number m/n, where m, n are integers, into itself. Write $m/n=x$. Then $m=nx$. Apply τ to get $m=n \cdot x'$, whence $x'=m/n$. Thus τ keeps the rational numbers fixed.

Next we prove that τ sends every positive number x into a positive number. In fact, if $x>0$, then x is a square, $x=a^2$. Hence, $\tau(x)= x'=a'^2$, where $a'=\tau(a)$. Since x' is a square, it must also be positive or zero. It could not be zero, since τ is univalent (one to one) and sends $0 \to 0$ and therefore does not send any x different from zero into zero. Thus $x'>0$.

Finally we prove $\tau : x \to x$ for every x. In fact, let $\tau : x \to x'$ and suppose $x' \neq x$. Let m/n be a rational number between x and x'. Then we have $x<m/n<x'$ or $x'<m/n<x$. We treat the first case, the second being similar. From $x<m/n$, we get $m/n-x>0$, whence applying τ, $m/n-x'>0$, so $m/n>x'$. This contradicts the assumption $m/n<x'$. Hence the assumption $x' \neq x$ is untenable, i.e., $\tau : x \to x$, every x. This proves that τ is the identity automorphism and completes the proof.

COROLLARY. *The only collineations (or automorphisms) of* PNK^2, *for* K *the field of real numbers, are the linear transformations.*

REMARK. A similar theorem for the complex field is false: there are infinitely many automorphisms of the complex field. In fact, there are infinitely many complex numbers satisfying no nontrivial polynomial relation over the rationals, and if x_1 and x_2 are two such quantities, then there is an automorphism of the complex field sending x_1 into x_2.

CHAPTER X

CONIC SECTIONS ANALYTICALLY TREATED

1. Derivation of equation of conic. Our object now is to study conics analytically, at least to the extent that we studied them synthetically. That is, we have already studied conics directly in terms of the primitive notions of *point*, *line*, and *incidence*; whereas now we will take advantage of the fact that our plane π has a coordinate system: the study shall now be in terms of the coordinates of points and of lines.

We start from the definition of the conic as the locus of the intersection of corresponding rays in a projectivity that is not a perspectivity. Let L, L' be two distinct points, σ a projectivity from the pencil of lines on L to the pencil of lines on L'; and σ not a perspectivity. If l is any line on L, then l and $(l)\sigma$ meet in just one point, $l \cdot l(\sigma)$; recall that $l = l(\sigma)$ is impossible, as $l = (l)\sigma$ would mean that LL' goes into $L'L$ under σ, and this would imply that σ is a perspectivity. The conic was defined as the locus of $l \cdot l(\sigma)$ as l varies through the pencil at L. Notation: $\Gamma = \Gamma(L, L'; \sigma)$.

Since our object is to study conics analytically, we should retain as little as possible from our synthetic study. Actually we will keep only the following: a line through L meets $\Gamma(L, L'; \sigma)$ in at most one point different from L; and similarly for L'; in particular, no point of LL' except L and L' are on the conic; and L and L' are on the conic. These facts follow directly from the definition.

Our first object is to get an algebraic condition (or conditions) on the coordinates of a point P_{x} that will be necessary and sufficient for the point to lie on the conic.

Let $\Gamma = \Gamma(L, L'; \sigma)$. We first study Γ in an especially convenient coordinate system, later in an arbitrary coordinate system. The coordinate system we have in mind will have L, L' as vertices. Let D_0, D_1, D_2, D be the vertices of a coordinate system with $D_1 = L$, $D_2 = L'$. Let P_{x} be a point on Γ; and let us first restrict ourselves to points not on LL' (Fig. 10.1). Let D_2P meet D_0D_1 in A; D_1P meet D_0D_2 in A'. Since $A \rightarrow AD_2$ is a perspectivity (from the range on

D_0D_1 to the pencil on D_2), and similarly $A'D \to A'$ is a perspectivity, and since $D_2A \to D_1A'$ is a projectivity, we see that the mapping $A \to A'$ is a projectivity. Let $P = P_{\mathbf{x}}$; then A has coordinates $\begin{pmatrix} x_0 \\ x_1 \\ 0 \end{pmatrix}$ and A' has coordinates $\begin{pmatrix} x_0 \\ 0 \\ x_2 \end{pmatrix}$. From our discussion of coordinate systems on a line, we know that we can take a coordinate system on D_0D_1 such that $\begin{pmatrix} x_0 \\ x_1 \end{pmatrix}$ are the coordinates of A in that coordinate system; and similarly on D_0D_2 we can take a coordinate system so that

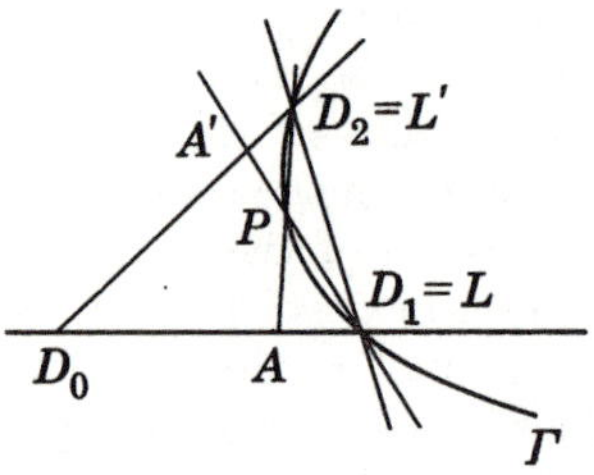

FIG. 10.1

$\begin{pmatrix} x_0 \\ x_2 \end{pmatrix}$ are the coordinates of A' in that system. (We write the first point as $A\begin{pmatrix} x_0 \\ x_1 \end{pmatrix}$, the second as $A'\begin{pmatrix} x_0 \\ x_2 \end{pmatrix}$.)

We start with some given coordinate system $\mathfrak{C}_1$ of π; every other coordinate system $\mathfrak{C}_2$ of π is related to $\mathfrak{C}_1$ by the formula $\mathfrak{C}_2 = \mathfrak{C}_1 L_A P^{\alpha}$. It is sufficient, however, to restrict ourselves to coordinate systems of the type $\mathfrak{C}_1 L_A$; and correspondingly, the coordinate systems on lines induced by these coordinate systems of the plane. In effect, this restriction amounts to considering only those coordinate systems discussed in Chapter VIII. As a result we know that the passage from $A'\begin{pmatrix} x_0 \\ x_2 \end{pmatrix}$ to $A\begin{pmatrix} x_0 \\ x_1 \end{pmatrix}$ is given by a matrix multiplication. Let $\begin{pmatrix} c_{00} & c_{01} \\ c_{10} & c_{11} \end{pmatrix}$ be the (nonsingular) matrix in question. Applying the matrix to $\begin{pmatrix} x_0 \\ x_2 \end{pmatrix}$, we get

$c_{00}x_0+c_{01}x_2$, $c_{10}x_0+c_{11}x_2$ as coordinates of $A\begin{pmatrix}x_0\\x_1\end{pmatrix}$. This does not mean that $x_0=c_{00}x_0+c_{01}x_2$, $x_1=c_{10}x_0+c_{11}x_2$, but only that

$$\begin{aligned}\rho x_0 &= c_{00}x_0+c_{01}x_2\\ \rho x_1 &= c_{10}x_0+c_{11}x_2\end{aligned}$$

for some $\rho\neq 0$; here ρ may depend on the point $P_{\mathbf{x}}$. (In fact, though this point need not enter the considerations, the number ρ definitely varies with $P_{\mathbf{x}}$; it could not be constant, for if it were, all the points of the conic (not on LL') would lie on the line $\rho x_0=c_{00}x_0+c_{01}x_2$, and this is not so.) Since we are considering first the points for which $x_0\neq 0$, we can write:

$$\text{(1)} \qquad \frac{x_1}{x_0} = \frac{c_{10}x_0+c_{11}x_2}{c_{00}x_0+c_{01}x_2}$$

whence

$$\text{(2)} \qquad x_1(c_{00}x_0+c_{01}x_2) = x_0(c_{10}x_0+c_{11}x_2).$$

Thus every point of the conic not on LL' satisfies the condition (2); the restriction to points not on LL' may now be removed, since one sees that $\begin{pmatrix}0\\0\\1\end{pmatrix}$ and $\begin{pmatrix}0\\1\\0\end{pmatrix}$ satisfy the condition.

Conversely, let $\mathbf{x}$ satisfy (2): to see that $P_{\mathbf{x}}$ lies on the conic. First consider the points for which $x_0\neq 0$. We have to see that $A'\begin{pmatrix}x_0\\x_2\end{pmatrix}$, $A\begin{pmatrix}x_0\\x_1\end{pmatrix}$ are corresponding points under σ; this amounts to establishing equation (1); and we have (2). We can divide (2) through by x_0, since $x_0\neq 0$; but we wish also to divide by $c_{00}x_0+c_{01}x_2$, and must see that this number is not zero. That is so, because if it were zero, we would also have $x_0(c_{10}x_0+c_{11}x_2)=0$, and since $x_0\neq 0$, $c_{11}x_0+c_{11}x_2=0$. Thus we would have

$$\begin{aligned}c_{00}x_0+c_{01}x_2 &= 0\\ c_{10}x_0+c_{11}x_2 &= 0,\end{aligned}$$

and since $c_{00}c_{11}-c_{10}c_{01}\neq 0$, the only solution of these equations is $x_0=0$, $x_1=0$; this is a contradiction, since $x_0\neq 0$. Hence we can divide both sides of (2) by $x_0(c_{00}x_0+c_{01}x_2)$ and get (1). This shows that $P_{\mathbf{x}}$ lies on the conic, at least if $x_0\neq 0$. It remains to consider the points satisfying (2) for which $x_0=0$. Here we note that $c_{01}\neq 0$; in fact, if $c_{01}=0$, then the transformation from D_0L to D_0L' defined by $\begin{pmatrix}c_{00} & c_{01}\\ c_{10} & c_{11}\end{pmatrix}$

would send $L = A\begin{pmatrix}0\\1\end{pmatrix}$ into $A'\begin{pmatrix}0\\1\end{pmatrix} = L'$, whence σ would send LL' into $L'L$, making σ a perspectivity; and this is not so. Hence in (2), if $x_0 = 0$, we find $x_1 = 0$ or $x_2 = 0$. There is only one point for which $x_0 = 0$, $x_1 = 0$, and this is L'; similarly, $x_0 = 0$, $x_2 = 0$ corresponds to L. Thus every point satisfying (2) lies on the given conic.

Summing up the considerations so far, we have the following theorem.

THEOREM. *Every conic is given in an appropriate coordinate system by an equation of the form* $x_1(c_{00}x_0 + c_{01}x_0) - x_0(c_{10}x_2 + c_{11}x_2) = 0$, *with* $(c_{00}c_{11} - c_{10}c_{01}) \cdot c_{01} \neq 0$.

In a previous chapter we saw that if a locus is given in one coordinate system by an equation of the form $F(x_0, x_1, x_2) = 0$, where $F(x_0, x_1, x_2)$ is a homogeneous polynomial of degree n, then it is given in any other coordinate system by an equation $G(x_0, x_1, x_2) = 0$, where G is a homogeneous polynomial of degree n. Thus in any coordinate system a conic is given by an equation of the form $G(x_0, x_1, x_2) = 0$, where G is a homogeneous polynomial of degree 2. Moreover if the conic is given by $G = 0$, with G homogeneous of degree 2, then G cannot be properly factored (i.e., into several factors at least two of which involve the x_i); *a priori* G could only factor into two homogeneous linear forms (and factors not involving the x_i). The easiest way to see that this cannot actually happen is to observe that if it did, the conic would consist of the points on one or two lines, in fact of one or two lines on the points L and L' (these would be the lines whose equations would be given by setting the factors equal to zero); this would imply that some line through L or L' meets the conic in at least three points, whereas a line through L meets the conic in at most one point different from L; and similarly for L'. Thus we have the following theorem:

THEOREM. *Every conic is given in every coordinate system by an equation* $Q(x_0, x_1, x_2) = 0$, *where* Q *is a homogeneous irreducible polynomial of degree* 2. (*Moreover the locus* $Q(x_0, x_1, x_2) = 0$ *has in it at least two points, since, in fact, the locus is the conic and a conic always has on it at least two points.*)

Let us consider the converse question, whether a locus given by an irreducible homogeneous equation of degree 2 is a conic. The answer to this depends on the field K; it is *yes* if K is the complex field, but *no* if K is the *real* field. Thus over the real field consider the locus given by the irreducible equation $x_0^2 + x_1^2 + x_2^2 = 0$; if real numbers x_0, x_1, x_2 satisfy this condition, then $x_0 = x_1 = x_2 = 0$, so there are no points on this locus, and hence it could not be a conic, which from the definition always contains at least two points. Another instructive example is the locus given by the equation $x_0^2+x_1^2=0$; over the real field, this equation is irreducible; and it has on it just the one point $x_0=0$, $x_1=0$, $x_2=1$. Clearly, in order that a locus $Q=0$ be a conic it is necessary that it contain at least two points. With this extra condition we get the converse of the last theorem. Thus:

THEOREM. *Let $Q(x_0, x_1, x_2)$ be a homogeneous polynomial of degree* 2 *with coefficients in a field K and let Q be irreducible over K* (i.e., *when the proposed factors have their coefficients restricted to K*). *If the locus $Q=0$ has in it at least two points, then the locus is a conic.*

PROOF. Our locus is given by $Q=0$ in some definite coordinate system. If we pass to a new coordinate system, the equation $Q=0$ transforms into an equation $Q'=0$, where Q' is homogeneous of degree 2. Moreover Q' is also irreducible, for if it factored, then returning to the old coordinates the factors of Q' would transform into factors of Q. Of course the locus of $Q'=0$ has in it at least two points, for this locus is just the locus $Q=0$ we started with. Thus we may change our coordinate system at will, without losing our hypotheses.

Let L, L' be two points of our locus and let us take them as two points D_1, D_2 of a coordinate system. Every homogeneous equation of degree 2 is of the form:

$$a_{00}x_0^2+a_{01}x_0x_1+a_{02}x_0x_2+a_{11}x_1^2+a_{12}x_1x_2+a_{22}x_2^2 = 0,$$

and in particular our locus is given by such an equation. Since the locus passes through $D_1=\begin{pmatrix}0\\1\\0\end{pmatrix}$, we see that $a_{11}=0$; and similarly, using D_2, that $a_{22}=0$. The equation can then be written:

$$x_1(a_{01}x_0+a_{12}x_2)-x_0(-a_{00}x_0-a_{02}x_2) = 0,$$

or, what is the same,

$$(3) \qquad x_1(c_{00}x_0+c_{01}x_2)-x_0(c_{10}x_0+c_{11}x_2) = 0,$$

where $a_{01}=c_{00}$, $a_{12}=c_{01}$, $-a_{00}=c_{10}$, $-a_{02}=c_{11}$. Compare this with equation (2) in the argument leading to the first theorem above. Here we also have $\begin{vmatrix}c_{00} & c_{01}\\ c_{10} & c_{11}\end{vmatrix}\neq 0$, as otherwise there exists a $\lambda\neq 0$ such that $c_{10}=\lambda c_{00}$, $c_{11}=\lambda c_{01}$ (or $c_{00}=\lambda c_{10}$, $c_{01}=\lambda c_{11}$); in either case one would see that the left-hand side of (3) has a linear factor, and this is not so. Also $c_{01}\neq 0$, as otherwise x_0 would be a factor of the left-hand side of (3). Since $\begin{vmatrix}c_{00} & c_{01}\\ c_{10} & c_{11}\end{vmatrix}\neq 0$, there is a projectivity $A\rightarrow A'$ from D_0D_1 to D_0D_2

defined by the matrix $\begin{pmatrix} c_{00} & c_{01} \\ c_{10} & c_{00} \end{pmatrix}$; and correspondingly, a projectivity $\sigma: D_1A' \to D_2A$ from the pencil at L to the pencil at L'. From the fact that $c_{01} \neq 0$, one sees that in the former projectivity, D_1 and D_2 do not correspond, so that under σ, D_1D_2 and D_2D_1 do not correspond, so σ is not a perspectivity. Now consider the conic $\Gamma(L, L'; \sigma)$. Its equation as deduced before is (2), which is the same as (3). Hence (3) is the locus of a conic, which was to be proved.

A definition is merely the starting point of a logical argument, and as such is somewhat arbitrary. For example, a set of points that is a conic can be characterized in several ways; and any one of these ways can equally well be taken as the definition. The intention is to come to a certain body of theorems, and the particular way of fixing the definition depends on how one proposes to come into possession of the theorems. Our definition of conic is a convenient one for a synthetic treatment; but if our considerations are to be analytic, it is convenient to start from a purely analytic definition. Thus in the case of a conic, we could have defined it as a locus containing at least two points which is given by an irreducible homogeneous equation of degree 2, and it would have been convenient to do so. Moreover, having decided on a definition, it is not necessary, within a given logical development, to compare it with other definitions; the reason for comparison is to show that two different logical developments amount to the same thing. Previously we did this for two definitions of the conic. Subsequently we are going to give new (analytical) definitions for terms already encountered, for example, *polar*. Having, say, defined polar, we do not have to compare this with our previous definition, but as the intention is that we are talking about the same thing previously called polar, the argument is developed at least to the point where the identity can be seen.

2. Uniqueness of the equation. Let a conic Γ be given by the homogeneous equation $F(\mathbf{x})=0$ of degree 2 in a given coordinate system $\mathfrak{C}$. Of course, the conic is also given by $cF(\mathbf{x})=0$, where c is any number different from zero. Since definitions will frequently be given in terms of an equation $F(\mathbf{x})=0$, it is important to know whether there are perhaps several equations for Γ. Actually, we shall show, *if* $G(\mathbf{x})=0$ *is also a homogeneous equation of degree* 2 *for* Γ (*in the given coordinate system* $\mathfrak{C}$), *then* $G(\mathbf{x})=cF(\mathbf{x})$ *for some* $c\neq 0$. We express this fact by saying that the equation of Γ is "essentially unique," where the word "essentially" refers to the fact that an arbitrary factor $c\neq 0$ can be introduced into the equation. Also we frequently leave out the word "essentially," supposing it then to be understood.

Let, then, $F(\mathbf{x})=0$, $G(\mathbf{x})=0$ be equations for Γ in the coordinate system $\mathfrak{C}$. Let D_0, D_1, D_2 be three noncollinear points on Γ. From $\mathfrak{C}$ go over by a matrix multiplication to a coordinate system $\mathfrak{C}_1$ having D_0, D_1, D_2 as vertices. Under $\mathbf{x}\to\mathbf{x}'=A\mathbf{x}$, $F(\mathbf{x})$ goes over into $F(A^{-1}\mathbf{x}')=F_1(\mathbf{x}')$ and $G(\mathbf{x})$ goes over into $G_1(\mathbf{x}')$. If we can show that $G_1(\mathbf{x}')=cF_1(\mathbf{x}')$, it will follow that $G(\mathbf{x})=cF(\mathbf{x})$.

As explained above, in $\mathfrak{C}_1$ we find

$$F_1(\mathbf{x}') = c_2x_0'x_1'+c_0x_1'x_2'+c_1x_2'x_0', \quad c_0c_1c_2\neq 0$$
$$G_1(\mathbf{x}') = d_2x_0'x_1'+d_0x_1'x_2'+d_1x_2'x_0', \quad d_0d_1d_2 \neq 0.$$

Hence we find that all points of Γ satisfy $d_1F_1(\mathbf{x}')-c_1G_1(\mathbf{x}')=0$; but $d_1F_1(\mathbf{x}')-c_1G_1(\mathbf{x}')=x_1'[d_1(c_2x_0'+c_0x_2')-c_1(d_2x_0'+d_0x_2')]$, so that (if $d_1F_1\neq c_1G_1$) all the points of Γ would lie on two lines. Suppose for a moment that on Γ there are at least 5 points; since no 3 are collinear, the 5 points could not distribute themselves on two lines, and the proof in this case is complete.

Can a conic contain less than 5 points? From the definition, we see that there is a one-to-one correspondence between the lines on L and the points on $\Gamma(L, L'; \sigma)$. Thus there are as many points on Γ as there are lines on a point, or points on a line. This leaves the theorem to be verified in the projective planes with 7 and 13 points, respectively; or what is the same thing, in PNK^2 for K the two element field and K the three element field, respectively. For the two element field, there is *a priori* only one possibility for $F_1(\mathbf{x}')$, namely, $x_0'x_1'+x_1'x_2'+x_2'x_0'$; and similarly for $G_1(\mathbf{x}')$, so $G_1(\mathbf{x}')=F_1(\mathbf{x}')$. For the three element field, extra considerations are necessary, and we leave this remaining case as an exercise.

3. Projective equivalence of conics. Let $\Gamma=\Gamma(L, L'; \sigma)$ be a conic. Let l be a line on L such that $l\neq LL'$, $(l)\sigma\neq L'L$; then $l\cdot(l)\sigma$ is a point of Γ not collinear with L, L'. Take three noncollinear points on Γ (say L, L', $l\cdot(l)\sigma$) and make them the vertices D_0, D_1, D_2 of a coordinate system. In this system Γ has the equation

$$a_{00}x_0^2+a_{01}x_0x_1+a_{02}x_0x_2+a_{11}x_1^2+a_{12}x_1x_2+a_{22}x_2^2 = 0;$$

but since D_0 lies on the conic, one finds $a_{00}\cdot 1^2=0$. Similarly because D_1, D_2 are on the conic, one finds $a_{11}=0$, $a_{22}=0$. Therefore the equation appears as:

$$a_{01}x_0x_1+a_{02}x_0x_2+a_{12}x_1x_2 = 0.$$

Here $a_{01}\neq 0$, as otherwise, x_2 would be a factor of the left-hand side. Similarly $a_{02}\neq 0$ and $a_{12}\neq 0$. Now consider a change of coordinates given by

$$x_0 = a_{12}x_0', \qquad x_1 = a_{02}x_1', \qquad x_2 = a_{01}x_2'.$$

Then in the new coordinate system, the conic is given by

$$x_0'x_1' + x_0'x_2' + x_1'x_2' = 0$$

If we refer only to this last coordinate system, there is no point, notationally, in keeping the primes; and one has the following theorem:

THEOREM. *Every conic* Γ *has, in an appropriate coordinate system, the equation*: $x_0x_1 + x_0x_2 + x_1x_2 = 0$.

Let now Γ_1, Γ_2 be two conics. Fix a coordinate system, with say E_0, E_1, E_2, E as vertices and unit point. Previously we took three points D_0, D_1, D_2 on Γ as vertices of a new coordinate system. We can regard this process as a linear transformation taking D_0, D_1, D_2, respectively, into E_0, E_1, E_2. Thus we can say that by a linear transformation one can send the conic Γ_1 into the conic $x_0x_1 + x_1x_2 + x_2x_0 = 0$. Similarly, by a linear transformation we can send Γ_2 into $x_0x_1 + x_1x_2 + x_2x_0 = 0$. The first of these transformations times the inverse of the second is a linear transformation taking Γ_1 into Γ_2. Hence we have:

THEOREM. *Any conic can be sent into any other by a linear transformation.*

Exercise. In the preceding theorem one can make the transformation take any three assigned points of the first conic into any three assigned points of the second; and this can be done in only one way.

4. Poles and polars. Let $F(\mathbf{x})$ be a homogeneous polynomial of degree 2 in the letters x_0, x_1, x_2. Then $F(\mathbf{u}+\mathbf{x})$ is a homogeneous polynomial of degree 2 in the six letters x_0, x_1, x_2, u_0, u_1, u_2. Each term of $F(\mathbf{u}+\mathbf{x})$ has two u's as factors, or two x's as factors, or one u and one x. Placing $\mathbf{x}=\mathbf{0}$, we see that $F(\mathbf{u})$ is the sum of the terms in $F(\mathbf{u}+\mathbf{x})$ which are of degree 2 in u_0, u_1, u_2; and similarly, by placing $\mathbf{u}=\mathbf{0}$, that $F(\mathbf{x})$ is the sum of the terms of degree 2 in the x's. Thus $G(\mathbf{u}, \mathbf{x}) = F(\mathbf{u}+\mathbf{x}) - F(\mathbf{u}) - F(\mathbf{x})$ is the sum of the terms in $F(\mathbf{u}+\mathbf{x})$ which are of degree 1 in the u's and of degree 1 in the x's; and hence $G(\mathbf{u}, \mathbf{x})$ is homogeneous of degree 1 in the u's and homogeneous of degree 1 in the x's. Let $\mathbf{u}$ be the coordinates of a point and let $\mathbf{x}$ be variable; then $G(\mathbf{u}, \mathbf{x})$ is a linear form in x_0, x_1, x_2.

DEFINITION. If the coefficients of $G(\mathbf{u}, \mathbf{x})$ are not all zero then $G(\mathbf{u}, \mathbf{x}) = 0$ is a line, called the *polar of the point* $P_{\mathbf{u}}$ *with respect* to $F(\mathbf{x})$. If $F(\mathbf{x}) = 0$ is a conic or one or two straight lines, we will also say that the polar is *with respect to the locus* $F(\mathbf{x}) = 0$.

To see that this definition is any good, or in other words, whether the polar is a geometric object, we have to consider a change of coordinates, compute the polar in the new system, and see that we get the same line.

Consider a change of coordinates given by the formula $\mathbf{x}=A\mathbf{x}'$. Then $F(\mathbf{x})$ goes over into the form $F(A\mathbf{x}')=F'(\mathbf{x}')$; and $F(\mathbf{u}+\mathbf{x})$ goes over into $F(A\mathbf{u}'+A\mathbf{x}')=F(A(\mathbf{u}'+\mathbf{x}'))=F'(\mathbf{u}'+\mathbf{x}')$; whence $G(\mathbf{u},\ \mathbf{x})$ goes over into $F'(\mathbf{u}'+\mathbf{x}')-F'(\mathbf{u}')-F'(\mathbf{x}')$. Call this $G'(\mathbf{u}',\mathbf{x}')$. Then $G'(\mathbf{u}',\mathbf{x}')=G(\mathbf{u},\mathbf{x})$. In particular, one sees that $G'(\mathbf{u}',\mathbf{x}')$, regarded as a linear form in x_0', x_1', x_2', does not have all its coefficients zero, as otherwise $G(\mathbf{u}',\mathbf{x}')$ would vanish for every point $\mathbf{x}'$, and hence $G(\mathbf{u},\mathbf{x})$ would vanish for every point $\mathbf{x}$. The locus $G(\mathbf{u}',\mathbf{x}')=0$ is the polar as computed in the new coordinates; and since $G(\mathbf{u}',\mathbf{x}')=G(\mathbf{u},\mathbf{x})$, this is the same as the locus $G(\mathbf{u},\mathbf{x})=0$, which is what was to be seen.

Exercise. Consider also a coordinate system change of the form $\mathbf{x}\to\mathbf{x}^{\alpha}$, where α is an automorphism of K.

THEOREM. *If $F(\mathbf{x})=0$ is the locus of a conic, then every point has a polar with respect to the locus $F(\mathbf{x})=0$ (unless $1+1=0$, in which case this is not so).*

PROOF. We take a coordinate system in such way that $F(\mathbf{x})=x_0x_1+x_1x_2+x_2x_0$. Computing $F(\mathbf{u}+\mathbf{x})-F(\mathbf{u})-F(\mathbf{x})$, we find

$$G(\mathbf{u},\mathbf{x}) = (u_1+u_2)x_0+(u_2+u_0)x_1+(u_0+u_1)x_2.$$

The question now is whether there is any point $\mathbf{u}$ such that the coefficients $u_1+u_2,\ u_2+u_0,\ u_0+u_1$ are all zero. We know that

$$\begin{aligned} u_1+u_2 &= 0\\ u_0 \quad + \quad u_2 &= 0\\ u_0+u_1 \qquad &= 0\end{aligned}$$

have a nontrivial solution if and only if $\begin{vmatrix}0&1&1\\1&0&1\\1&1&0\end{vmatrix}=0$. Computing the determinant, we find its value to be 2. Thus if $1+1\neq 0$, every point has a polar with respect to every conic.

If $1+1=0$, then the point $(1, 1, 1)$ does not have a polar with respect to the above conic. Thus *the case in which $1+1=0$ in K forms an exception and will be excluded from our further considerations on conics.* Recall that in the synthetic treatment we also made this restriction.

For computational purposes we seek a good expression for $G(\mathbf{u},\mathbf{x})$. Recall that in calculus one defines derivatives in such way that it turns out that the derivative of x^n is nx^{n-1}. There the definition depends on limit concepts, but if here we define the derivative of the polynomial $ax^n+bx^{n-1}+cx^{n-2}+\cdots$, where $a, b, c\cdots$ are in K, to be $nax^{n-1}+(n-1)bx^{n-2}+(n-2)cx^{n-3}+\cdots$, then one finds that the familiar rules of computation obtain, in particular, $(f(x)g(x))' = f'(x)g(x) + g'(x)f(x)$ and $(f(x)+g(x))' = f'(x)+g'(x)$, where $'$ stands for derivative. One also proves that $f(x+h) = f(x) + f'(x)h +$ terms of degree ≥ 2 in h; this is a special case of Taylor's theorem—we will need the result only for polynomials of degree ≤ 2, where the result is easily checked directly. In the case of a polynomial f in several variables, say $f = f(x_0, x_1, x_2)$ one finds similarly that

$$f(x_0+u_0, x_1+u_1, x_2+u_2) = f(x_0, x_1, x_2) + \frac{\partial f}{\partial x_0} u_0 + \frac{\partial f}{\partial x_1} u_1 + \frac{\partial f}{\partial x_2} u_2$$
$$+ \text{terms of degree } \geq 2 \text{ in } u_0, u_1, u_2.$$

Here $\frac{\partial f}{\partial x_0}$ means the derivative of $f(x_0, x_1, x_2)$ relative to the letter x_0 (i.e., we think of $f(x_0, x_1, x_2)$ as a polynomial in the one letter x_0, and take the derivative as explained). We can also write

$$f(x_0+u_0, x_1+u_1, x_2+u_2) = f(u_0, u_1, u_2) + \frac{\partial f}{\partial u_0} x_0 + \frac{\partial f}{\partial u_1} x_1 + \frac{\partial f}{\partial u_2} x_2$$
$$+ \text{terms of degree } \geq 2 \text{ in } x_0, x_1, x_2.$$

Here $\frac{\partial f}{\partial u_0}$ means the derivative of $f(u_0, u_1, u_2)$ relative to the letter u_0; sometimes we like to think of u_0, u_1, u_2 as fixed and in that case could describe $\frac{\partial f}{\partial u_0}$ as the result of substituting u_0 for x_0, u_1 for x_1, u_2 for x_2 in $\frac{\partial f}{\partial x_0}$. The assertions above will only be needed for $f(x_0, x_1, x_2)$ of degree ≤ 2 in x_0, x_1, x_2 and can be verified easily by a direct computation.

Let now $F(x_0, x_1, x_2)$ be homogeneous of degree 2:

$$F(x_0, x_1, x_2) = a_{00}x_0^2 + a_{01}x_0x_1 + a_{02}x_0x_2 + a_{11}x_1^2 + a_{12}x_1x_2 + a_{22}x_2^2.$$

Since $G(\mathbf{x}, \mathbf{u}) = F(\mathbf{x}+\mathbf{u}) - F(\mathbf{x}) - F(\mathbf{u})$ is the sum of the linear terms in $\mathbf{x}$ of $F(\mathbf{x}+\mathbf{u})$, we find:

$$G(\mathbf{x}, \mathbf{u}) = \frac{\partial F}{\partial u_0} x_0 + \frac{\partial F}{\partial u_1} x_1 + \frac{\partial F}{\partial u_2} x_2.$$

Here

$$\frac{\partial F}{\partial u_0} = 2a_{00}u_0 + a_{01}u_1 + a_{02}u_2$$
$$\frac{\partial F}{\partial u_1} = a_{01}u_0 + 2a_{11}u_1 + a_{12}u_2$$
$$\frac{\partial F}{\partial u_2} = a_{02}u_0 + a_{12}u_1 + 2a_{22}u_2.$$

Applying the criterion for the linear equations $\frac{\partial F}{\partial u_0}=0$, $\frac{\partial F}{\partial u_1}=0$, $\frac{\partial F}{\partial u_2}=0$ to have a nontrivial solution, we get the following theorem.

THEOREM. *A necessary and sufficient condition for the locus $F=0$ to have a polar with respect to every point u is that*

$$\Delta = \begin{vmatrix} 2a_{00} & a_{01} & a_{02} \\ a_{01} & 2a_{11} & a_{12} \\ a_{02} & a_{12} & 2a_{22} \end{vmatrix} \neq 0.$$

The following is a corollary of the last two theorems.

COROLLARY. *If $F(\mathbf{x})$ is irreducible, homogeneous, and of degree 2, and the locus of $F=0$ contains at least two points, then $\Delta \neq 0$.*

Note that we can write

$$\left(\frac{\partial F}{\partial u_0}, \frac{\partial F}{\partial u_1}, \frac{\partial F}{\partial u_2}\right) = (u_0, u_1, u_2)\begin{pmatrix} 2a_{00} & a_{01} & a_{02} \\ a_{01} & 2a_{11} & a_{12} \\ a_{02} & a_{12} & 2a_{22} \end{pmatrix} = \mathbf{u}^T A,$$

where $\mathbf{u}^T$ stands for the row matrix (u_0, u_1, u_2), and A stands for the matrix exhibited. The matrix A is *symmetric*, i.e., the entry in the ith row and jth column is the same as that in the jth row and ith column. Letting A^T designate the matrix obtained from an arbitrarily given matrix by interchange of rows and columns—first row of A becoming first column of A^T, second row of A becoming second column of A^T, etc.—the *symmetry* of A *can be expressed by saying* $A=A^T$.

Exercise. Show that $(AB)^T=B^TA^T$.

We can write: $G(\mathbf{x}, \mathbf{u})=\mathbf{u}^TA\mathbf{x}$. Replacing $\mathbf{u}$ by $\mathbf{x}$, we see that $G(\mathbf{x}, \mathbf{x})=\mathbf{x}^TA\mathbf{x}$; moreover, $G(\mathbf{x}, \mathbf{x})=F(2\mathbf{x})-F(\mathbf{x})-F(\mathbf{x})=4F(\mathbf{x})-2F(\mathbf{x})=2F(\mathbf{x})$. Thus:

$$2F(\mathbf{x}) = \mathbf{x}^TA\mathbf{x}$$

or

$$F(\mathbf{x}) = \mathbf{x}^TB\mathbf{x},$$

where

$$B = \tfrac{1}{2}A = \begin{pmatrix} a_{00} & a_{01}/2 & a_{02}/2 \\ a_{01}/2 & a_{11} & a_{12}/2 \\ a_{02}/2 & a_{12}/2 & a_{22} \end{pmatrix}.$$

Thus we see that any homogeneous polynomial $F(\mathbf{x})$ can be written in the form $\mathbf{x}^T C\mathbf{x}$ in at least one way. Actually this can be done in several ways. Expanding

$$(x_0, x_1, x_2)\begin{pmatrix} c_{00} & c_{01} & c_{02} \\ c_{10} & c_{11} & c_{12} \\ c_{20} & c_{21} & c_{22} \end{pmatrix}\begin{pmatrix} x_0 \\ x_1 \\ x_2 \end{pmatrix}, \text{ we get}$$

$$\mathbf{x}^T C\mathbf{x} = c_{00}x_0{}^2 + (c_{01}+c_{10})x_0x_1 + (c_{02}+c_{20})x_0x_2 + c_{11}x_1{}^2 + (c_{12}+c_{21})x_1x_2 + c_{22}x_2{}^2,$$

and comparing this with

$$F(\mathbf{x}) = a_{00}x_0{}^2 + a_{01}x_0x_1 + a_{02}x_0x_2 + a_{11}x_1{}^2 + a_{12}x_1x_2 + a_{22}x_2{}^2,$$

we see that $\mathbf{x}^T C\mathbf{x} = F(\mathbf{x})$ if and only if $c_{00} = a_{00}$, $c_{01}+c_{10} = a_{01}$, $c_{02}+c_{20} = a_{02}$, $c_{11} = a_{11}$, $c_{12}+c_{21} = a_{12}$, $c_{22} = a_{22}$. There are several ways to pick the c_{ij} so that these conditions obtain, but we restore uniqueness by requiring C to be symmetric, and then $C = B$.

Recalling that we are excluding the case that $1+1 = 0$, we can summarize the above computations as follows:

THEOREM. *Every homogeneous polynomial* $F(\mathbf{x})$ *of degree* 2 *can be written in the form* $\mathbf{x}^T A\mathbf{x}$, *with* $A^T = A$, *in one and only one way; the polar of* $P_{\mathbf{u}}$ *with respect to* $F(\mathbf{x})$ *is (if it exists)* $\mathbf{u}^T A\mathbf{x} = 0$.

Exercise. If the polar of P lies on Q, then the polar of Q lies on P.

THEOREM. *Every point* $P_{\mathbf{u}}$ *has a polar with respect to* $\mathbf{x}^T A\mathbf{x}$ *if and only if* $\det A \neq 0$. *(In this statement, as in other similar ones, we tacitly assume* A *to be symmetric.)*

PROOF. The polar, if it exists, is given by $\mathbf{u}^T A\mathbf{x} = 0$; and it exists if and only if $\mathbf{u}^T A \neq 0$. To say $\mathbf{u}^T A \neq 0$ for every $\mathbf{u}$ $(\neq 0)$ comes to saying that the rows of A are linearly independent, and this is so if and only if $\det A \neq 0$.

COROLLARY. *If the locus of* $\mathbf{x}^T A\mathbf{x} = 0$ *is a conic, then* $\det A \neq 0$.

The corollary follows from the theorem and a previous theorem to the effect that every point has a polar with respect to every conic. One can see this point anew as follows. Making a coordinate change $\mathbf{x} = B\mathbf{x}'$, the form $\mathbf{x}^T A\mathbf{x}$ goes over into $\mathbf{x}'^T B^T A B\mathbf{x}'$, so that the matrix of the form $\mathbf{x}^T A\mathbf{x}$ in the new system is $B^T AB$: note that $B^T AB$ is symmetric, as $(B^T AB)^T$ = product of transposes in reverse order = $B^T A^T B^{TT} = B^T AB$. The determinant of the matrix does, in general, change with a change of coordinate system, in fact, $\det B^T AB = (\det B)^2 \det A$; but whether the determinant is zero or not does not change. Since any two conics are projectively equivalent, the corollary will hold in general if it holds for some conic. It is thus sufficient to check the corollary for some conic, say $x_0x_1 + x_1x_2 + x_2x_0 = 0$.

Exercises

1. Using the idea of the last paragraph, show that if $\mathbf{x}^T A\mathbf{x}=0$ is one or two lines, then det $A=0$.

2. If $\mathbf{x}^T A\mathbf{x}=0$ consists of two distinct lines, then rank A (=number of linear independent columns of A)=2; if $\mathbf{x}^T A\mathbf{x}=0$ consists of just one line, then rank $A=1$.

3. Two points are called conjugate with respect to a conic if each is on the polar of the other. Show that if two pairs of opposite vertices of a complete quadrilateral are conjugate with respect to a conic Γ, then so are the remaining pair of vertices.

Over the real field one can separate the irreducible polynomials $\mathbf{x}^T A\mathbf{x}$ into three categories: (i) those for which the locus $\mathbf{x}^T A\mathbf{x}=0$ has at least two points, hence is a conic, (ii) those for which the locus $\mathbf{x}^T A\mathbf{x}=0$ has no points, (iii) those for which the locus $\mathbf{x}^T A\mathbf{x}=0$ has just one point. Over the complex field the division into cases disappears, since there are always plenty of points satisfying $\mathbf{x}^T A\mathbf{x}=0$, as we see from the following theorem.

THEOREM. *Over the complex field, every locus* $\mathbf{x}^T A\mathbf{x}=0$ *has a point on every line. In fact, if* $F(\mathbf{x})$ *is any homogeneous polynomial, then every line meets the locus* $F(\mathbf{x})=0$.

PROOF. Let l be a given line, and take a coordinate system in which l has the equation $x_0=0$. In effect, then, the theorem has only to be proved for the line $x_0=0$.

Let $\mathbf{x}^T A\mathbf{x}=a_{00}x_0^2+a_{01}x_0x_1+a_{02}x_0x_2+a_{11}x_1^2+a_{12}x_1x_2+a_{22}x_2^2$. We seek the intersection of $x_0=0$ with $\mathbf{x}^T A\mathbf{x}=0$. Placing $x_0=0$ in $\mathbf{x}^T A\mathbf{x}$, we have to see for which values of x_1, x_2 we have $a_{11}x_1^2+a_{12}x_1x_2+a_{22}x_2^2=0$. If $a_{22}=0$, then $x_1=0$, $x_2=1$ is a solution. If $a_{22}\neq 0$, note first that in no solution can we have $x_1=0$, for then $a_{22}x_2^2=0$ and $x_2=0$; this gives (0, 0, 0) as a "solution," which does not count, however, since it does not represent a point. Since any solution $(0, x_1, x_2)$ need only be determined within a factor, we seek for $(0, 1, x_2/x_1)$. The condition on x_2/x_1 is obtained from $a_{11}x_1^2+a_{12}x_1x_2+a_{22}x_2^2=0$ by dividing this equation by x_1^2. This yields the equation $a_{11}+a_{12}(x_2/x_1)+a_{22}(x_2/x_1)^2=0$. Over the reals, such an equation need not have a solution, but over the complexes there always exists at least one solution. Thus every line meets the locus $\mathbf{x}^T A\mathbf{x}=0$. A similar argument holds for any homogeneous $F(\mathbf{x})$.

COROLLARY. *On every locus* $F(\mathbf{x})=0$ *there are, over the complex field, infinitely many points.*

PROOF. Let $P_1,\cdots, P_s$ be any finite number of points on $F(\mathbf{x})=0$; we show that there is at least one more point on $F(\mathbf{x})=0$. In the complex projective plane there are infinitely many points on a line, hence infinitely many lines on every point. Let A be a point different from the P_i; and let l be a line through A different from the lines AP_i. Let l meet $F(\mathbf{x})=0$ in a point Q. Then Q cannot be one of the points P_i, say P_1, as otherwise l would be the line AP_1, which is not so.

THEOREM. *Over the complex field, the polynomial* $\mathbf{x}^T A\mathbf{x}$ (*assumed* $\neq 0$) *is either reducible, in which case* $\det A=0$ *and* $\mathbf{x}^T A\mathbf{x}=0$ *represents one or two lines; or it is irreducible, in which case* $\det A\neq 0$ *and* $\mathbf{x}^T A\mathbf{x}$ *represents a conic.*

This is a corollary to the previous theorem, since over the complexes a locus $\mathbf{x}^T A\mathbf{x}=0$ cannot be empty or even contain a sole point.

Exercises

1. Let A be a symmetric matrix with real entries. If the locus $\mathbf{x}^T A\mathbf{x}=0$ has only one point (over the reals), then over the complex field $\mathbf{x}^T A\mathbf{x}=0$ consists of two distinct lines. On the other hand, if $\mathbf{x}^T A\mathbf{x}=0$ is empty (over the reals), then over the complex field $\mathbf{x}^T A\mathbf{x}=0$ is a conic.

2. Let A be a symmetric matrix with entries from a field K. Consider the form $\mathbf{x}^T A\mathbf{x}$. Show that rank A is a geometric object. For K the complex field, give a synthetic interpretation of the rank.

We are now going to verify that the previous definition of polar and the present one coincide. The verification is straightforward. There are two cases: (i) the point $\mathbf{a}$ is not on the conic $\mathbf{x}^T A\mathbf{x}=0$, (ii) the point $\mathbf{a}$ is on the conic. We will consider case (i), leaving (ii) as an exercise.

Let $\mathbf{u}$ be a point on $\mathbf{x}^T A\mathbf{x}=0$ and let us ask where the line joining $\mathbf{a}$ and $\mathbf{u}$ meets the conic: analytically we ask for the values of λ such that $\lambda\mathbf{a}+\mathbf{u}$ is on the conic, i.e., $(\lambda\mathbf{a}^T+\mathbf{u}^T)A(\lambda\mathbf{a}+\mathbf{u})=0$. Since $\mathbf{u}^T A\mathbf{u}=0$, this comes to solving $\lambda^2(\mathbf{a}^T A\mathbf{a})+2\lambda\mathbf{u}^T A\mathbf{a}=0$. One root of this equation is $\lambda=0$, which corresponds to the fact that $\mathbf{u}$ is on the conic. The other root is $\lambda=-2\mathbf{u}^T A\mathbf{a}/\mathbf{a}^T A\mathbf{a}$, and one concludes that for this value of λ, the point $\lambda\mathbf{a}+\mathbf{u}$ is on the conic.

Now we compute the fourth harmonic $\mu\mathbf{a}+\mathbf{u}$ of $\mathbf{a}$ with respect to $\mathbf{u}$ and $\lambda\mathbf{a}+\mathbf{u}$. We require then, that $R(\mathbf{a}, \mu\mathbf{a}+\mathbf{u};\ \mathbf{u}, \lambda\mathbf{a}+\mathbf{u})=-1$, or, what is the same thing, $R(\mathbf{a}, \mathbf{u};\ \mu\mathbf{a}+\mathbf{u}, \lambda\mathbf{a}+\mathbf{u})=2$. This gives $(1/\mu)/(1/\lambda)=2$, or $\mu=\lambda/2$.

We now have to verify that $(-\mathbf{u}^T A\mathbf{a}/\mathbf{a}^T A\mathbf{a})\mathbf{a}+\mathbf{u}$ lies on the polar of $\mathbf{a}$; i.e., that $\mathbf{a}^T A\{[-\mathbf{u}^T A\mathbf{a}/\mathbf{a}^T A\mathbf{a}]\mathbf{a}+\mathbf{u}\}=0$. This is immediate and completes the verification.

5. Polarities and conics. We recall that a projective collineation of PNK is an automorphism of the plane that induces a projectivity on the ranges of corresponding lines. Every automorphism is of the form $L_A P^\alpha$, where L_A is a linear transformation and P^α is an automorphism of PNK induced by an automorphism α of K. Since projectivities preserve cross-ratio, so do projective collineations. Let $\mathbf{x}$, $\mathbf{y}$, $\lambda\mathbf{x}+\mu\mathbf{y}$, $\rho\mathbf{x}+\sigma\mathbf{y}$ be four collinear points, of cross-ratio $(\mu/\lambda)/(\sigma/\rho)$. Then $L_A P^\alpha$ sends them into four collinear points of cross-ratio $[(\mu/\lambda)/(\sigma/\rho)]^\alpha$. Hence if $L_A P^\alpha$ is a projective collineation, we must have $\alpha=1$. Conversely, the linear transformations are projective collineations.

Bringing in the lines on a level with the points, we have seen that a linear transformation, or what is the same thing, a projective collineation, has equations of the form

$$\begin{aligned} \mathbf{x} \to \mathbf{x}' &= A\mathbf{x} \\ \mathbf{u} \to \mathbf{u}' &= \mathbf{u}A^{-1}. \end{aligned}$$

Hence a correlation has equations of the form

$$\begin{aligned} \mathbf{x} \to \mathbf{u}' &= (A\mathbf{x})^T = \mathbf{x}^T A^T \\ \mathbf{u} \to \mathbf{x}' &= (\mathbf{u}A^{-1})^T = (A^{-1})^T \mathbf{u}^T = (A^T)^{-1}\mathbf{u}^T. \end{aligned}$$

For the correlation to be a polarity, the line $\mathbf{x}^T A^T$, into which $\mathbf{x}$ goes, itself must go into $\mathbf{x}$: in other words, $(A^T)^{-1}(\mathbf{x}^T A^T)^T = \rho\mathbf{x}$, for some ρ. This yields $(A^T)^{-1}A\mathbf{x}=\rho\mathbf{x}$. Placing $\mathbf{x}$ successively equal to (1, 0, 0), (0, 1, 0), (0, 0, 1) one sees that $(A^T)^{-1}A = \begin{pmatrix} \rho & & \\ & \rho & \\ & & \rho \end{pmatrix}$. Thus we get $A = \rho A^T$. Applying T to both sides, we get $A^T=\rho A$; and from this and $A=\rho A^T$ we get $A=\rho^2 A$. Since A is nonsingular, it has a non-zero entry, from which we conclude that $\rho^2=1$, hence $\rho=+1$ or -1. The value $\rho=-1$ is excluded, for if $\rho=-1$, we get that $A^T=-A$; and, since A is 3×3, that $\det A^T=(-1)^3 \det A$; whence $\det A=\det A^T= -\det A$ and $\det A=0$. This is a contradiction, since A is nonsingular. Thus, in every polarity, $A=A^T$. Conversely, every correlation with $A=A^T$ is a polarity. To sum up:

THEOREM. *The polarities are the correlations*

$$\begin{aligned} \mathbf{x} \to \mathbf{u}' &= \mathbf{x}^T A \\ \mathbf{u} \to \mathbf{x}' &= A^{-1}\mathbf{u}^T \end{aligned}$$

with $A=A^T$.

Thus every polarity is associated with a nonsingular symmetric matrix A. It is also true that every conic is associated with a nonsingular symmetric matrix A: namely, the conic $\mathbf{x}^T A\mathbf{x}=0$ is associated with the matrix A. Through A, the conic is associated with a polarity —the matrix ρA gives the same polarity. Moreover we see that the polarity associated with the conic $\mathbf{x}^T A\mathbf{x}=0$ sends each point into its pole with respect to the conic and each line into its polar.

The question now is whether every polarity (of matrix A) arises in this way from a conic. This amounts to asking whether the locus $\mathbf{x}^T A\mathbf{x}=0$ is a conic. The answer is: not necessarily so. Thus for $A=\begin{pmatrix}1 & & \\ & 1 & \\ & & 1\end{pmatrix}$, the transformation $\mathbf{x}\to\mathbf{x}^T$, $\mathbf{u}\to\mathbf{u}^T$ is a polarity, but, over the real field, the locus $\mathbf{x}^T\mathbf{x}=0$, i.e., $x_0^2+x_1^2+x_2^2=0$ is empty, since the only numbers for which $x_0^2+x_1^2+x_2^2=0$ are $x_0=x_1=x_2=0$, and (0, 0, 0) is not a point. We can prove, however, that if $\mathbf{x}^T A\mathbf{x}=0$ is not empty, then it is a conic. Of course, if $\mathbf{x}^T A\mathbf{x}=0$ has two points (at least), then we know it is a conic (since $\det A\neq 0$, $\mathbf{x}^T A\mathbf{x}$ is irreducible). Thus we have to show that if $\mathbf{x}^T A\mathbf{x}=0$ is not empty, then it contains at least two points.

First a lemma.

LEMMA. *By a change of coordinates, any quadratic form*

$$a_{00}x_0^2+2a_{01}x_0x_1+2a_{02}x_0x_2+a_{11}x_1^2+2a_{12}x_1x_2+a_{22}x_2^2$$

can be brought to the form $ax_0'^2+bx_1'^2+cx_2'^2$.

PROOF. If all the a_{ij} are zero, then the quadratic is already in the desired form. Assume, then, that some $a_{ij}\neq 0$. If a_{00}, a_{11}, a_{22} are not all zero, one of them, a_{00} say, is not zero: if a_{00}, a_{11}, a_{22} are all zero, then a_{01}, a_{02}, a_{12} are not all zero, say $a_{01}\neq 0$; in this case we make the substitution $x_0=x_0'+x_1'$, $x_1=x_0'-x_1'$, $x_2=x_2'$ and in the resulting form find the coefficient of $x_0'^2$ to be different from zero. Thus we may suppose $a_{00}\neq 0$. The quadratic may now be written as

$$a_{00}\left(x_0+\frac{a_{01}}{a_{00}}x_1\right)^2+2a_{02}\left(x_0+\frac{a_{01}}{a_{00}}x_1\right)x_2+\left(a_{11}-\frac{a_{01}^2}{a_{00}}\right)x_1^2$$
$$+2\left(a_{12}-\frac{a_{02}a_{01}}{a_{00}}\right)x_1x_2+a_{22}x_2^2,$$

and making the substitution $x_0+\dfrac{a_{01}}{a_{00}}x_1=x_0'$, $x_1=x_1'$, $x_2=x_2'$, one finds in the resulting expression that the coefficient of $x_0'x_1'$ is zero and the coefficient of $x_0'^2$ is not zero. Thus we may assume the quadratic already to be in the form

$$a_{00}x_0^2+2a_{02}x_0x_2+a_{11}x_1^2+2a_{12}x_1x_2+a_{22}x_2^2$$

with $a_{00}\neq 0$. By a second application of the above argument, we may further suppose $a_{02}=0$, or, in other words, that the given quadratic is in the form

$$a_{00}x_0^2+a_{11}x_1^2+2a_{12}x_1x_2+a_{22}x_2^2.$$

By a repetition of the argument on the sum $a_{11}x_1^2+2a_{12}x_1x_2+a_{22}x_2^2$, we may first suppose $a_{11}\neq 0$ and then $a_{12}=0$. This completes the proof.

THEOREM. *If $A^T=A$, $\det A\neq 0$, and $\mathbf{x}^TA\mathbf{x}=0$ is not empty, then $\mathbf{x}^TA\mathbf{x}=0$ is a conic.*

PROOF. By the lemma, we may suppose $A=\begin{pmatrix} a & & \\ & b & \\ & & c \end{pmatrix}$, with $abc\neq 0$.

Suppose now that (x_0, x_1, x_2) satisfies $ax_0^2+bx_1^2+cx_2^2=0$, and that $x_2\neq 0$. Then first we observe that not both $x_0=0$ and $x_1=0$; otherwise $cx_2^2=0$, $x_2\neq 0$, implies $c=0$, and this is not so. Next we observe that $(x_0, x_1, -x_2)$ also satisfies $ax_0^2+bx_1^2+cx_2^2=0$. Since (x_0, x_1, x_2) and $(x_0, x_1, -x_2)$ are distinct points, the locus $\mathbf{x}^TA\mathbf{x}=0$ has on it at least two points. This, together with the fact that $\det A\neq 0$, establishes that $\mathbf{x}^TA\mathbf{x}=0$ is a conic.

The points $\mathbf{x}$ and $\mathbf{y}$ are conjugates in a polarity if each lies on the polar of the other. The points $\mathbf{x}$ is self-conjugate if it lies on its own polar. In the polarity $\mathbf{x}\to\mathbf{x}^TA$, $\mathbf{u}\to A^{-1}\mathbf{u}^T$, the point $\mathbf{x}$ is self-conjugate if and only if $\mathbf{x}^TA\mathbf{x}=0$. Thus this section can be summarized as follows:

A conic can be defined as the set of self-conjugate points of a polarity having at least one self-conjugate point.

APPENDIX TO CHAPTER X

A1. Factorization of linear transformations into polarities. In this section we will show that every linear transformation in the plane is the product of two polarities. As always when speaking of polarities, we assume that the base field K is commutative and that in it $1+1\neq 0$. Also at a certain point we shall assume K to be infinite, or, at any rate, that it has at least eight elements.

First we will prove an analogue of the stated theorem for linear transformations (=projectivities) on a line. We define an *involution* to be a linear transformation I such that $I\cdot I$ is the identity transformation. If I is an involution and $I:A\to A'$, then $I:A'\to A$. Thus an involution pairs the points on a line into so-called conjugate points, which are interchanged by I. Conversely, if I is a linear transformation that interchanges pairs, then I is an involution.

LEMMA. *If a linear transformation T on a line l interchanges a single pair A, A', $A\neq A'$, of points, then T is an involution.*

PROOF. Let B be another point on l and let $T:B\to B'$. To show that $T:B'\to B$. This is immediate if $B=B'$, so we assume $B\neq B'$. By an early theorem we know that there is a linear transformation U interchanging A, A', B, B'. By the fundamental theorem, there is only one linear transformation such that $A \to A'$, $A' \to A$, $B \to B'$. Hence $U = T$; whence $T: B' \to B$. Q.E.D.

THEOREM. *Every linear transformation T on a line l is either an involution or is the product of two involutions.*

PROOF. If T leaves every point on l fixed, then it is the identity, and hence is the product of any involution by itself. Assume, then, that T is not the identity. By the fundamental theorem, then, T can have at most two fixed points (points X such that $X \to X$). Let A be not fixed under T; and let $T: A \to A'$ and $A' \to A''$. If $A = A''$, then by the lemma T is an involution; suppose $A \neq A''$. Let I be a linear transformation such that $I: A' \to A'$, $A \to A''$, $A'' \to A$. Such a transformation exists by the fundamental theorem and is an involution by the lemma. We now show that $TI = J$ is an involution: in fact, $J(A) = (A)TI = A'I = A'$ and $J(A') = (A')TI = A''I = A$, so J interchanges A and A' and thus is an involution. Hence $T = TII = JI$ is the product of two involutions. Q.E.D.

For the plane it will also be convenient to have some lemmas.

LEMMA 1. *Let ABC be a triangle. A correlation that sends each vertex into the opposite side is a polarity.*

PROOF. Let A, B, C be chosen as the vertices of a coordinate system: the points are therefore $\begin{pmatrix}1\\0\\0\end{pmatrix}$, $\begin{pmatrix}0\\1\\0\end{pmatrix}$, $\begin{pmatrix}0\\0\\1\end{pmatrix}$; and the opposite sides are, respectively, $(1, 0, 0)$, $(0, 1, 0)$, $(0, 0, 1)$. A matrix M must take $\begin{pmatrix}1\\0\\0\end{pmatrix}$ into $(1, 0, 0)$ or a multiple $(a, 0, 0)$ thereof. Hence the first column of M is $\begin{pmatrix}a\\0\\0\end{pmatrix}$. Arguing similarly for the other vertices, we find that M is of the form $\begin{pmatrix}a & 0 & 0\\0 & b & 0\\0 & 0 & c\end{pmatrix}$. Hence, $M = M^T$, so the correlation is a polarity.

LEMMA 2. *Let ABC be a triangle, D a point not on the sides thereof, d a line through D not on the vertices thereof. Then there is a polarity having D and d as self-conjugate elements ($D \to d \to D$) and sending the vertices into the opposite sides.*

PROOF. Let A, B, C have, respectively, the coordinates $\begin{pmatrix}1\\0\\0\end{pmatrix}$, $\begin{pmatrix}0\\1\\0\end{pmatrix}$, $\begin{pmatrix}0\\0\\1\end{pmatrix}$. Let $D: (p_0, p_1, p_2)$; $d: (q_0, q_1, q_2)$. Then $p_0p_1p_2 \neq 0$, since D is not on any of the sides of triangle ABC; and $q_0q_1q_2 \neq 0$ since d is not on A, B, or C. Hence $M = \begin{pmatrix}q_0/p_0 & & \\ & q_1/p_1 & \\ & & q_2/p_2\end{pmatrix}$ defines a polarity of the desired kind.

THEOREM. *Every linear transformation T in the plane is the product of two polarities.*

PROOF. Let A be a point, a a line on A. Consider the transforms under $T: A \to A' \to A''$ and $a \to a' \to a''$. In order to get at one main idea of the proof, let us assume that A, A', A'' are not collinear, a, a', a'' not concurrent, and that each of the lines a, a', a'' is on just one of A, A', A'' (namely, A, A', A'', respectively).

We have

$$T: AA'aa' \to A'A''a'a'',$$

and claim there is a polarity P such that

$$P: A'A''a'a'' \to a'aA'A,$$

namely, the polarity, given by Lemma 2, which sends each vertex of the triangle AA'' $(a \cdot a'')$ into its opposite side and sends A' into a'. Then

$$TP: AA'aa' \to a'aA'A.$$

Thus $TP = Q$ sends each vertex of triangle $AA'(a \cdot a')$ into its opposite side, hence, by Lemma 1, is a polarity. Therefore, $T = T \cdot P \cdot P = Q \cdot P$, which shows that T is a product of two polarities.

It is possible that A, A', A'' are collinear for every choice of A; let us suppose that this is so. We can also suppose that T does not have three noncollinear fixed points, for if it does, then taking them as the three vertices of a coordinate system, T is affected by a matrix of the form $\begin{pmatrix} a & & \\ & b & \\ & & c \end{pmatrix}$, for which the theorem is readily verified. Thus, if there are fixed points for T we may suppose them on some line l. Let A be a point not on l. Then $A \neq A'$; and since A'' is collinear with A, A', T takes line AA' into itself. Let now B be a point not on l and not on AA'. Then BB' is also invariant under T (where $B' = T(B)$). So $O = AA' \cdot BB'$ is a fixed point of T. Now let C be any further point. If C is not fixed for T, then CC' must pass through O, as otherwise the vertices of the triangle of sides AA', BB', CC' would be fixed points. Let now $C = AB \cdot A'B'$. Then C is fixed under T, since C must go into a point on $A'B'$ and into a point of OC, i.e., it must go into $OC \cdot A'B'$, which is C. Now take $a = AB$. Then a, a', a'' are concurrent at C. Using the argument of the proof of Lemma 2, one can find a polarity P such that $P: A'A''a'a'' \to a'aA'A$. The conclusion of the proof (in the presently supposed case) is now exactly as in the last paragraph.

Going back to the case that A, A', A'' are not always collinear, let A, A' A'' be noncollinear. For a we must avoid the lines AA', AA''. For a' we must avoid $A'A$ and $A'A''$ and therefore for a we must avoid the lines which under T go into $A'A$ and $A'A''$. Similarly for a'' we must avoid $A''A$, $A''A'$, and this comes to avoiding two other lines through A. Thus, if six lines through A are avoided, one will find that of the lines a, a', a'' at most one goes through each of A, A', A'' (namely a, a', a'', respectively). We also wish to have a, a', a'' not concurrent. Suppose a, a', a'' are concurrent for three choices of a. The point of concurrency of a, a', a'' is fixed for T, and so we get three fixed points. Assuming as we may by an argument of the last paragraph that T does not have three noncollinear fixed points, we get three collinear points, and hence a whole line of fixed points. The case of a line of fixed points is dual to the case treated in the last paragraph. Therefore we may assume that for three choices of a, not all three times are a, a', a'' concurrent. Thus if we avoid eight lines through A in choosing a, we shall have the situation supposed in the first paragraph, and the proof is complete.

DEFINITION. A plane figure consisting of points and lines is called *self-dual* if there is a one-to-one mapping of points to lines and lines to points preserving incidence.

COROLLARY TO THE THEOREM. *The set of fixed points and fixed lines of a linear transformation forms a self-dual figure.*

¶ *Note:* The above corollary holds with no restriction on the base field K. As the case that K is infinite has already been covered, we may assume K finite. In the case K is finite, one can show that any automorphism of PK^2 has the same number of fixed points as fixed lines: see M. Hall, *The Theory of Groups*, p. 400. Now if A and B are two fixed points of an automorphism of PK^2, the line AB is also fixed; similarly, if a and b are fixed lines, then ab is a fixed point. Thus the set of fixed points and lines satisfy our axioms 1, 2, and 3 (p. 43) for the set to be a plane. If there are four fixed points, no three of which are collinear, then our axiom 4* (p. 55) is satisfied, and the set of fixed points and lines forms a projective plane, hence certainly a self-dual figure. If axiom 4* does not obtain, the given set of points and lines (which satisfy axioms 1, 2, 3) is called a *degenerate plane*. A simple analysis shows that a degenerate plane having the same finite number of points as lines is a self-dual figure.

PROOF. Let the given linear transformation T be written as the product of two polarities: $T = PQ$. Let A be a fixed point of T. Then $(A)P = a$ is a fixed line of T: for $(a)Q = A$, whence $(a)PQ = (A)Q = a$. Similarly, if a is a fixed line of T, then $(a)P = A$ is a fixed point of T. Thus the polarity P sends the figure of fixed points and lines into itself.

Exercises

1. Find the different possible fixed figures of points and lines for linear transformations. Excepting the identity, the figures are:

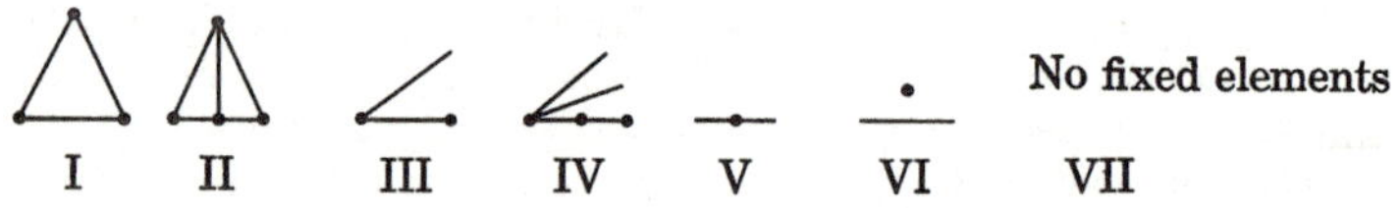

For K the complex field, the type VI cannot occur, though for K real it occurs. Type VII cannot occur for K real or complex.

2. Write down equations for a linear transformation of each of the types mentioned in the previous exercise.

3. A translation in the Euclidean plane induces in the associated projective plane a linear transformation of type IV.

CHAPTER XI

COORDINATES ON A CONIC

1. Coordinates on a conic. We have considered *ranges of points* and *pencils of lines* and now will consider along with them the *set of points on a conic* and the set of *tangents to a conic.* To have a common name for these sets we will call them *one-dimensional forms.* Previously certain simple one-to-one correspondences between ranges and pencils were called perspectivities. Now we extend the term *perspectivity* to include certain simple one-to-one correspondences between one-dimensional forms. Thus let Γ be a conic, A a point on Γ, and make a line on A correspond to its other intersection with Γ: this correspondence we will call a perspectivity. Dually we get a perspectivity between the tangents to Γ and the points on some tangent a to Γ. A *projectivity* is again defined as a product of perspectivities: one has to be careful, however, to check that this word has the same meaning as before when applied to correspondences between ranges of points and pencils of lines. This is essentially accomplished in the following theorem.

THEOREM. *A correspondence between two ranges of points that can be written as a product of perspectivities between one-dimensional forms can also be written as a product of perspectivities from ranges to pencils and from pencils to ranges.*

PROOF. Let $\gamma = \pi_1\pi_2\cdots\pi_s$ be the correspondence written as a product of perspectivities between one-dimensional forms. Let π_i be the first that is to a conic Γ, say to the points of Γ—the case that π_i is to the lines of Γ is treated dually. Then π_i is from a pencil of center A, with A on Γ; and π_{i+1} is from Γ to a pencil of center B on Γ. Thus $\pi_i\pi_{i+1}$ is a projectivity (in the earlier sense), and hence can be rewritten in terms of perspectivities between ranges and pencils. In this way we rewrite γ, perhaps several times, eliminating all perspectivities involving conics.

Exercise. Let Γ be a conic. The correspondence that associates to a point A of Γ the tangent a at A to Γ is a projectivity.

In order to deal analytically with the one-dimensional form of points on a conic Γ, we introduce homogeneous coordinates $\begin{pmatrix} x_0 \\ x_1 \end{pmatrix}$ for the points on Γ as follows. Let A be a point on Γ and fix coordinates in the pencil at A. This comes to choosing three lines $\mathbf{a}_0$, $\mathbf{a}_1$, $\mathbf{a}_2$ through A with $\mathbf{a}_2 = \mathbf{a}_0 + \mathbf{a}_1$, writing any other line through A as $x_0\mathbf{a}_0 + x_1\mathbf{a}_1$, and then ascribing to this line the coordinates $\begin{pmatrix} x_0 \\ x_1 \end{pmatrix}$. Let this line cut Γ in A and X. By definition we now ascribe the coordinates $\begin{pmatrix} x_0 \\ x_1 \end{pmatrix}$ to X.

If coordinates $\begin{pmatrix} x_0' \\ x_1' \end{pmatrix}$ are ascribed to X through a pencil at point B on Γ, then the passage from $\begin{pmatrix} x_0 \\ x_1 \end{pmatrix}$ to $\begin{pmatrix} x_0' \\ x_1' \end{pmatrix}$ is effected by a nonsingular 2×2 matrix, since that is the way one passes from coordinates of AX to coordinates of the projectively related BX. On the basis of this remark one can survey the totality of coordinate systems on Γ. The results are just like those for coordinate systems on a line. For example, any three points on Γ can be assigned the coordinates $\begin{pmatrix} 1 \\ 0 \end{pmatrix}$, $\begin{pmatrix} 0 \\ 1 \end{pmatrix}$, $\begin{pmatrix} 1 \\ 1 \end{pmatrix}$, respectively, and this assignment determines uniquely a coordinate system.

It is immediate that any projectivity $\begin{pmatrix} x_0 \\ x_1 \end{pmatrix} \rightarrow \begin{pmatrix} x_0' \\ x_1' \end{pmatrix}$ between one-dimensional forms is given by the formulas:

$$\begin{aligned} x_0' &= ax_0 + bx_1 \\ x_1' &= cx_0 + dx_1, \end{aligned} \qquad \begin{vmatrix} a & b \\ c & d \end{vmatrix} \neq 0,$$

and conversely, such formulas determine a projectivity.

Introducing the nonhomogeneous coordinates $\theta = x_1/x_0$, $\theta' = x_1'/x_0'$, the formulas read: $\theta' = \dfrac{c + d\theta}{a + b\theta}$, or $b\theta\theta' + a\theta' - d\theta - c = 0$, with $ad - bc \neq 0$. The nonhomogeneous coordinates are frequently convenient; one will, however, have to introduce the symbol ∞ and compute with it in the standard way. The number θ is then called the *parameter* for the element whose coordinate is θ.

Exercise. Show that it is possible to introduce coordinates for the points of a conic Γ and for the lines of Γ in such a way that each point A of Γ has the same coordinates as the tangent a at A.

Consider the conic $x_0x_2 = x_1^2$. Placing $x_0 = 1$, $x_1 = \theta$, we see that the point $(1, \theta, \theta^2)$ is on the conic; and that conversely every point on the conic for which $x_0 \neq 0$, can be written in the form $(1, \theta, \theta^2)$. Moreover, coordinates in the pencil of center $\begin{pmatrix}0\\0\\1\end{pmatrix}$ can be so chosen that the line $x_1 - \theta x_0 = 0$ has coordinate θ. Since any conic is projectively equivalent to $x_0x_2 - x_1^2 = 0$, we can sum up the argument as follows:

THEOREM. *Given a conic* Γ, *one can choose coordinates in the plane and coordinates on* Γ *so that the point on* Γ *of parameter* θ *has coordinates* $(1, \theta, \theta^2)$. (*In homogeneous form: the point of conical coordinates* $\begin{pmatrix}\theta_0\\\theta_1\end{pmatrix}$ *has plane coordinates* $(\theta_0^2, \theta_0\theta_1, \theta_1^2)$.)

2. Projectivities on a conic. Let Γ be a conic, P a point not on Γ. Let X be a point on Γ, and let X' be the other intersection of PX with Γ ($X' = X$ if PX is tangent to Γ). The question is: Is the mapping $X \to X'$ a projectivity?

That the answer is *yes*, at least for the complex field, can be made plausible as follows. Introduce coordinates on Γ and let θ, θ' be the coordinates of X, X', respectively. Given θ, we can, by algebraic formulas that mirror the geometric facts, find θ'. We find, or *would* find, that θ, θ' are related by a formula $f(\theta, \theta') = 0$, where f is a polynomial. We may suppose f free of multiple factors, since, if there are multiple factors, we suppress the multiplicity—i.e., for example, if we had, say, $f = g^r h$, with $r > 1$, we simply replace f by gh. Supposing now that $f(\theta, \theta')$ is free of multiple factors, it is not difficult to show that for a general value of θ the polynomial $f(\theta, \theta')$ has no multiple factors (as a polynomial in θ'), although for special values of θ there may well be multiple factors. Thus, we see (or may suppose) that the degree of $f(\theta, \theta')$ is the number of (distinct) roots θ' of $f(\theta, \theta') = 0$ for general θ. Since X determines X' uniquely, the degree of $f(\theta, \theta')$ in θ' is 1. By the same argument, degree of $f(\theta, \theta')$ in θ is 1. Thus, θ, θ' are related by a formula: $a\theta\theta' + b\theta + c\theta' + d = 0$ or $\theta' = \dfrac{-b\theta - d}{a\theta + c}$. Here $bc \neq ad$; otherwise θ' would be constant. Hence θ, θ' are projectively related.

What we have made plausible is that, if two variables θ, θ' are algebraically related and if the correspondence $\theta \to \theta'$ is one-to-one, then θ, θ' are projectively related. This statement can be made precise and given a rigorous proof—but not with the limited amount of algebra presupposed for the book. Meanwhile the ideas expressed can be used heuristically to guide us to some of the geometric facts.

Let us now proceed to a bona fide proof. In order to introduce a simplifying line of thought, we will first suppose K is the field of complex numbers. In this case the polar p of P meets Γ in two points. Let A be one of them. We may suppose the coordinates on Γ to be established via the pencil of center A. We have $H(AP, p; AX, AX')$. By an early theorem, then, $AX \to AX'$ is a projectivity; hence also $X \to X'$ is a projectivity.

For K the field of reals we can get a proof in the following instructive manner. We consider simultaneously PNK and $PN\bar{K}$, where $\bar{K}$ is the field of complex numbers. We think of PNK as a subplane of $PN\bar{K}$: PNK consist of the points in $PN\bar{K}$ that have real coordinates. The conic Γ is given by some quadratic equation: we use this quadratic equation to define a conic $\bar{\Gamma}$ in $PN\bar{K}$. The conic Γ then consists of the points on $\bar{\Gamma}$ having real coordinates. The correspondence $X \to X'$ on Γ is part of a similar correspondence on $\bar{\Gamma}$. Applying the result for $\bar{\Gamma}$, we have $\theta' = \dfrac{\alpha\theta + \beta}{\gamma\theta + \delta}$, with α, β, γ, δ complex. Write $\alpha = a + ia'$, where a, a' are real and $i = \sqrt{-1}$; and similarly for β, γ, δ. Then we have $\theta' = \dfrac{a\theta + b + i(a'\theta + b')}{c\theta + d + i(c'\theta + d')}$. Consider this equation for real θ only; then θ' is also real, since $(1, \theta', \theta'^2)$ is a point in PNK. Cross-multiplying and comparing real parts of both sides, we get $\theta' = \dfrac{a\theta + b}{c\theta + d}$; here $ad \neq bc$, since θ' is not constant. Hence θ, θ' are projectively related, which completes the proof for K real.

(For arbitrary K, at least if $1 + 1 \neq 0$ in K, a similar line of argument leads to the desired conclusion. Here again one would have to make use of certain advanced, although simple, algebraic notions not presupposed for the book; and therefore we will not pursue this point further. Besides we claim that the correspondence $X \to X'$ is projective in every case, even if $1 + 1 = 0$: we have made use of the inequality $1 + 1 \neq 0$ in bringing in the theory of polars.)

For a completely general proof, we follow up our initial suggestion of finding θ' when θ is given; we do this directly, and therefore have no need for the heuristic remarks above. In order to bring these computations within an expedient radius, we choose coordinate systems in the plane and on the conic in such way that the computations are simplified. According to the last theorem, we take Γ to be the conic $x_0 x_2 = x_1^2$, with θ as parameter, and $\{(1, \theta, \theta^2)\}$ as the points of Γ. Let $P: (a, b, c)$. Then we have:

$$\begin{vmatrix} a & b & c \\ 1 & \theta & \theta^2 \\ 1 & \theta' & \theta'^2 \end{vmatrix} = 0.$$

Subtracting the second row from the third, we get

$$\begin{vmatrix} a & b & c \\ 1 & \theta & \theta^2 \\ 0 & \theta'-\theta & \theta'^2-\theta^2 \end{vmatrix} = 0, \qquad (\theta'-\theta)\cdot\begin{vmatrix} a & b & c \\ 1 & \theta & \theta^2 \\ 0 & 1 & \theta'+\theta \end{vmatrix} = 0.$$

This determinantal equation expresses the condition that the point X' (of coordinate θ') be simultaneously collinear with P and X and on Γ: there are two points X' that satisfy this condition, namely, X itself satisfies the condition, and there is one "other" point X' that satisfies it. The fact that there are "two" values of X' is reflected in the fact that the equation is of degree 2 in θ'. The equation has $\theta'-\theta$ as a factor. We get the "two" points X' by setting the factors separately equal to zero. The equation $\theta'-\theta=0$ gives one value of X', namely X. The equation $\begin{vmatrix} a & b & c \\ 1 & \theta & \theta^2 \\ 0 & 1 & \theta'+\theta \end{vmatrix}=0$ gives another value of X' (which may coincide with X). It is this second value of X' that is originally put into correspondence with X. Therefore $\begin{vmatrix} a & b & c \\ 1 & \theta & \theta^2 \\ 0 & 1 & \theta'+\theta \end{vmatrix}=0$ gives the desired relation between θ and θ'.

Expanding we find: $a\theta\theta'-b(\theta'+\theta)+c=0$, or $\theta'=\dfrac{b\theta-c}{a\theta-b}$; and $b^2\neq ac$, as $P:(a, b, c)$ is not on $\Gamma: x_0x_2=x_1^2$. Hence θ and θ' are projectively related. This completes the proof in the most general case.

Defining an *involution* to be a projectivity whose square is the identity, we can sum the above and its converse in the following theorem.

THEOREM. *Let Γ be a conic, P a point not on Γ. Let X be a point on Γ and let PX cut Γ again in X'. Then the mapping $X\to X'$ is an involution. Conversely, every involution can be obtained this way.*

PROOF. For the converse, let $X_1\leftrightarrow X_1'$, $X_2\leftrightarrow X_2'$ in some involution. Consider the involution determined on Γ by the pencil of center $P=X_1X_1'\cdot X_2X_2'$. Both involutions coincide on X_1, X_1', X_2, hence are identical.

Exercises

1. Find the equation of the involution sending $\infty\to\infty$, $0\to\theta_0$.

2. Show that the following is a construction for $\theta_1+\theta_2$ on a conic. Join θ_1 and θ_2 and let this line cut the tangent at ∞ at P. Then the line joining P to 0 intersects the conic again at $\theta_1+\theta_2$. Show that the following is a construction for $\theta_1\cdot\theta_2$. Join θ_1 and θ_2 and let this cut the join of 0 and ∞ at P. Then the line joining P to 1 cuts the conic again at $\theta_1\theta_2$.

3. Consider the Euclidean plane, and on its line at infinity establish a pairing of points $X \leftrightarrow X'$ by making X and X' correspond if lines through X and X' (and entering the Euclidean plane) are perpendicular. Show that $X \to X'$ is an involution. Hence, show that if Γ is a conic and A a point thereon, then the chords BC (B on Γ, C on Γ) which subtend a right angle at A (i.e., for which BAC is right) all pass through a point on the normal to Γ at A (normal at A = line perpendicular to tangent at A).

4. By a change of coordinates, bring the conic $3x_0^2+x_1^2-5x_2^2-2x_1x_2+x_0x_2=0$ into the form $x_1'^2=x_0'x_2'$. Hence, find a parametrization of the given conic of the form:

$$x_0 = a_{00}\theta^2+a_{01}\theta+a_{02}, \qquad x_1 = a_{10}\theta^2+a_{11}\theta+a_{12}, \qquad x_2 = a_{20}\theta^2+a_{21}\theta+a_{22}.$$

Here det $(a_{ij})\neq 0$; otherwise the points (x_0, x_1, x_2) would all lie on a line.

5. If θ varies over K, the point $(a_{00}\theta^2+a_{01}\theta+a_{02},\ a_{10}\theta^2+a_{11}\theta+a_{12},\ a_{20}\theta^2+a_{21}\theta+a_{22})$ traces out a conic, provided det $(a_{ij})\neq 0$.

6. Let $\theta \to \theta'$ be a projectivity on a conic Γ. If $\theta \to \theta'$ is an involution, then the line joining θ and θ' passes through a fixed point; if $\theta \to \theta'$ is not an involution, then the line joining θ and θ' is tangent to a fixed conic.

7. Let Γ, Γ' be two conics. Any linear transformation taking Γ into Γ' induces a projectivity between Γ and Γ'. Conversely, every projectivity between Γ, Γ' can be obtained this way. Recall a previous exercise to the effect that there is one and only one linear transformation taking Γ into Γ' and three given points A, B, C of Γ, respectively, into three given points A', B', C' of Γ'.

CHAPTER XII

PAIRS OF CONICS

1. Pencils of Conics.† Let $\mathbf{x}^T A\mathbf{x}=0$ and $\mathbf{x}^T B\mathbf{x}=0$ be conics ($A^T=A$, $B^T=B$). By a pencil of conics we mean all the conics of the form $\mathbf{x}^T(\lambda A+\mu B)\mathbf{x}=0$. We can fix (λ, μ) as coordinates of a conic in the pencil: introducing λ/μ, instead of λ, μ, one frequently writes the pencil in the form $\mathbf{x}^T(\lambda A+B)\mathbf{x}=0$, where $\lambda=\infty$ comes into consideration.

The locus $\mathbf{x}^T(\lambda A+B)\mathbf{x}=0$ need not be a conic: necessary and sufficient for it to be a conic is that $|\lambda A+B|\neq 0$. The condition $|\lambda A+B|=0$ is a cubic in λ; therefore, in K there is at least one solution; and there are exactly three if each root λ_0 is counted to the number of times $\lambda-\lambda_0$ is a factor of the polynomial $|\lambda A+B|$. For any such root λ_0, the locus $\mathbf{x}^T(\lambda_0 A+B)\mathbf{x}=0$ consists of two lines, or just one if $\mathbf{x}^T(\lambda_0 A+B)\mathbf{x}$ is a perfect square, in which case we say the line is counted twice. A locus consisting of two lines, or one counted twice, is conveniently referred to as a *degenerate conic*; and the degenerate conics are also conveniently included in the pencil $\mathbf{x}^T(\lambda A+B)\mathbf{x}=0$ (i.e., in order to include degenerate conics, we change slightly the definition of pencil given in the previous paragraph).

Let A, B, C, D be four points, no three of which are collinear. Then the conics through A, B, C, D (including the degenerate ones) form a pencil. In fact, let $f=0$, $g=0$ be two nondegenerate conics through A, B, C, D, where f, g are polynomials of degree 2. Then $\lambda f+g=0$ is also a conic, possibly degenerate, through A, B, C, D, as one immediately checks. Conversely, any conic, possibly degenerate, through A, B, C, D, is of the form $\lambda f+g=0$. To see this, first consider a nondegenerate conic $h=0$ through A, B, C, D. We can suppose that h is not a multiple of f, as in this case we get h from $\lambda f+g$ by placing $\lambda=\infty$. Let E be a point on the conic $h=0$ not on the conic $f=0$ (any point E on $h=0$ distinct from A, B, C, D will have this property). We now substitute the coordinates of E into $\lambda f+g=0$ and solve for λ. This

† In the present chapter, the base field K is taken to be the complex number field.

gives a locus $\lambda f+g=0$ which passes through A, B, C, D, E. This locus is nondegenerate, since no three of A, B, C, D, E are collinear. Thus the conics $h=0$ and $\lambda f+g=0$ have five points, no three of which are collinear, in common, and therefore coincide. This completes the proof for $h=0$ nondegenerate, and the degenerate case is handled similarly.

Exercises

1. Prove that a pencil of conics is a geometric object.

2. Let $f=0$, $g=0$ be two conics (not necessarily meeting in four distinct points). Let $h=0$ be a conic in the pencil $\lambda f+\mu g=0$ different from $f=0$. Then $f=0$ and $h=0$ determine the same pencil as $f=0$ and $g=0$.

2. Intersection multiplicities. Previously we defined an algebraic curve to be a locus given by a homogeneous polynomial equation: $f(x_0, x_1, x_2)=0$. The degree of $f=0$ is called the order of the curve. Let $f=0$, $g=0$ be two curves having only a finite number of points in common. There is a famous theorem, the theorem of Bezout, that states that there are certain geometric objects associated with the points of intersection of the curves, certain positive integers, the so-called intersection multiplicities, such that the sum of the intersection multiplicities equals $(\deg f)\cdot(\deg g)$.

The establishment of Bezout's Theorem, though perhaps possible on the basis of the algebra presupposed, lies beyond the scope of this text. Instead, we will make an *ad hoc* attempt to understand Bezout's Theorem for the case of two conics. The main point is to define intersection multiplicity at a point P of two conics passing through P.

We make some heuristic remarks. Above we saw that the conics through four points A, B, C, D, no three of which are collinear, form a pencil. Let $f=0$, $g=0$ be two conics through A, B, C, D. Since by Bezout's Theorem, the sum of the intersection multiplicities is 4, this shows us that we should define the intersection multiplicity at A, B, C, D to be 1 each. Other cases of pencils can heuristically be thought of as arising by our letting some of the points A, B, C, D come together; if m points come together at A, we should define the intersection multiplicity at A as m. If we consider the pencil $\lambda f+g=0$, then the points A, B, C, D are the same for any two members of the pencil. Hence to define the intersection multiplicity of $f=0$, $g=0$ at A, a point of intersection, we may (or should be able to) replace f, g by any two members of the pencil $\lambda f+g=0$. Let us replace them by f and one of the degenerate members of the pencil. The degenerate conic consists of two lines, l_1, l_2. It is clear how we should define the intersection multiplicity of l_1 with $f=0$ at A—namely, it should be 0 if l_1 does not pass through A; it should be 1 if l_1 passes through A but is not tangent to $f=0$; and it should be 2 if l_1 passes through A but is tangent to $f=0$. This should be true also for l_2; and the intersection multiplicity at A of l_1+l_2 should be the sum of the individual multiplicities.

There is one difficulty in the above suggestion for defining the intersection multiplicity of $f=0$ and $g=0$ at A; namely, in the pencil $\lambda f+g=0$, there may be two distinct degenerate conics l_1+l_2 and l_3+l_4; and conceivably the intersection multiplicity as defined by l_1+l_2 is not the same as that defined by l_3+l_4. Conjecturing that this is not so, we are led to the following theorem:

THEOREM. *Sum of intersection multiplicities of l_1 and of l_2 at $A=$ sum of intersection multiplicities of l_3 and of l_4 at A.*

Before giving the proof, we classify the possible ways a pair of lines can meet a conic. We take into consideration (a) whether the lines are distinct, (b) if they are, whether their intersection is on the conic, and (c) whether one or more of the lines is tangent to the conic.

CASE (i): the lines are distinct, do not meet on the conic, and neither is tangent to the conic.

CASE (ii): the lines are distinct, do not meet on the conic, and just one of them is tangent to the conic.

CASE (iii): the lines are distinct, do not meet on the conic, and both of them are tangent to the conic.

CASE (iv): the lines are distinct, meet on the conic, and neither is tangent to the conic.

CASE (v): the lines are distinct, meet on the conic, and just one of them is tangent to the conic.

CASE (vi): the lines consist of a line not tangent to the conic counted twice.

CASE (vii): the lines consist of a tangent counted twice.

PROOF OF THE THEOREM. Let l_1+l_2 present case (i) relative to $f=0$. Then l_1 and l_2 intersect $f_1=0$ in four distinct points A, B, C, D, with, say $l_1=AB$, $l_2=CD$. The possibilities for l_3, l_4 are $l_3=AC$, $l_4=BD$ and $l_3=AD$, $l_4=BC$. In every case one checks the theorem: both sides of the equality are equal to 1.

Let l_1+l_2 present case (ii) relative to $f=0$. Let l_1 be tangent to the conic at A and l_2 cut the conic in BC ($B\neq C$). Introduce nonhomogeneous coordinates by taking a line at infinity not through A, B, C; taking A to be the origin; and taking l_1 to be the x-axis ($y=0$). Then f has the form $ax+by+Q(x, y)$, where $Q(x, y)$ is the sum of quadratic terms. We know how to compute the tangent at $A=(0, 0)$ to $f=0$; and doing this, we find that it is $ax+by=0$. Since we took $y=0$ to be the tangent, f has the form $y+Q(x, y)$ (or a constant $c\neq 0$ times this, but we may take $c=1$). Let $ux+vy+w=0$ be equation of l_2; $w\neq 0$ since l_2 does not pass through A. Let $h=y(ux+vy+w)$. Since $h=rf+sg$, r, s constants, $s\neq 0$, the linear combinations of f and g are the same as the linear combinations of f and h; therefore, in looking for the degenerate members $\lambda f+\mu g$, we may as well look for the degenerate members $\lambda f+\mu h$. Suppose, then, that $\lambda f+\mu h=0$ consists of two straight lines. If both lines pass through A, then these lines must be AB and AC (since $\lambda f+\mu h$ vanishes at B and at C); and in this case the theorem can be verified at once. If just one of the lines passes through A, then $\lambda f+\mu h=(ax+by)(cx+dy+1)$; therefore the linear terms of $\lambda f+\mu h$ are $ax+by$: but the linear terms of f and of h are y; therefore $\lambda f+\mu h=by(cx+dy+1)$. Of the two lines $\lambda f+\mu h=0$, then, one must be $y=0$; and since $\lambda f+\mu h$ vanishes at B and at C, the other line must be BC. In this case, the theorem is trivially verified.

Let l_1+l_2 present case (iii), and let l_1 be tangent at A and l_2 tangent at C. Let l_3+l_4 be a degenerate member of the pencil $\lambda f+\mu g=0$. If just one of l_3, l_4 passes through A, then by the argument of the last paragraph this must be the tangent at A; in that event, only one of l_3, l_4 passes through C, and, by the same argument, it is the tangent at C. Thus $l_1+l_2=l_3+l_4$ in this case, and the theorem is verified. The remaining case is that both l_3, l_4 pass through both A and C. In this case $l_3=l_4$ and again the theorem is immediately checked.

Let l_1+l_2 present case (iv), l_1, l_2 meeting at A and cutting the conic at B and C. Introduce nonhomogeneous coordinates with a line at infinity not through A, B, or C; with A at the origin; and with $y=0$ as tangent to $f=0$ at origin. We want now to consider the degenerate conics of the form $\lambda f+\mu h=0$, where $h=0$ is equation of l_1+l_2 (and hence h has no linear terms; $f=y+$quadratic terms). Requiring $\lambda f+\mu h=0$ to pass through a point D on BC, $D\neq B$, $D\neq C$, makes $\lambda f+\mu h=0$ break up into two lines, one of which is BC. The other must pass through A. Moreover, this other one is $y=0$, since the linear terms of $\lambda f+\mu h$ are λy. Thus, we are in case (ii), and the theorem is proved in this case.

Let l_1+l_2 present case (v). Let them meet at A. Introduce nonhomogeneous coordinates with A the origin, $y=0$ the tangent, and $x=0$ the other line (so that $xy=0$ is the equation for l_1+l_2; and $f=y+$quadratic terms); let this other line cut the conic again at B. We first observe that both lines l_3, l_4 of a degenerate conic $\lambda f+\mu xy=0$ must pass through A. If only one did, then by a previous argument one sees that this must be $y=0$; say $y=0$ is l_3. The other line, l_4, then would pass through B; and since the total intersection of l_3+l_4 with $f=0$ is $A+B$, l_4 would have to be the tangent at B. But then l_3+l_4 presents case (iii), and we saw under case (iii) that if one of the degenerate conics is in case (iii), the only other one is in case (vi), whereas it is in case (v). Thus both l_3, l_4 must pass through A. Therefore, the equation of l_3+l_4 must have no linear terms. Now $\lambda f+\mu xy=0$ is an equation for l_3+l_4. Therefore, $\lambda=0$ (as there must be no linear terms). Hence $l_3+l_4=l_1+l_2$ and the theorem is verified in case (v). (Note that in case (i) there are three distinct degenerate members in the pencil $\lambda f+\mu g=0$; in cases (ii), (iii), (iv) there are just two degenerate members; whereas in case (v) there is just one degenerate member.)

Let l_1+l_2 present case (vi), so that $l_1=l_2$ and l_1 cuts $f=0$ in points A and B. Introduce nonhomogeneous coordinates with a line at infinity not through A or B, with A the origin, $y=0$ as tangent at A to $f=0$, and $x=0$ as line AB (so that l_1+l_2 has the equation $x^2=0$; $f=y+ax^2+bxy+cy^2$). In the pencil $\lambda f+\mu x^2=0$, consider $y+ax^2+bxy+cy^2-ax^2=0$. This is a degenerate conic: $y(1+bx+cy)=0$. One of these lines, $y=0$, passes through A and is tangent there—therefore does not pass through B; hence the other line $1+bx+cy=0$ passes through B. The conic $y(1+bx+cy)=0$ cuts $f=0$ in $A+B$, and therefore $1+bx+cy=0$ cuts $f=0$ in at most $A+B$; but since this line does not pass through A, it must cut $f=0$ only in B, hence is the tangent at B. Thus we come back to case (iii), and the theorem is verified in this case.

Finally, let l_1+l_2 present case (vii). Let l_1+l_2 cut $f=0$ at A. The point A is the total intersection of $f=0$ and $g=0$; hence also the total intersection of $f=0$ and any degenerate conic l_3+l_4 in the pencil $\lambda f+\mu g=0$. Hence $l_3=l_4=$ (tangent to $f=0$ at A) $=l_1=l_2$. In this case the theorem is, of course, verified at once.

To sum up: we define the intersection multiplicity of $f=0$ and $g=0$ at P to be sum of intersection multiplicity of l_1 with $f=0$ at P and intersection multiplicity of l_2 with $f=0$ at P, where l_1+l_2 is any degenerate member of the pencil $\lambda f+\mu g=0$. With this definition, Bezout's Theorem is verified for conics.

COROLLARY. *Let $h=0$ be any conic in the pencil $\lambda f+\mu g=0$. Then the intersection multiplicity of $h=0$ with $f=0$ at P is the same as that of $g=0$ with $f=0$ at P.*

PROOF. Both intersection multiplicities are defined via the same degenerate member of the pencil $\lambda f+\mu g=0$.

Exercises

1. Let P be a point not on the conic $\Gamma: f=0$. Let $l=0$ be the polar of P with respect to Γ. Then $l^2(P)f-f(P)l^2=0$ is the equation for the pair of tangents from P to Γ.

2. Two conics $f=0$, $g=0$ meeting at P have intersection multiplicity 1 there if and only if the tangents at P to $f=0$ and $g=0$ are distinct.

CHAPTER XIII

QUADRIC SURFACES

1. Projectivities between pencils of planes. In this chapter we will work in a PNK^3, with K, at least at first, an arbitrary commutative field. By a *pencil of planes* we mean the set of planes on a line. Let $a_0x_0+a_1x_1+a_2x_2+a_3x_3=0$, $b_0x_0+b_1x_1+b_2x_2+b_3x_3=0$ be two (distinct) planes: we refer to the planes $\mathbf{a}$ and $\mathbf{b}$, where $\mathbf{a}$ abbreviates (a_0, a_1, a_2, a_3). Clearly every plane $\lambda\mathbf{a}+\mu\mathbf{b}$ passes through the intersection of $\mathbf{a}$ and $\mathbf{b}$, since $\mathbf{ax}=0$, $\mathbf{bx}=0$ implies $(\lambda\mathbf{a}+\mu\mathbf{b})\mathbf{x}=\lambda\mathbf{ax}+\mu\mathbf{bx}=0+0=0$. Conversely, every plane through the intersection of $\mathbf{a}$ and $\mathbf{b}$ is of the form $\lambda\mathbf{a}+\mu\mathbf{b}$. In fact, let $\mathbf{c}$ be such a plane and let $\mathbf{x}$ be a point of $\mathbf{c}$ not on the intersection of $\mathbf{a}$ and $\mathbf{b}$. Solving $\lambda\mathbf{ax}+\mu\mathbf{bx}=0$ for λ, μ, we get a plane passing through the intersection of $\mathbf{a}$ and $\mathbf{b}$ and through $\mathbf{x}$, hence coinciding with $\mathbf{c}$.

We establish *coordinates in the pencil of planes* on a line l as follows. Take three planes $\pi_{\mathbf{a}}$, $\pi_{\mathbf{b}}$, $\pi_{\mathbf{c}}$ on l. Then $\mathbf{c}=\lambda\mathbf{a}+\mu\mathbf{b}$. Replacing $\mathbf{a}$ by $\lambda\mathbf{a}$, $\mathbf{b}$ by $\mu\mathbf{b}$, we adjust the 4-tuples $\mathbf{a}$ and $\mathbf{b}$ so that $\mathbf{c}=\mathbf{a}+\mathbf{b}$. Then any plane $\mathbf{d}$ on l is such that $\mathbf{d}=\rho\mathbf{a}+\sigma\mathbf{b}$ and (ρ, σ) are called the coordinates of $\mathbf{d}$ in the coordinate system $(\pi_{\mathbf{a}}, \pi_{\mathbf{b}}, \pi_{\mathbf{c}})$.

In a similar way, in fact dually, we establish coordinates in a range of points.

Now let p be a pencil of planes on a line l; and let m be a line not meeting l. Then we define a *perspectivity* from p to m as follows: a plane π of p goes into the point of intersection $\pi \cap m$. Similarly we define a perspectivity from m to p. A *projectivity* is then, by definition, a product of perspectivities. (In defining projectivity we presumably ought also to allow the perspectivities between ranges of points and (plane) pencils of lines: this is so, but the above procedure will be sufficient for our immediate purposes.)

On a line, let $(P_{\mathbf{a}}, P_{\mathbf{b}}, P_{\mathbf{a}+\mathbf{b}})$, $(P_{\mathbf{a}'}, P_{\mathbf{b}'}, P_{\mathbf{a}'+\mathbf{b}'})$ be two coordinate systems. Let $\mathbf{a}=\lambda\mathbf{a}'+\mu\mathbf{b}'$, $\mathbf{b}=\rho\mathbf{a}'+\sigma\mathbf{b}'$. Then $x_0\mathbf{a}+x_1\mathbf{b}=(\lambda x_0+\rho x_1)\mathbf{a}'+$

$(\mu x_0+\sigma x_1)\mathbf{b}'$, whence $\begin{pmatrix} x_0' \\ x_1' \end{pmatrix} = \begin{pmatrix} \lambda & \rho \\ \mu & \sigma \end{pmatrix} \begin{pmatrix} x_0 \\ x_1 \end{pmatrix}$; moreover $\begin{vmatrix} \lambda & \rho \\ \mu & \sigma \end{vmatrix} \neq 0$. Thus new coordinates = nonsingular 2×2 matrix times old coordinates. Conversely, any nonsingular 2×2 matrix yields a coordinate change. Similar considerations hold for pencils of planes.

Now we obtain equations for a perspectivity between a pencil p and a range m. Let p be determined by two planes $\mathbf{a}, \mathbf{b}$; and let $\pi_{\mathbf{a}}, \pi_{\mathbf{b}}, \pi_{\mathbf{a}+\mathbf{b}}$ determine a coordinate system in p. Let m intersect $\mathbf{a}$, $\mathbf{b}$ in $\mathbf{x}$, $\mathbf{y}$, respectively; and adjust $\mathbf{x}, \mathbf{y}$ so that m intersects $\pi_{\mathbf{a}+\mathbf{b}}$ in $\mathbf{x}+\mathbf{y}$. Then $\mathbf{ax}=0$, $\mathbf{by}=0$, $(\mathbf{a}+\mathbf{b})\ (\mathbf{x}+\mathbf{y})=0$, whence $\mathbf{ay}+\mathbf{bx}=0$. Then $\pi_{\lambda\mathbf{a}+\mu\mathbf{b}}$ intersects m in $P_{\lambda\mathbf{x}+\mu\mathbf{y}}$: in fact, $(\lambda\mathbf{a}+\mu\mathbf{b})(\lambda\mathbf{x}+\mu\mathbf{y}) = \lambda^2\mathbf{ax}+\lambda\mu(\mathbf{ay}+\mathbf{bx})+\mu^2\mathbf{by}=0+0+0=0$. Hence, if the perspectivity is $\begin{pmatrix} \lambda \\ \mu \end{pmatrix} \to \begin{pmatrix} \lambda' \\ \mu' \end{pmatrix}$, then, in the coordinate systems determined by $(\pi_{\mathbf{a}}, \pi_{\mathbf{b}}, \pi_{\mathbf{a}+\mathbf{b}})$ and $(P_{\mathbf{x}}, P_{\mathbf{y}}, P_{\mathbf{x}+\mathbf{y}})$, the transformation is given by $\begin{pmatrix} \lambda' \\ \mu \end{pmatrix} = \begin{pmatrix} 1 & 0 \\ 0 & 1 \end{pmatrix} \begin{pmatrix} \lambda \\ \mu \end{pmatrix}$, i.e., the transformation is effected by a matrix multiplication (namely, by $\begin{pmatrix} 1 & 0 \\ 0 & 1 \end{pmatrix}$). From our knowledge of all coordinate systems in p and in m, we see that in any coordinate systems, the perspectivity is effected by a 2×2 matrix. Hence also *any projectivity is given by a* 2×2 (*nonsingular*) *matrix*. Since the analytic expressions for the projectivities here considered are the same as those previously considered between lines in a plane, the same, or corresponding, theorems of existence and uniqueness hold here as before. In particular we have the following theorem.

THEOREM. *There is one and only one projectivity sending three given planes of one pencil respectively into three given planes of another pencil.*

Exercises

1. Establish this theorem in a purely synthetic way, assuming that the fundamental theorem holds in some plane.

2. Define (as before) a perspectivity between a range of points and a pencil of lines; and dually define perspectivity between a pencil of planes and a pencil of lines. Let these additional perspectivities intervene in a new definition of projectivity between two pencils of planes. Show that the new definition is equivalent with the old. Use analytic methods if this seems preferable.

THEOREM. *Let p, p' be two pencils of planes on two skew lines l, l'. Let m be a line skew to both l, l'. Let π be a (variable) plane on l. Then $\pi \to \pi' = (\pi \cap m) \cdot l'$ is a projectivity between p and p'. Conversely, every projectivity between p and p' arises in this way.*

PROOF. That $\pi \to \pi' = (\pi \cap m) \cdot l'$ is a projectivity is immediate from the definition. Conversely, let τ be a projectivity between p and p'. Let π_1, π_2, π_3 be three planes on l; and let π_1', π_2', π_3' be their images under τ. Let $\pi_1 \cap \pi_1' = n_1$, $\pi_2 \cap \pi_2' = n_2$, $\pi_3 \cap \pi_3' = n_3$. The three lines n_1, n_2, n_3, meet l, l' and are distinct from them (for example, $n_1 \neq l$, since n_1 is in a plane on l' but l is not). The lines n_1, n_2, n_3 are mutually skew: in fact, n_1, n_2 could not meet on l', as then π_1 would equal π_2; and similarly n_1, n_2 do not meet on l; hence they do not meet at all, as otherwise by Pasch's axiom l, l' would meet. Now let Q_1 be any point of n_1; then through Q_1 there is one and only one line meeting n_1, n_2, n_3. Certainly at most one line, otherwise n_2, n_3 would be coplanar with these lines, and this is not so. To get the one line, join Q_1 to n_2 and let this plane cut n_3 in Q_3. Then Q_1Q_3 cuts n_2 in a point Q_2; and $Q_1Q_2Q_3$ is the desired line. Take Q_1 not on l or l' and call the resulting line m. Then m is skew to l, otherwise at least two of n_1, n_2, n_3 are coplanar; and m is similarly skew to l'. Let $\sigma : \pi \to \pi' = (\pi \cap m) \cdot l'$. Then σ and τ coincide at three elements of p, i.e., at π_1, π_2, π_3. Hence, $\sigma = \tau$ by the previous theorem. Q.E.D.

2. Reguli and quadric surfaces. By a *regulus* one means the set of lines meeting three mutually skew lines. By a *quadric surface* one means the set of points lying on the lines of a regulus.

NOTATION. The quadric surface consisting of the points on the lines meeting three mutually skew lines l, l', m is designated $Q(l, l', m)$. If $\tau : \pi \to \pi' = (\pi \cap m) \cdot l'$ (where π is on l), then the surface is also designated $Q(l, l'; \tau)$.

The regulus of lines meeting l, l', m is called the *first regulus* of $Q(l, l', m)$.

Exercise. Any two lines of a regulus are skew to each other.

THEOREM. *If a line meets three lines of a regulus, in particular, the first regulus of $Q(l, l', m)$, then it meets them all. Hence the lines meeting all the lines of the first regulus themselves form a regulus, called the second regulus of $Q(l, l', m)$. The first regulus consists of all lines meeting all lines of the second regulus. Through any point of any line of the first regulus goes one and only one line of the second; and through any point of a line of the second regulus goes one and only one line of the first.*

PROOF. Let $\overline{m}$ meet three lines l_1, l_2, l_3 of the first regulus. By a notation considered above, let $Q(l, l', m) = Q(l, l'; \tau)$; and similarly, let $Q(l, l', \overline{m}) = Q(l, l'; \overline{\tau})$. Both τ and $\overline{\tau}$ send the planes ll_1, ll_2, ll_3 into $l'l_1$, $l'l_2$, $l'l_3$, respectively. Hence $\tau = \overline{\tau}$; and so the first regulus of $Q(l, l', m)$ (which consists of the intersections of corresponding elements under τ) coincides with the first regulus of $Q(l, l', \overline{m})$. This proves the first point of the theorem. The lines that meet all the lines of the first regulus are the lines that meet some specified three of them, hence form a regulus: this proves the second point. The set of lines meeting the second regulus is, by what has already been proved, the same as the set meeting any three of them, in particular, l, l', m, and thus is the first regulus: this proves the third point. Let P be a point on a line of the first regulus. Through P take a line meeting two other lines of the first regulus; this line will belong to the second regulus; similarly if P is on a line of the second regulus. The proof is now complete.

COROLLARY. *Let l_1, l_1', m_1 be distinct lines of either the first or of the second regulus of $Q(l, l', m)$. Then $Q(l, l', m) = Q(l_1, l_1', m_1)$.*

THEOREM. *If a line meets a quadric in three points, then the line belongs either to the first or to the second regulus.*

PROOF. Let n meet the quadric in P_1, P_2, P_3. Suppose n does not belong to the first regulus. Then n meets the lines through P_1, P_2, P_3 that do belong to the first regulus. Hence n belongs to the second regulus.

Exercises

1. No two lines of the same regulus meet. Every line of the first regulus meets every line of the second. No plane can lie entirely in a quadric.

2. On a quadric there are no reguli except the two already accounted for.

THEOREM. *If a plane cuts a quadric in one line, it cuts it in a second line; the two lines belong to the two reguli.*

PROOF. Let π go through l; and let l belong, say, to the first regulus. Let l' be another line of the first regulus, and write the quadric in the form $Q(l, l'; \tau)$. Let $\tau(\pi) = \pi'$. Then $\pi \cap \pi' = m$ is a line of the second regulus, and π passes through m.

DEFINITION. Let P be any point of a quadric and let l, m be the lines of the first and second reguli through P. Then plane (l, m) is called the *tangent plane to the quadric at* P.

THEOREM. *If a plane is not tangent to a quadric, then it cuts the quadric in a (nondegenerate) conic.*

PROOF. The plane π^* contains no line in the quadric. Let the quadric be $Q(l, l'; \tau)$. Let $\tau : \pi \to \pi'$. Then $\pi \cap \pi^* \to \pi' \cap \pi^*$ is a projectivity, not a perspectivity, as otherwise the fixed line of the perspectivity would be in π^* and on the quadric. The section $\pi^* \cap Q(l, l'; \tau)$ is the set of intersections of corresponding elements of this projectivity, and so is a conic. Q.E.D.

THEOREM. *Every quadric surface has in some coordinate system the equation* $x_0x_3 = x_1x_2$.

PROOF. Let l_1, l_2, l_3 be three lines of the first regulus, m_1, m_2, m_3 three lines of the other. Let $l_1 \cdot m_1 = A$, $l_3 \cdot m_1 = B$, $l_1 \cdot m_3 = C$, $l_3 \cdot m_3 = D$, $l_2 \cdot m_2 = E$. No three of A, B, C, D, E are collinear, so they may be given coordinates

$$\begin{pmatrix}1\\0\\0\\0\end{pmatrix}, \begin{pmatrix}0\\1\\0\\0\end{pmatrix}, \begin{pmatrix}0\\0\\1\\0\end{pmatrix}, \begin{pmatrix}0\\0\\0\\1\end{pmatrix}, \begin{pmatrix}1\\1\\1\\1\end{pmatrix},$$

respectively. Let $l_2 \cdot m_1 = F$, $l_1 \cdot m_2 = G$, $l_3 \cdot m_2 = H$, $l_2 \cdot m_3 = I$. The points F, E, I are collinear: hence the coordinates of E are a linear combination of those of F and I. The first two coordinates of I are zero. Hence the first two coordinates of F are proportional to the first two of E; hence equal. Thus the coordinates of F are $\begin{pmatrix}1\\1\\0\\0\end{pmatrix}$. Similarly

$$G: \begin{pmatrix}1\\0\\1\\0\end{pmatrix}, \quad H: \begin{pmatrix}0\\1\\0\\1\end{pmatrix}, \quad I: \begin{pmatrix}0\\0\\1\\1\end{pmatrix}. \quad \text{Let } P: \begin{pmatrix}x_0\\x_1\\x_2\\x_3\end{pmatrix}$$

be on the quadric. Let the line of the first regulus through P cut m_1, m_3 in R, U, respectively; let the line of the second regulus through P cut l_1, l_3 in S, T, respectively. Then by an argument of the type just employed for F, one finds the coordinates of R, S, T, U to be, respectively,

$$\begin{pmatrix}x_0\\x_1\\0\\0\end{pmatrix}, \begin{pmatrix}x_0\\0\\x_2\\0\end{pmatrix}, \begin{pmatrix}0\\x_1\\0\\x_3\end{pmatrix}, \begin{pmatrix}0\\0\\x_2\\x_3\end{pmatrix}.$$

Let ST meet l_2 in V; then $V: \rho\begin{pmatrix}x_0\\0\\x_2\\0\end{pmatrix} + \sigma\begin{pmatrix}0\\x_1\\0\\x_3\end{pmatrix}$, where $\rho \neq 0$, $\sigma \neq 0$.

This point is a linear combination of F and I; hence the first two coordinates are equal, and so are the last two. Therefore, $\rho x_0 = \sigma x_1$, $\rho x_2 = \sigma x_3$, whence

$$\rho\sigma(x_1x_2 - x_0x_3) = 0 \quad \text{and} \quad x_1x_2 - x_0x_3 = 0.$$

Conversely, suppose a point satisfies the condition $x_1x_2-x_0x_3=0$. If one of the coordinates, say x_0, is zero, then also one of x_1, x_2, say $x_1=0$; the point is then on line CD, hence on the surface. Therefore, we may suppose $x_0x_1x_2x_3\neq 0$. Let

$$\begin{pmatrix} x_0\\ x_1\\ 0\\ 0\end{pmatrix},\quad \begin{pmatrix} x_0\\ 0\\ x_2\\ 0\end{pmatrix},\quad \begin{pmatrix} 0\\ x_1\\ 0\\ x_3\end{pmatrix},\quad \begin{pmatrix} 0\\ 0\\ x_2\\ x_3\end{pmatrix}$$

be labeled R, S, T, U, respectively. The point

$$x_1\begin{pmatrix} x_0\\ 0\\ x_2\\ 0\end{pmatrix}+x_0\begin{pmatrix} 0\\ x_1\\ 0\\ x_3\end{pmatrix}=\begin{pmatrix} x_1x_0\\ x_1x_0\\ x_1x_2\\ x_0x_3\end{pmatrix}$$

is on FI. Hence ST is a line of the second regulus, and on it lies $\begin{pmatrix} x_0\\ x_1\\ x_2\\ x_3\end{pmatrix}$,

i.e., this point is on the surface. Q.E.D.

COROLLARY. *Any two quadric surfaces are projectively equivalent: any one can be sent into any other by a linear transformation.*

COROLLARY. *In any coordinate system a quadric is given by a homogeneous quadratic equation.*

3. Quadric surfaces over the complex field. Let

$$\begin{aligned} f(x_0, x_1, x_2, x_3) &= a_{00}x_0^2+a_{01}x_0x_1+a_{02}x_0x_2+a_{03}x_0x_3\\ &\quad +a_{11}x_1^2+a_{12}x_1x_2+a_{13}x_1x_3+a_{22}x_2^2+a_{23}x_2x_3+a_{33}x_3^2\end{aligned}$$

be an arbitrary quadratic form in four variables. As in the case of three variables, we can write f in one and only one way in the form $\mathbf{x}^TA\mathbf{x}$ with $A^T=A$.

Starting with PNK^3, we consider the coordinate changes of Chapter VIII. If a locus in one coordinate system is given by $\mathbf{x}^TA\mathbf{x}=0$, with $A^T=A$, then in any other coordinate system it is given by $\mathbf{x}'^TB^TAB\mathbf{x}'=0$. Here, if $\mathbf{x}=B\mathbf{x}'$, one finds $\mathbf{x}^TA\mathbf{x}=\mathbf{x}'^TB^TAB\mathbf{x}'$. Note that

$$(B^TAB)^T=B^TA^TB^{TT}=B^TAB.$$

THEOREM. *If* $\mathbf{x}^TA\mathbf{x}=0$, *with* $A^T=A$, *is a quadric, then* $\det A\neq 0$.

PROOF. Change of coordinates brings $\mathbf{x}^T A\mathbf{x}$ into the form $x_1'x_2' - x_0'x_3' = 0$. The matrix for $x_1'x_2' - x_0'x_3'$ is

$$C = \begin{pmatrix} 0 & 0 & 0 & -1/2 \\ 0 & 0 & 1/2 & 0 \\ 0 & 1/2 & 0 & 0 \\ -1/2 & 0 & 0 & 0 \end{pmatrix}$$

i.e., $x_1'x_2' - x_0'x_3' = \mathbf{x}'^T C\mathbf{x}'$. Hence, for some matrix B, $B^T AB = C$. Since $\det C = (\det B)^2 \cdot \det A$, and $\det C \neq 0$, we have $\det A \neq 0$. Q.E.D.

Conversely, for the complex number field, we have:

THEOREM. *If* $\det A \neq 0$, *then* $\mathbf{x}^T A\mathbf{x} = 0$, *with* $A^T = A$, *is a quadric.*

PROOF. Given any homogeneous quadratic form $f(x_0, x_1, x_2, x_3)$ we can by a nonsingular linear transformation bring it to the form $ax_0^2 + bx_1^2 + cx_2^2 + dx_3^2 = 0$. The method of doing this has already been employed in the case of conics. If $f = \mathbf{x}^T A\mathbf{x}$ and $ax_0^2 + bx_1^2 + cx_2^2 + dx_3^2 = \mathbf{x}^T B^T AB\mathbf{x}$, then $\det B^T AB = (\det B)^2 \det A \neq 0$. Moreover, $\det B^T AB = abcd$. Therefore $a \neq 0$, $b \neq 0$, $c \neq 0$, $d \neq 0$. Making the transformation $x_0' = \sqrt{a}x_0$, $x_1' = \sqrt{b}x_1$, $x_2' = \sqrt{c}x_2$, $x_3' = \sqrt{d}x_3$, one can bring $f = 0$ to the form $x_0^2 + x_1^2 + x_2^2 + x_3^2 = 0$. Thus every locus $\mathbf{x}^T A\mathbf{x} = 0$ is projectively equivalent with any other locus $\mathbf{x}^T \bar{A}\mathbf{x} = 0$ provided $\det A \neq 0$ and $\det \bar{A} \neq 0$. In particular every such locus is projectively equivalent with $x_1x_3 - x_0x_2 = 0$; i.e., it is projectively equivalent to a quadric, and so it is a quadric. Q.E.D.

Exercise. The theorem is false for K = real field.

4. Some properties of the sphere. In this section we start with the Euclidean 3-space (over the reals) and assume an elementary knowledge of Euclidean geometry. Just how much is assumed is intentionally being left a little vague, as we wish to make some illustrations without committing ourselves either to a detailed development of the subject or to a linking of it with projective geometry.

First, some properties of circles in the plane will be considered. The plane is the Euclidean plane, but we adjoin the line at infinity as explained earlier.

So far in studying projective geometry we have always posited a base field K. Now, however, we will employ a useful concept, namely, that of *extending the base field.* Let L be a field containing K and let an algebraic curve in PNK be given by an equation $f(x_0, x_1, x_2) = 0$ with coefficients in K—we will say the curve is *over* K or *defined over* K. The idea is to study $f = 0$ over L and from the study get information on the curve over K. Below we always take K = reals, L = complexes.

Let $f^*(x_0, x_1, x_2) = 0$ be an algebraic curve in the plane, where f^* is homogeneous of degree n; and assume that f^* is not divisible by x_0. Placing $x_0=1$, $x_1=x$, $x_2=y$, we get an equation $f^*(1, x, y)=0$, where $f^*(1, x, y)=f(x, y)$ is a polynomial of degree n in x, y. Conversely, given any polynomial $f(x, y)$ of degree n, there is, as one easily sees, one and only one homogeneous polynomial $f^*(x_0, x_1, x_2)$ of degree n such that $f^*(1, x, y)=f(x, y)$; moreover f^* is not divisible by x_0. The loci $f^*(x_0, x_1, x_2)=0$ and $f(x, y)=0$ are not necessarily the same—usually $f^*(x_0, x_1, x_2)=0$ cuts l_∞, while $f(x, y)=0$ is in the affine portion of the plane—but all the same we refer to $f^*(x_0, x_1, x_2)=0$ and $f(x, y)=0$ as the same curve: this is justified by the fact that $f^*(x_0, x_1, x_2)=0$ and $f(x, y)=0$ coincide in the affine portion of the plane, and from the fact that we have a standard way of passing from f^* to f and from f to f^*.

To illustrate, let $f=xy-1$. Then $f^*=x_1x_2-x_0^2$. The locus $f^*=0$ contains just two more points than $f=0$, namely, (0, 1, 0) and (0, 0, 1). Because we call $f=0$ and $f^*=0$ the same curve, we also call $xy=1$ a conic.

Consider $f=x^2+y^2-1$. Then $f^*=x_1^2+x_2^2-x_0^2$. Over the real field $f=0$ and $f^*=0$ have the same locus. The locus $f^*=0$ over the complex field is a much larger set than $f=0$ over the real field. Nonetheless, we study $f^*=0$ over the complex field to get information on $f=0$ over the real field.

Consider $f=x^2+y^2+1$. Then $f^*=x_1^2+x_2^2+x_0^2$. Over the complex field, both loci are conics. Over the real field, both loci are empty, so neither is a conic.

THEOREM. *All the circles in the Euclidean plane cut the line at infinity in the same two points, I, J. Conversely, any conic defined over the reals that passes through I, J is a circle.*

PROOF. Every circle has equation of the form $x^2+y^2+ax+by+c=0$, or, in homogeneous coordinates, $x_1^2+x_2^2+ax_1x_0+bx_2x_0+cx_0^2=0$. Placing $x_0=0$, we get $x_1^2+x_2^2=(x_1+ix_2)(x_1-ix_2)$, from which one sees that all circles pass through the point $(0, 1, i)$ and $(0, 1, -i)$. These are the points I, J.

Conversely, suppose $dx^2+exy+fy^2+ax+by+c=0$ passes through I, J. Then one finds $d+ei+fi^2=0$ and $d-ei+fi^2=0$, i.e., $d-f+ei=0$ and $d-f-ei=0$. Adding and subtracting these equations, one finds $e=0$ and $d-f=0$, or $d=f$. Here $d\neq 0$, as otherwise the given equation is not that of a conic. Supposing, as we may, that $d=1$, the given conic has equation $x^2+y^2+ax+by+c=0$. This represents a circle if it is not empty; and it is not empty since a conic by definition is not empty.

DEFINITION. The lines through a point P joining I, J are called the *isotropic lines* through P.

THEOREM. *The cross-ratio that two lines, defined over the reals and meeting in P, make with the isotropic lines (in one order) is $(1-i\tan\theta)/(1+i\tan\theta)$, where θ is the angle between the lines.*

PROOF. Take P as origin and let the lines be $y=m_1x$, $y=m_2x$. The isotropic lines are $y=ix$, $y=-ix$. These lines meet l_∞ at $(0, 1, m_1)$, $(0, 1, m_2)$ $(0, 1, i)$, $(0, 1, -i)$. The cross-ratio is $\dfrac{m_1-i}{m_1+i}\Big/\dfrac{m_2-i}{m_2+i}$, which simplifies to the expression given in the theorem.

COROLLARY. *The angles between lines l_1, l_2 and $\bar{l}_1$, $\bar{l}_2$ are equal if and only if the cross-ratios between them and the isotropic lines through their intersections are equal. (If the isotropic lines through $P=l_1\cdot l_2$ are taken in the order PI, PJ, then the isotropic lines through $\bar{P}=\bar{l}_1\cdot\bar{l}_2$ are to be taken in the order $\bar{P}I$, $\bar{P}J$.)*

REMARK. In a systematic linking up of projective with Euclidean geometry, the content of the above theorem can be taken as a basis on which to define angle between two lines.

THEOREM. *Parallel planes have the same I, J points.*

PROOF. Take a circle Γ in one of the planes. A transformation $\tau: x'=x+a$, $y'=y+b$, $z'=z+c$ takes the first plane π into the other plane π' and the circle Γ into a circle Γ'. If we identify the planes π and π' with their projective extensions, and the transformation τ with the transformation $\tau^*: x_1'=x_1+ax_0$, $x_2'=x_2+bx_0$, $x_3'=x_3+cx_0$, then the line at infinity of π goes into the line at infinity in π'. Hence I_π, J_π go into $I_{\pi'}$, $J_{\pi'}$: this assertion would hold for any congruent transformation τ, not merely for those of the special kind being considered. Now we make use of the fact that τ, or τ^* rather, leaves every point at infinity fixed. Hence, $\{I_\pi, J_\pi\}=\{I_{\pi'}, J_{\pi'}\}$, Q.E.D.

Let S be a sphere: it has equation $x^2+y^2+z^2+ax+by+cz+d=0$, a, b, c, d real. We regard S as the real part of a locus $\bar{S}$ given by the same equation over the complex field. $\bar{S}$ is a quadric surface (as one easily checks). Let P be a point of S. Then P is a point of $\bar{S}$, and $\bar{S}$ has, according to our discussion of quadrics, a tangent plane at P. We want to verify that this tangent plane (or its real part, rather) is what ordinarily would be called the tangent plane in Euclidean geometry.

Take coordinates so that P is at the origin and the center of the sphere is on the z-axis. Then S has equation $x^2+y^2+z^2-az=0$. The intersection with $z=0$ is $\begin{cases} x^2+y^2=0 \\ z=0 \end{cases}$, or, from the point of view of the geometry of the z-plane, simply: $x^2+y^2=0$. Over the complex field, $x^2+y^2=(x+iy)(x-iy)$, i.e., the intersection is the set of isotropic lines through P, whence $z=0$ is the tangent to the sphere in the sense of our discussion on quadrics. We can put the matter thus:

THEOREM. *The plane π through a point P of a sphere S perpendicular to the radius through P intersects S in the isotropic lines of π through P.*

Exercise. A plane section of a sphere, if not empty, is a circle.

DEFINITION. Let N, S be diametrically opposite points of a sphere $\mathscr{S}$, and let π be the tangent plane to $\mathscr{S}$ at S. Let P be a (variable) point of $\mathscr{S}$, $P \neq N$. The mapping $P \to P' = NP \cap \pi$ is called a *stereographic projection.* The point N is called the *center* of the projection.

THEOREM. *A stereographic projection maps circles not through its center into circles.*

PROOF. Let Γ be a circle on $\mathscr{S}$ not through N. Using the definition of conic, one proves immediately that the projection of Γ onto π, the tangent plane at the antipode S of N, is a conic Γ'. It is now a matter of seeing that Γ' passes through the I, J points of π. Let Γ be cut out on $\mathscr{S}$ by the plane σ. This plane cuts the lines NI_π and NJ_π. Let π' be the tangent plane to $\mathscr{S}$ at N. Then $\{I_\pi, J_\pi\} = \{I_{\pi'}, J_{\pi'}\}$, so NI_π, NJ_π are the isotropic lines of π' through N, hence lie on the sphere by the last theorem. Hence, $\sigma \cap NI_\pi$ lies on Γ, as does $\sigma \cap NJ_\pi$. Hence the projections onto π of these two points lie on the projection of Γ, whence Γ' passes through I_π, J_π and so is a circle. Q.E.D.

THEOREM. *A stereographic projection from a sphere $\mathscr{S}$ to a plane π maps the isotropic lines through any point P of the tangent plane at P into the isotropic lines through P' of the plane π, where P' is the image of P.*

PROOF. If one designates by Γ the pair of isotropic lines at P, and their projection in π by Γ', the proof of the present theorem is just like the proof of the last theorem.

Exercise. By the angle between two curves (say circles) on a sphere $\mathscr{S}$ at a point of intersection P of the curves one means the angle between the tangents to the curves at P. A similar definition holds for plane curves. Prove that stereographic projection preserves angles.

CHAPTER XIV

THE JORDAN CANONICAL FORM

Let A be an $(n+1)\times(n+1)$ nonsingular matrix defining a linear transformation $\mathbf{x}\to\mathbf{y}=A\mathbf{x}$ in a projective n-space S over the complex field K. If $\mathbf{x}'=B\mathbf{x}$ is a change of coordinates, then $\mathbf{x}'\to\mathbf{y}'=B\mathbf{y}=BA\mathbf{x}=BAB^{-1}\mathbf{x}'$, so that in the new coordinate system the transformation is effected by the matrix BAB^{-1}. By varying B, the matrix BAB^{-1} can be given various, and hence simple, forms; and from the forms various information on the transformation can be read off.

By a *fixed space* of a linear transformation T one means a subspace F of S such that $T(P)$ is in F for every point P of F. Suppose S can be written as the join of two fixed spaces F, G that are skew to each other. Then $\dim F+\dim G=n-1$. Let $\dim F=r$, $\dim G=n-r-1$. Let $P_0,\cdots,P_r$ be a basis for F; $P_{r+1},\cdots,P_n$ a basis for G. Choose a coordinate system such that

$$P_0:\begin{pmatrix}1\\0\\0\\\vdots\\0\end{pmatrix}\quad P_1:\begin{pmatrix}0\\1\\0\\\vdots\\0\end{pmatrix},\ \text{etc.}$$

and let T be effected by A in this coordinate system. The points $P_{r+1},\cdots,P_n$ go into points of G, hence into points whose first $r+1$ coordinates are zero; and since the images of $P_{r+1},\cdots,P_n$ are given by the $(r+1)$th, $\cdots$, nth columns of A, we conclude that the first $r+1$ rows of the last $n-r$ columns of A are all zeros. Similarly the last $n-r$ rows of the first $r+1$ columns are all zeros. Schematically A has the following appearance:

$$\begin{array}{c} \\ r+1 \\ n-r \end{array}\begin{array}{c} \begin{array}{cc} r+1 & n-r \end{array} \\ \left(\begin{array}{c|c} C & O \\ \hline O & D \end{array}\right) \end{array}$$

The transformation T induces a linear transformation on F, and in fact this is given, as one easily sees, by the matrix C in a coordinate system having $P_0, \cdots, P_r$ as its vertices; and similarly for G and D. The study of A thus reduces to the study of two smaller matrices, C and D.

Referring to the above situation, we say "A has the blocks C, D along the diagonal and zeros elsewhere." Note that by a change of the subscripts on the P_i, the positions of C and D can be interchanged.

We write $G = F_1 + F_2 + \cdots + F_t$ if each F_i is a fixed space of A and is skew to the join of the other F_j: and we say that G is the *direct sum* of $F_1, \cdots, F_t$. If a fixed space F can be written as $F_1 + F_2$, with F_1, F_2 nonempty, then F is called *reducible*; otherwise it is called *irreducible*. [N.B. All definitions are with respect to a given matrix A or the corresponding transformation T.]

If $G = F_1 + \cdots + F_t$ and $F_t = F_t' + F_t''$, then

$$G = F_1 + \cdots + F_{t-1} + F_t' + F_t''.$$

as one easily checks. To simplify A, then, we first write S as a direct sum of irreducible spaces: this gives A the form of a number of blocks along the diagonal with zeros elsewhere; and the problem reduces to simplifying the appearance of the blocks, which operate on irreducible spaces.

Now we get some information on the fixed spaces.

THEOREM. *Every linear transformation has a fixed point; and, dually, a fixed hyperplane.*

PROOF. We have to see that there is an $\mathbf{x}$ such that $A\mathbf{x} = \lambda\mathbf{x}$ for some λ. This is the same as saying that $(A - \lambda E)\mathbf{x} = 0$ for some λ. And this system will have a solution $\mathbf{x}(\neq \mathbf{0})$ if $\det (A - \lambda E) = 0$. Since we are supposing K to be the complex field, the Fundamental Theorem of Algebra assures us of a solution of the equation $\det (A - \lambda E) = 0$; and hence of an $\mathbf{x}$ and λ such that $A\mathbf{x} = \lambda\mathbf{x}$.

COROLLARY. *Every fixed k-space contains a fixed $(k-1)$-space.*

The polynomial $\det (A - \lambda E)$ is called the *characteristic polynomial* of A. Note that BAB^{-1} has the same characteristic polynomial, since

$$BAB^{-1} - \lambda E = BAB^{-1} - \lambda BEB^{-1} = B(A - \lambda E)B^{-1},$$

and, applying the rule $\det UV = \det U \cdot \det V$,

$$\begin{aligned} \det (BAB^{-1} - \lambda E) &= \det B \cdot \det (A - \lambda E) \cdot \det B^{-1} \\ &= \det (A - \lambda E) \cdot \det(B \cdot B^{-1}) = \det (A - \lambda E). \end{aligned}$$

Thus, in changing coordinate systems we do not change the characteristic polynomial.

A root of $\det (A - \lambda E) = 0$ is called a *characteristic root*. If $A\mathbf{x} = \lambda\mathbf{x}$, then λ is called the associated characteristic root of the fixed point $\mathbf{x}$. To each fixed point there is associated one and only one characteristic root. The characteristic roots are necessarily distinct from zero since det A, the constant term in det $(A-\lambda E)$, is different from zero.

THEOREM. *The characteristic roots of A^{-1} are the reciprocals of the characteristic roots of A.*

PROOF. Let det $(A^{-1}-\mu E)=0$. Then $\mu \neq 0$ and

$$\left(\frac{1}{\mu}\right)^{n+1} \det A \cdot \det (A^{-1}-\mu E) = \det \left(\frac{1}{\mu} E - A\right),$$

so $1/\mu$ is a characteristic root of A. Similarly, the reciprocal of a characteristic root of A is a characteristic root of A^{-1}

If $\mathbf{x} \to \mathbf{y} = A\mathbf{x}$, then the induced transformation on the hyperplanes is $\mathbf{u} \to \mathbf{v} = \mathbf{u}A^{-1}$.

THEOREM. *If T has at least two fixed points, then it has at least two fixed hyperplanes; and conversely by duality.*

PROOF. If A has two distinct characteristic roots, then also A^{-1} has two distinct characteristic roots, and these yield distinct associated hyperplanes. Assume, then, that A has only one characteristic root, λ_0. Then $(A-\lambda_0 E)\mathbf{x}=0$ has at least two linearly independent solutions. This conclusion can be translated into a statement on the rank of $A-\lambda_0 E$ (namely, rank $(A-\lambda_0 E) \leq n-1$). Recall that, for any appropriate matrices U, V, rank $UV \leq$ rank V, and hence in particular that, if U is square and det $U \neq 0$, then rank $UV =$ rank V. Hence rank $(A-\lambda_0 E) =$ rank $(A^{-1}-\lambda_0^{-1}E)$, and we get in turn two linearly independent solutions to $\mathbf{u}(A^{-1}-\lambda_0^{-1}E)=0$.

We next want to prove that an irreducible subspace has only one fixed point, but for purposes of induction we make a slightly stronger statement.

THEOREM. *If S has more than one fixed point, then it is reducible. Moreover if P is a fixed point, then S can be written as $S=F_1+F_2$, with F_1 containing P.*

PROOF. Let Q be a second fixed point. By the previous theorem there are two distinct fixed hyperplanes g, h. If one of g, h does not pass through P, say g does not, then $P+g$ is a desired decomposition. Thus we may suppose both g, h pass through P. If now one does not pass through Q, say g, then $Q+g$ is a desired decomposition. Thus we may suppose both g, h pass through P and Q.

Consider now the space S' of lines through P, i.e., a point of S' is a line of S through P, and a line of S' is a plane of S through P. If S is n-dimensional, then S' is $(n-1)$-dimensional: we are going to make an induction on n. The spaces g, h represent for S' two hyperplanes. The transformation T induces in S' a linear transformation, and g and h are fixed in this transformation. Hence S' has two fixed points (one of which is the line PQ), and so, by induction on n, S' is the direct sum of two spaces, F_1', F_2', one of which, say F_1', contains PQ. Thus $S' = F_1' + F_2'$. In terms of S, one can say that S is the join of two fixed spaces F_1, F_2 with P, Q in F_1 and $F_1 \cap F_2 = P$.

In F_1, the transformation T induces a linear transformation having P, Q as fixed points, hence by induction $F_1 = F_3 + F_4$, with P in F_3. Applying the rule $\dim L + \dim M = \dim (L \cap M) + \dim [L, M]$, one shows that F_4 is skew to the join $[F_2, F_3]$. Hence $S = [F_2, F_3] + F_4$, and the proof is complete.

We now study A under the assumption that S is irreducible for A.

Since S is irreducible, there is only one fixed point, F_0, and hence only one fixed hyperplane F_{n-1}. The transformation induced in F_{n-1} by T has only F_0 as fixed point; therefore in F_{n-1} there is only one fixed $(n-2)$-space F_{n-2}; etc. Thus for each k there is one and only one fixed k-space F_k; and S contains F_{n-1}, which contains $F_{n-2}, \cdots$, which contains F_0.

Take vertices $E_0, E_1, \cdots, E_n$ of a coordinate system as follows. Let $E_0 = F_0$; E_1 any other point in F_1; E_2, any point in F_2 not collinear with E_0E_1; etc. Consider the appearance of A in this coordinate system. Since E_0 goes into itself, the first column of A must be zeros except for the first element; since E_1 goes into a linear combination of

$$\begin{pmatrix} 1 \\ 0 \\ 0 \\ \vdots \\ 0 \end{pmatrix} \quad \text{and} \quad \begin{pmatrix} 0 \\ 1 \\ 0 \\ \vdots \\ 0 \end{pmatrix},$$

the second column of A consists of zeros below the second element; etc. In this way one sees that A has zeros below the main diagonal. Such a matrix is called *triangular*. If $\lambda_0, \cdots, \lambda_n$ are the diagonal elements, then $\prod(\lambda_i - \lambda)$ is the characteristic polynomial, hence the elements in the diagonal are the characteristic roots; and since there is only one characteristic root, A is triangular with equal elements in the diagonal.

Now we are going to show that, by an appropriate choice of the vertices $E_0, \cdots, E_n$ and unit point G, A can be made to take the form

$$\begin{pmatrix} \lambda_1 & 1 & & & \\ & \lambda_1 & 1 & & \\ & & \lambda_1 & & \\ & & & \ddots & 1 \\ & & & & \lambda_1 \end{pmatrix},$$

i.e., the characteristic root λ_1 of A down the main diagonal, ones along the diagonal just above it, and zeros elsewhere. This is the so-called

Jordan canonical form for an *irreducible matrix* (i.e., S is irreducible under T_A).

Let $E_0, E_1, \cdots, E_n$ be taken as above. Then T induces a transformation on F_{n-1} given by a matrix B, where B is obtained from A by deleting the last row and last column of A. By induction on n we may suppose vertices $E_0, \cdots, E_{n-1}$ and unit point G' taken so that B is in Jordan canonical form. Only the last column of A will perhaps not now be in the desired form, and we will show that it takes the desired form upon appropriate choice of the unit point of the coordinate system.

Note that F_{n-1} has equation $x_n = 0$. We take E_n arbitrarily not in $x_n = 0$. That this can be done is an additional assertion, but we assert it and make it part of our induction. Let $E_n' = T(E_n)$. Then $E_n \neq E_n'$. The line E_nE_n' cuts $x_n = 0$ in a point V. The point V could not be in F_{n-2}, otherwise the join of E_n and F_{n-2} would be a fixed hyperplane different from F_{n-1}, and there is only one fixed hyperplane. By induction, then, we may suppose $V = E_{n-1}$.

We have

$$E_n: \begin{pmatrix} 0 \\ \vdots \\ 0 \\ 0 \\ 1 \end{pmatrix}, \quad E_{n-1}: \begin{pmatrix} 0 \\ \vdots \\ 0 \\ 1 \\ 0 \end{pmatrix}, \quad G': \begin{pmatrix} 1 \\ 1 \\ \vdots \\ 1 \\ 0 \end{pmatrix}.$$

We can choose

$$G: \begin{pmatrix} 1 \\ 1 \\ \vdots \\ 1 \end{pmatrix}$$

arbitrarily on E_nG_n', except that $G \neq G_n'$, $G \neq E_n$. Let

$$H: \begin{pmatrix} 1 \\ 1 \\ \vdots \\ 1 \\ 0 \\ 0 \end{pmatrix}.$$

Then HG meets E_nE_{n-1} in

$$K: \begin{pmatrix} 0 \\ 0 \\ \vdots \\ 0 \\ 1 \\ 1 \end{pmatrix}.$$

Thus K can be taken arbitrary on E_nE_{n-1}, except $K \neq E_n$, $K \neq E_{n-1}$. Let us take K so that $R(E_{n-1}, E_n; E_n', K) = \lambda_1$. Then one verifies that

$$E_n' : \begin{pmatrix} 0 \\ \vdots \\ 0 \\ 1 \\ \lambda_1 \end{pmatrix}.$$

Thus the last column of A is a multiple of

$$\begin{pmatrix} 0 \\ \vdots \\ 0 \\ 1 \\ \lambda_1 \end{pmatrix};$$

but since the last entry must be the characteristic root λ_1 itself, the last column is precisely

$$\begin{pmatrix} 0 \\ \vdots \\ 0 \\ 1 \\ \lambda_1 \end{pmatrix}.$$

Thus A is in Jordan canonical form.

Relaxing the condition that A be irreducible, we get the following theorem. First, however, we need a definition.

DEFINITION. A matrix A is said to be in *Jordon canonical form* if A consists of a number of blocks along the diagonal, each of which is in Jordan canonical form for an irreducible matrix.

THEOREM. *Every nonsingular matrix A can be brought to Jordan canonical form by a change of coordinates* $\mathbf{x}' = B\mathbf{x}$, i.e., *for some B, BAB^{-1} is in Jordan canonical form.*

Given a matrix in Jordan canonical form, we can change the order of the blocks at will. Each block belongs to some characteristic root (the number in the diagonal). Let us agree to put first all the blocks belonging to some characteristic root λ_1, then all those belonging to a root $\lambda_2 (\lambda_2 \neq \lambda_1)$, etc. Several blocks can belong to one characteristic root λ_1: we agree to put the smaller ones first. Let then A be in Jordan canonical form and let $A_1, \cdots, A_m$ be the blocks belonging to λ_1. Let A_i be $r_i \times r_i$; and let $r_1 \leq r_2 \leq \cdots \leq r_m$.

THEOREM. *$r_1, \cdots, r_m$ can be computed in terms of the ranks of the matrices $(A-\lambda_1 E)^\rho$, $\rho = 1, 2, \cdots$.*

PROOF. If a matrix A has blocks $C, D, \cdots$ along the diagonal and zeros elsewhere, then rank A = rank C + rank $D + \cdots$, and rank A^ρ = rank C^ρ + rank $D^\rho + \cdots$. Thus to compute the rank of $(A-\lambda_1 E)^\rho$, we must pay attention to the ranks of the blocks in $A-\lambda_1 E$ and their powers. The first block in $A-\lambda_1 E$ has the form

$$C = \begin{pmatrix} 0 & 1 & & & \\ & 0 & 1 & & \\ & & 0 & & \\ & & & \ddots & 1 \\ & & & & 0 \end{pmatrix},$$

where C is $r_1 \times r_1$. One sees that C is of rank $r_1 - 1$. We have

$$C^2 = \begin{pmatrix} 0 & 0 & 1 & & & \\ & 0 & 0 & 1 & & \\ & & \cdot & \cdot & \cdot & \\ & & & 0 & 0 & 1 \\ & & & & 0 & 0 \\ & & & & & 0 \end{pmatrix}$$

and C^2 is of rank $r_1 - 2$; similarly for higher powers of C. Assume now that $r_1 = r_2 = \cdots = r_k < r_{k+1}$. Then rank A − rank $(A-\lambda_1 E) = m$, rank $(A-\lambda_1 E)$ − rank $(A-\lambda_1 E)^2 = m, \cdots,$ rank $(A-\lambda_1 E)^{r_1-1}$ − rank $(A-\lambda_1 E)^{r_1} = m$, rank $(A-\lambda_1 E)^{r_1}$ − rank $(A-\lambda_1 E)^{r_1+1} = m-k$. Thus there is a shift in the differences of the ranks at the (r_1+1)th stage: from this fact one can compute r_1. Moreover, the shift is k, so also k can be computed from the ranks of $(A-\lambda_1 E)^\rho$, $\rho = 1, 2, \cdots$. Suppose now that $r_{k+1} = r_{k+2} = \cdots = r_l < r_{l+1}$. By application of the same ideas one computes r_{k+1} and l; and similarly for all the r_i. This completes the proof.

THEOREM. *Let $A_0 = BAB^{-1}$ and $\bar{A}_0 = \bar{B}A\bar{B}^{-1}$ be Jordan canonical forms for A. Then A_0 and $\bar{A}_0$ have the same blocks on the diagonal: shifting the blocks in A_0 appropriately, we shall have $A_0 = \bar{A}_0$. In other words, the Jordan canonical form of a matrix A is essentially uniquely determined.*

PROOF. The characteristic roots of A, A_0, $\bar{A}_0$ are all the same. Let λ_1 be a common characteristic root. From the previous theorem it will be sufficient to show that rank $(A_0-\lambda_1 E)^\rho$ = rank $(\bar{A}_0-\lambda_1 E)^\rho$; and this equality will follow by showing that each side is rank $(A-\lambda_1 E)^\rho$.

In fact, recalling that rank $UV \leq$ rank V for any matrices U, V, we obtain

$$\begin{aligned}\text{rank } (A_0 - \lambda_1 E)^\rho &= \text{rank } (BAB^{-1} - \lambda_1 E)^\rho = \text{rank } [B(A - \lambda_1 E)B^{-1}]^\rho \\ &= \text{rank } B(A - \lambda_1 E)^\rho B^{-1} \leq \text{rank } (A - \lambda_1 E)^\rho;\end{aligned}$$

and by a similar computation rank $(A - \lambda_1 E)^\rho \leq$ rank $(A_0 - \lambda_1 E)^\rho$. Thus rank $(A_0 - \lambda_1 E)^\rho =$ rank $(A - \lambda_1 E)^\rho$; and similarly rank $(\bar{A}_0 - \lambda_1 E)^\rho$ $=$ rank $(A - \lambda_1 E)^\rho$. This completes the proof.

By the *Segre symbol* of a matrix A in canonical form, or of the corresponding transformation, is meant the following. If a characteristic root λ_1 is associated with blocks of length $p, q, \cdots$ one writes $p, q, \cdots$ in round parenthesis; one does this for each characteristic root; and then places these parentheses in square brackets. Thus the Segre symbol of

$$\begin{pmatrix} \lambda_1 & 1 & & \\ & \lambda_1 & & \\ & & \lambda_1 & \\ & & & \lambda_2 \end{pmatrix},$$

where $\lambda_1 \neq \lambda_2$, is [(21)(1)]. The *type* of a linear transformation is its Segre symbol.

Exercises

1. Write down the possible types of linear transformations in the plane. Write down the corresponding canonical forms. For each type describe the configuration of fixed points and lines.

2. Find the Jordan canonical forms for the following matrices

$$\begin{pmatrix} 0 & 1 \\ 1 & 0 \end{pmatrix}, \quad \begin{pmatrix} 1 & 1 & 1 \\ 0 & 1 & 1 \\ 0 & 0 & 1 \end{pmatrix}, \quad \begin{pmatrix} 1 & 1 & 0 \\ 0 & 1 & 0 \\ 0 & 1 & 1 \end{pmatrix}.$$

3. Simplification of a singular $(n+1) \times (n+1)$ matrix A can be accomplished as follows. To A one adds an xE such that det $(A + xE) \neq 0$, simplifies $A + xE$, i.e., brings it to Jordan canonical form, and then subtracts xE. Defining Jordan canonical form for singular matrices just as above for non-singular, find canonical forms for

$$\begin{pmatrix} 0 & 0 & 1 \\ 0 & 0 & 1 \\ 0 & 0 & 1 \end{pmatrix} \quad \text{and for} \quad \begin{pmatrix} 0 & 1 & 1 \\ 0 & 0 & 1 \\ 0 & 0 & 1 \end{pmatrix}.$$

What significance if any do the characteristic roots of A have for $T = T_A$? Since ρA determines the same transformation as A (but the characteristic roots of ρA are ρ times those of A), the roots cannot have a significance: but the quotients, two by two, do have a significance. This is shown by the following theorem:

THEOREM. *Let P, Q be distinct fixed points for A and let λ, μ be the corresponding associated characteristic roots. Let S be a third point on PQ and let $A: S \to S'$. Then $R(P, Q; S, S') = \lambda/\mu$.*

PROOF. Let P:x, Q:y. Then $A\mathbf{x} = \lambda\mathbf{x}$, $A\mathbf{y} = \mu\mathbf{y}$. Let x, y be chosen so that S:x+y. Then $A(\mathbf{x}+\mathbf{y}) = \lambda\mathbf{x} + \mu\mathbf{y}$ and $R(P, Q; S, S') = 1/1/\mu/\lambda = \lambda/\mu$. Q.E.D.

Exercises

1. Let $\mathcal{S} = F + F'$ be a decomposition of space for a transformation T. Let P, Q be fixed points of F, F', respectively. Let U be a point on PQ and $T: U \to V$. Then T is uniquely determined by the transformations induced in F and in F' and by $R(P,Q; U, V)$.

2. For each type of linear transformation in the plane describe the configuration of fixed points and lines.

3. Use the theorem developed above to show that every involution I on the line different from the identity is given in some coordinate system by the matrix $\begin{pmatrix} 1 & 0 \\ 0 & -1 \end{pmatrix}$; hence, that it has two distinct fixed points P, Q; and that if $I: U \to V$, then U, V separate P, Q harmonically.

BIBLIOGRAPHICAL NOTE

An excellent work on projective geometry is O. Veblen and J. W. Young, *Projective Geometry*, 2 vols., Boston (1910, 1918). See especially Volume II for a systematic linking up on an axiomatic basis of Euclidean (and non-Euclidean) geometry with projective geometry.

For non-Euclidean geometry, see H. Busemann and P. J. Kelly, *Projective Geometry and Projective Metrics*, New York (1953).

For projective geometry without Desargues' Theorem, see M. Hall, Jr., *Projective Planes and Related Topics*, California Institute of Technology (1954); and his book *The Theory of Groups*, New York (1959), especially pp. 346–420. See also G. Pickert, *Projektive Ebenen*, Berlin (1955), which contains an extensive bibliography.

On the introduction of coordinates, see E. Artin, *Geometric Algebra*, New York (1957).

For a development of geometry on an axiomatic, but not projective, basis, see D. Hilbert, *Grundlagen der Geometrie*, 7th ed., Berlin (1930). In English translation there is only the first edition of this work: *Foundations of Geometry*, Illinois (1938).

Once Desargues' Theorem is assumed, a field K makes its appearance and the powerful methods of linear algebra are available. For this, see R. Baer, *Linear Algebra and Projective Geometry*, New York (1952).

To some extent nonlinear geometry can be studied over a noncommutative field. For example, B. Segre, "Elementi di geometria non lineare sopra un corpo sghembo," *Rend. Circ. Mat. Palermo*, Vol. 7 (1958), pp. 81–122, defines a quadric surface as the set of points swept out by the lines (called generators) which meet three mutually skew lines in 3-space, and a conic to be the intersection of a quadric with a plane not through a generator. For the most part, however, the commutativity of K is essential in nonlinear algebro-geometric developments.

With K commutative, one is on the road to algebraic geometry. For an elementary text which continues the study of algebraic curves and surfaces to those of order three, see J. G. Semple and G. T. Kneebone, *Algebraic Projective Geometry*, Oxford (1952).

For algebraic curves, see R. J. Walker, *Algebraic Curves*, Princeton (1950). For algebraic varieties of arbitrary dimension, an excellent textbook is B. L. van der Waerden's, *Einführung in die algebraische Geometrie*, New York (1945). See also W. V. D. Hodge and D. Pedoe, *Methods of Algebraic Geometry*, 3 vols., Cambridge (1954); S. Lang, *Introduction to Algebraic Geometry*, New York (1958); S. Lefschetz, *Algebraic Geometry*, Princeton (1953); J. G. Semple and L. Roth, *Introduction to Algebraic Geometry*, Oxford (1949).

Algebraic geometry is closely associated with commutative algebra. For this, see O. Zariski and P. Samuel, *Commutative Algebra*, Vols. 1 and 2, D. Van Nostrand Co., Inc., Princeton, N.J. (1958, 1960).

For the geometric derivation of the Jordan canonical form, see St. Cohn-Vossen, *Mathematische Annalen*, Vol. 115 (1939), pp. 80–86. For an algebraic derivation, see B. L. van der Waerden, *Modern Algebra*, Vol. 2, pp. 119–122, New York (1950).

INDEX

Affine line, 31
Affine plane, 31, 52
Affine space, 162
Affine 3-space, 103
Algebraic curve, 172
Alias transformation, 153
Alibi transformation, 153
Analytic model, 72, 137
$ANK^n(= AK^n)$, 163
Associativity, 68 f.
Automorphism, 121, 167
Axially perspective, 9
Axioms, 42
  of alignment, 43, 56, 96, 117, 126
Axis, 47
Basis, 124
Betweenness, 7
Bezout's Theorem, 204
Bounded figure, 6
Boundedness, 6
Brianchon's Theorem, 120
Center, of a perspectivity, 12, 47
  of a projection, 3
Centrally perspective, 9
Characteristic root, 219
Class, 37, 140, 142
Collineation, 64, 121
Commutativity, 68 f., 82
Complete 4-line, 63
Complete quadrangle or 4-point, 17, 117
Configuration, 65
Confined, 65
Conics, 106 ff., 177 ff.
  degenerate, 115, 203
  point conic, 120
  line conic, 120
Conjugate, 121
Consistency, 49 ff., 71
Coordinates, 34, 36, 68 ff., 73, 143, 162, 165, 208
  on a conic, 197 ff.
Coordinate systems, 141 ff., 165 ff.
Correlation, 121
Cross-ratio, 21 ff., 27 ff., 159
Dependent, 123, 126, 142
Desargues' Theorem, 9, 56, 60 ff., 75, 88, 89, 102
Determinant, 150
Diagonal point, 17
Dimension, 129
Direction, 8
Distributivity, 69, 81
Division of a segment in a ratio, 8
Duality, 43, 46, 134 ff.
Equivalent, 124, 141, 183
Euclidean plane, 31
Exchange Axiom, 122
Exchange Theorem, 124
Field, 68 ff.
Fixed space, 218
Four-line, 63
Four-point, 17, 117
Fourth harmonic, 15, 17
Free plane, 65
Fundamental Theorem, 25, 89, 94, 95
Generate, 129
Geometric object, 170
Harmonic tetrad, 15, 19, 20 f., 28, 116
Higher-dimensional space, 96 ff., 122 ff.
Homogeneous coordinates, 36
Hyperplane, 133
Ideal points, 31
Identity Axiom, 122
Image, 3
Incidence, 55
Independence, 53 ff.
Independent, 123, 142
Inner product, 145
Intersection multiplicity, 204
Involution, 121, 201
Irreducible space, 219
Isomorphism, 55
Isotropic lines, 215
Join, of two points, 4
  of two spaces, 129
Jordan canonical form, 218 ff.
Left multiplication, 82
Linear dependence, 142
Linear equations, 38
Linear transformations, 141 ff., 153, 155, 160, 163
Matrix, 145
Model, 49, 51, 72
Mystic hexagon, 111
Neutral element, 69, 71
Noncommutative, 69, 86 ff.

Notation, for a conic, 112
for a cross-ratio, 22
for a harmonic tetrad, 17
for the intersection of two spaces, 129
for the join of two spaces, 129
for an n-space over a field, 137
for a perspectivity, 12
for a projectivity, 13
for a quadrangular hexad, 60
for a quadric surface, 210
for a 3-space, 137
Onto, 154
Opposite sides, of a complete quadrangle, 17
of a hexagon, 110
Orientation, 8, 22
Pappus' Theorem, 11, 26, 56, 76, 82, 88, 89
Partial plane, 65
Pascal's Theorem, 110
Pasch's Axiom, 96
Pencil, of lines, 31
of conics, 203
of planes, 208
Perspective, 9
Perspectivity, 12, 46, 47 f., 161, 197, 208
PNK ($= PNK^2$), 165
PNK^n ($= PK^n$), 102, 137
right, 139
left, 139
Point triple, 60
Polar, 115 ff., 184
Polarity, 121, 190
Pole, 119, 184
Polynomial, 172
Polynomial equation, 172
Product of transformations, 13
Projection, 3
Projective invariants, 3 ff.
Projective line, 31
Projective plane, 30, 34, 43, 56, 97
Projective theorems, 1, 11, 18
Projectivity, 12, 46, 157, 158, 197, 208
Quadrangular hexad, 60
Quadric surfaces, 208 ff.
Range, 47
Rank, 125, 140
Regulus, 210
Representative, 36, 73, 141
Right multiplication, 82
Scalar product, 145
Segment, 7
Segre symbol, 225
Self-conjugate, 121
Self-dual, 44, 196
Semilinear, 169
Solvability, 68, 69
Space, 128
3-dimensional, 96 ff.
4-dimensional, 102
Span, 129
Sphere, 214
Stereographic projection, 217
Symmetric, 187
System of type Σ, 55
Tangent, 7, 106, 115, 211
Theorem A, 25, 89, 95
Theory of Dependence, 122
Three-space, 96, 100
Transformation, 12 f.
incidence-preserving, 155
linear, 153, 155, 157 ff., 162, 163
nonsingular, 153
one-to-one, 153
onto, 154
semilinear, 169
singular, 153
univalent, 153
Transpose, 151
Triangle, 17
triple, 60
Trivial solution, 36
Type of a linear transformation, 225
Undefined terms, 42, 43
Unit point, 144, 156
Unproved propositions, 42, 43
Vanishing line, 5
Vanishing points, 4
Vector space, 163
Vertices of a complete 4-point, 17
of a coordinate system, 144, 156
Zero, 70

A CATALOG OF SELECTED

DOVER BOOKS

IN SCIENCE AND MATHEMATICS

Astronomy

BURNHAM'S CELESTIAL HANDBOOK, Robert Burnham, Jr. Thorough guide to the stars beyond our solar system. Exhaustive treatment. Alphabetical by constellation: Andromeda to Cetus in Vol. 1; Chamaeleon to Orion in Vol. 2; and Pavo to Vulpecula in Vol. 3. Hundreds of illustrations. Index in Vol. 3. 2,000pp. 6⅛ x 9¼.
Vol. I: 0-486-23567-X
Vol. II: 0-486-23568-8
Vol. III: 0-486-23673-0

EXPLORING THE MOON THROUGH BINOCULARS AND SMALL TELESCOPES, Ernest H. Cherrington, Jr. Informative, profusely illustrated guide to locating and identifying craters, rills, seas, mountains, other lunar features. Newly revised and updated with special section of new photos. Over 100 photos and diagrams. 240pp. 8¼ x 11. 0-486-24491-1

THE EXTRATERRESTRIAL LIFE DEBATE, 1750–1900, Michael J. Crowe. First detailed, scholarly study in English of the many ideas that developed from 1750 to 1900 regarding the existence of intelligent extraterrestrial life. Examines ideas of Kant, Herschel, Voltaire, Percival Lowell, many other scientists and thinkers. 16 illustrations. 704pp. 5⅜ x 8½. 0-486-40675-X

THEORIES OF THE WORLD FROM ANTIQUITY TO THE COPERNICAN REVOLUTION, Michael J. Crowe. Newly revised edition of an accessible, enlightening book recreates the change from an earth-centered to a sun-centered conception of the solar system. 242pp. 5⅜ x 8½. 0-486-41444-2

A HISTORY OF ASTRONOMY, A. Pannekoek. Well-balanced, carefully reasoned study covers such topics as Ptolemaic theory, work of Copernicus, Kepler, Newton, Eddington's work on stars, much more. Illustrated. References. 521pp. 5⅜ x 8½.
0-486-65994-1

A COMPLETE MANUAL OF AMATEUR ASTRONOMY: TOOLS AND TECHNIQUES FOR ASTRONOMICAL OBSERVATIONS, P. Clay Sherrod with Thomas L. Koed. Concise, highly readable book discusses: selecting, setting up and maintaining a telescope; amateur studies of the sun; lunar topography and occultations; observations of Mars, Jupiter, Saturn, the minor planets and the stars; an introduction to photoelectric photometry; more. 1981 ed. 124 figures. 25 halftones. 37 tables. 335pp. 6½ x 9¼. 0-486-40675-X

AMATEUR ASTRONOMER'S HANDBOOK, J. B. Sidgwick. Timeless, comprehensive coverage of telescopes, mirrors, lenses, mountings, telescope drives, micrometers, spectroscopes, more. 189 illustrations. 576pp. 5⅜ x 8¼. (Available in U.S. only.)
0-486-24034-7

STARS AND RELATIVITY, Ya. B. Zel'dovich and I. D. Novikov. Vol. 1 of *Relativistic Astrophysics* by famed Russian scientists. General relativity, properties of matter under astrophysical conditions, stars, and stellar systems. Deep physical insights, clear presentation. 1971 edition. References. 544pp. 5⅜ x 8¼. 0-486-69424-0

Chemistry

THE SCEPTICAL CHYMIST: THE CLASSIC 1661 TEXT, Robert Boyle. Boyle defines the term "element," asserting that all natural phenomena can be explained by the motion and organization of primary particles. 1911 ed. viii+232pp. 5⅜ x 8½. 0-486-42825-7

RADIOACTIVE SUBSTANCES, Marie Curie. Here is the celebrated scientist's doctoral thesis, the prelude to her receipt of the 1903 Nobel Prize. Curie discusses establishing atomic character of radioactivity found in compounds of uranium and thorium; extraction from pitchblende of polonium and radium; isolation of pure radium chloride; determination of atomic weight of radium; plus electric, photographic, luminous, heat, color effects of radioactivity. ii+94pp. 5⅜ x 8½. 0-486-42550-9

CHEMICAL MAGIC, Leonard A. Ford. Second Edition, Revised by E. Winston Grundmeier. Over 100 unusual stunts demonstrating cold fire, dust explosions, much more. Text explains scientific principles and stresses safety precautions. 128pp. 5⅜ x 8½. 0-486-67628-5

THE DEVELOPMENT OF MODERN CHEMISTRY, Aaron J. Ihde. Authoritative history of chemistry from ancient Greek theory to 20th-century innovation. Covers major chemists and their discoveries. 209 illustrations. 14 tables. Bibliographies. Indices. Appendices. 851pp. 5⅜ x 8½. 0-486-64235-6

CATALYSIS IN CHEMISTRY AND ENZYMOLOGY, William P. Jencks. Exceptionally clear coverage of mechanisms for catalysis, forces in aqueous solution, carbonyl- and acyl-group reactions, practical kinetics, more. 864pp. 5⅜ x 8½. 0-486-65460-5

ELEMENTS OF CHEMISTRY, Antoine Lavoisier. Monumental classic by founder of modern chemistry in remarkable reprint of rare 1790 Kerr translation. A must for every student of chemistry or the history of science. 539pp. 5⅜ x 8½. 0-486-64624-6

THE HISTORICAL BACKGROUND OF CHEMISTRY, Henry M. Leicester. Evolution of ideas, not individual biography. Concentrates on formulation of a coherent set of chemical laws. 260pp. 5⅜ x 8½. 0-486-61053-5

A SHORT HISTORY OF CHEMISTRY, J. R. Partington. Classic exposition explores origins of chemistry, alchemy, early medical chemistry, nature of atmosphere, theory of valency, laws and structure of atomic theory, much more. 428pp. 5⅜ x 8½. (Available in U.S. only.) 0-486-65977-1

GENERAL CHEMISTRY, Linus Pauling. Revised 3rd edition of classic first-year text by Nobel laureate. Atomic and molecular structure, quantum mechanics, statistical mechanics, thermodynamics correlated with descriptive chemistry. Problems. 992pp. 5⅜ x 8½. 0-486-65622-5

FROM ALCHEMY TO CHEMISTRY, John Read. Broad, humanistic treatment focuses on great figures of chemistry and ideas that revolutionized the science. 50 illustrations. 240pp. 5⅜ x 8½. 0-486-28690-8

Engineering

DE RE METALLICA, Georgius Agricola. The famous Hoover translation of greatest treatise on technological chemistry, engineering, geology, mining of early modern times (1556). All 289 original woodcuts. 638pp. 6¾ x 11. 0-486-60006-8

FUNDAMENTALS OF ASTRODYNAMICS, Roger Bate et al. Modern approach developed by U.S. Air Force Academy. Designed as a first course. Problems, exercises. Numerous illustrations. 455pp. 5⅜ x 8½. 0-486-60061-0

DYNAMICS OF FLUIDS IN POROUS MEDIA, Jacob Bear. For advanced students of ground water hydrology, soil mechanics and physics, drainage and irrigation engineering and more. 335 illustrations. Exercises, with answers. 784pp. 6⅛ x 9¼. 0-486-65675-6

THEORY OF VISCOELASTICITY (Second Edition), Richard M. Christensen. Complete consistent description of the linear theory of the viscoelastic behavior of materials. Problem-solving techniques discussed. 1982 edition. 29 figures. xiv+364pp. 6⅛ x 9¼. 0-486-42880-X

MECHANICS, J. P. Den Hartog. A classic introductory text or refresher. Hundreds of applications and design problems illuminate fundamentals of trusses, loaded beams and cables, etc. 334 answered problems. 462pp. 5⅜ x 8½. 0-486-60754-2

MECHANICAL VIBRATIONS, J. P. Den Hartog. Classic textbook offers lucid explanations and illustrative models, applying theories of vibrations to a variety of practical industrial engineering problems. Numerous figures. 233 problems, solutions. Appendix. Index. Preface. 436pp. 5⅜ x 8½. 0-486-64785-4

STRENGTH OF MATERIALS, J. P. Den Hartog. Full, clear treatment of basic material (tension, torsion, bending, etc.) plus advanced material on engineering methods, applications. 350 answered problems. 323pp. 5⅜ x 8½. 0-486-60755-0

A HISTORY OF MECHANICS, René Dugas. Monumental study of mechanical principles from antiquity to quantum mechanics. Contributions of ancient Greeks, Galileo, Leonardo, Kepler, Lagrange, many others. 671pp. 5⅜ x 8½. 0-486-65632-2

STABILITY THEORY AND ITS APPLICATIONS TO STRUCTURAL MECHANICS, Clive L. Dym. Self-contained text focuses on Koiter postbuckling analyses, with mathematical notions of stability of motion. Basing minimum energy principles for static stability upon dynamic concepts of stability of motion, it develops asymptotic buckling and postbuckling analyses from potential energy considerations, with applications to columns, plates, and arches. 1974 ed. 208pp. 5⅜ x 8½. 0-486-42541-X

METAL FATIGUE, N. E. Frost, K. J. Marsh, and L. P. Pook. Definitive, clearly written, and well-illustrated volume addresses all aspects of the subject, from the historical development of understanding metal fatigue to vital concepts of the cyclic stress that causes a crack to grow. Includes 7 appendixes. 544pp. 5⅜ x 8½. 0-486-40927-9

ROCKETS, Robert Goddard. Two of the most significant publications in the history of rocketry and jet propulsion: "A Method of Reaching Extreme Altitudes" (1919) and "Liquid Propellant Rocket Development" (1936). 128pp. 5⅜ x 8½. 0-486-42537-1

STATISTICAL MECHANICS: PRINCIPLES AND APPLICATIONS, Terrell L. Hill. Standard text covers fundamentals of statistical mechanics, applications to fluctuation theory, imperfect gases, distribution functions, more. 448pp. 5⅜ x 8½. 0-486-65390-0

ENGINEERING AND TECHNOLOGY 1650–1750: ILLUSTRATIONS AND TEXTS FROM ORIGINAL SOURCES, Martin Jensen. Highly readable text with more than 200 contemporary drawings and detailed engravings of engineering projects dealing with surveying, leveling, materials, hand tools, lifting equipment, transport and erection, piling, bailing, water supply, hydraulic engineering, and more. Among the specific projects outlined-transporting a 50-ton stone to the Louvre, erecting an obelisk, building timber locks, and dredging canals. 207pp. 8⅜ x 11¼. 0-486-42232-1

THE VARIATIONAL PRINCIPLES OF MECHANICS, Cornelius Lanczos. Graduate level coverage of calculus of variations, equations of motion, relativistic mechanics, more. First inexpensive paperbound edition of classic treatise. Index. Bibliography. 418pp. 5⅜ x 8½. 0-486-65067-7

PROTECTION OF ELECTRONIC CIRCUITS FROM OVERVOLTAGES, Ronald B. Standler. Five-part treatment presents practical rules and strategies for circuits designed to protect electronic systems from damage by transient overvoltages. 1989 ed. xxiv+434pp. 6⅛ x 9¼. 0-486-42552-5

ROTARY WING AERODYNAMICS, W. Z. Stepniewski. Clear, concise text covers aerodynamic phenomena of the rotor and offers guidelines for helicopter performance evaluation. Originally prepared for NASA. 537 figures. 640pp. 6⅛ x 9¼. 0-486-64647-5

INTRODUCTION TO SPACE DYNAMICS, William Tyrrell Thomson. Comprehensive, classic introduction to space-flight engineering for advanced undergraduate and graduate students. Includes vector algebra, kinematics, transformation of coordinates. Bibliography. Index. 352pp. 5⅜ x 8½. 0-486-65113-4

HISTORY OF STRENGTH OF MATERIALS, Stephen P. Timoshenko. Excellent historical survey of the strength of materials with many references to the theories of elasticity and structure. 245 figures. 452pp. 5⅜ x 8½. 0-486-61187-6

ANALYTICAL FRACTURE MECHANICS, David J. Unger. Self-contained text supplements standard fracture mechanics texts by focusing on analytical methods for determining crack-tip stress and strain fields. 336pp. 6⅛ x 9¼. 0-486-41737-9

STATISTICAL MECHANICS OF ELASTICITY, J. H. Weiner. Advanced, self-contained treatment illustrates general principles and elastic behavior of solids. Part 1, based on classical mechanics, studies thermoelastic behavior of crystalline and polymeric solids. Part 2, based on quantum mechanics, focuses on interatomic force laws, behavior of solids, and thermally activated processes. For students of physics and chemistry and for polymer physicists. 1983 ed. 96 figures. 496pp. 5⅜ x 8½. 0-486-42260-7

Mathematics

FUNCTIONAL ANALYSIS (Second Corrected Edition), George Bachman and Lawrence Narici. Excellent treatment of subject geared toward students with background in linear algebra, advanced calculus, physics and engineering. Text covers introduction to inner-product spaces, normed, metric spaces, and topological spaces; complete orthonormal sets, the Hahn-Banach Theorem and its consequences, and many other related subjects. 1966 ed. 544pp. 6⅛ x 9¼. 0-486-40251-7

ASYMPTOTIC EXPANSIONS OF INTEGRALS, Norman Bleistein & Richard A. Handelsman. Best introduction to important field with applications in a variety of scientific disciplines. New preface. Problems. Diagrams. Tables. Bibliography. Index. 448pp. 5⅜ x 8½. 0-486-65082-0

VECTOR AND TENSOR ANALYSIS WITH APPLICATIONS, A. I. Borisenko and I. E. Tarapov. Concise introduction. Worked-out problems, solutions, exercises. 257pp. 5⅝ x 8¼. 0-486-63833-2

AN INTRODUCTION TO ORDINARY DIFFERENTIAL EQUATIONS, Earl A. Coddington. A thorough and systematic first course in elementary differential equations for undergraduates in mathematics and science, with many exercises and problems (with answers). Index. 304pp. 5⅜ x 8½. 0-486-65942-9

FOURIER SERIES AND ORTHOGONAL FUNCTIONS, Harry F. Davis. An incisive text combining theory and practical example to introduce Fourier series, orthogonal functions and applications of the Fourier method to boundary-value problems. 570 exercises. Answers and notes. 416pp. 5⅜ x 8½. 0-486-65973-9

COMPUTABILITY AND UNSOLVABILITY, Martin Davis. Classic graduate-level introduction to theory of computability, usually referred to as theory of recurrent functions. New preface and appendix. 288pp. 5⅜ x 8½. 0-486-61471-9

ASYMPTOTIC METHODS IN ANALYSIS, N. G. de Bruijn. An inexpensive, comprehensive guide to asymptotic methods–the pioneering work that teaches by explaining worked examples in detail. Index. 224pp. 5⅜ x 8½ 0-486-64221-6

APPLIED COMPLEX VARIABLES, John W. Dettman. Step-by-step coverage of fundamentals of analytic function theory–plus lucid exposition of five important applications: Potential Theory; Ordinary Differential Equations; Fourier Transforms; Laplace Transforms; Asymptotic Expansions. 66 figures. Exercises at chapter ends. 512pp. 5⅜ x 8½. 0-486-64670-X

INTRODUCTION TO LINEAR ALGEBRA AND DIFFERENTIAL EQUATIONS, John W. Dettman. Excellent text covers complex numbers, determinants, orthonormal bases, Laplace transforms, much more. Exercises with solutions. Undergraduate level. 416pp. 5⅜ x 8½. 0-486-65191-6

RIEMANN'S ZETA FUNCTION, H. M. Edwards. Superb, high-level study of landmark 1859 publication entitled "On the Number of Primes Less Than a Given Magnitude" traces developments in mathematical theory that it inspired. xiv+315pp. 5⅜ x 8½. 0-486-41740-9

TENSOR CALCULUS, J.L. Synge and A. Schild. Widely used introductory text covers spaces and tensors, basic operations in Riemannian space, non-Riemannian spaces, etc. 324pp. 5⅜ x 8¼. 0-486-63612-7

ORDINARY DIFFERENTIAL EQUATIONS, Morris Tenenbaum and Harry Pollard. Exhaustive survey of ordinary differential equations for undergraduates in mathematics, engineering, science. Thorough analysis of theorems. Diagrams. Bibliography. Index. 818pp. 5⅜ x 8½. 0-486-64940-7

INTEGRAL EQUATIONS, F. G. Tricomi. Authoritative, well-written treatment of extremely useful mathematical tool with wide applications. Volterra Equations, Fredholm Equations, much more. Advanced undergraduate to graduate level. Exercises. Bibliography. 238pp. 5⅜ x 8½. 0-486-64828-1

FOURIER SERIES, Georgi P. Tolstov. Translated by Richard A. Silverman. A valuable addition to the literature on the subject, moving clearly from subject to subject and theorem to theorem. 107 problems, answers. 336pp. 5⅜ x 8½. 0-486-63317-9

INTRODUCTION TO MATHEMATICAL THINKING, Friedrich Waismann. Examinations of arithmetic, geometry, and theory of integers; rational and natural numbers; complete induction; limit and point of accumulation; remarkable curves; complex and hypercomplex numbers, more. 1959 ed. 27 figures. xii+260pp. 5⅜ x 8½. 0-486-63317-9

POPULAR LECTURES ON MATHEMATICAL LOGIC, Hao Wang. Noted logician's lucid treatment of historical developments, set theory, model theory, recursion theory and constructivism, proof theory, more. 3 appendixes. Bibliography. 1981 edition. ix + 283pp. 5⅜ x 8½. 0-486-67632-3

CALCULUS OF VARIATIONS, Robert Weinstock. Basic introduction covering isoperimetric problems, theory of elasticity, quantum mechanics, electrostatics, etc. Exercises throughout. 326pp. 5⅜ x 8½. 0-486-63069-2

THE CONTINUUM: A CRITICAL EXAMINATION OF THE FOUNDATION OF ANALYSIS, Hermann Weyl. Classic of 20th-century foundational research deals with the conceptual problem posed by the continuum. 156pp. 5⅜ x 8½. 0-486-67982-9

CHALLENGING MATHEMATICAL PROBLEMS WITH ELEMENTARY SOLUTIONS, A. M. Yaglom and I. M. Yaglom. Over 170 challenging problems on probability theory, combinatorial analysis, points and lines, topology, convex polygons, many other topics. Solutions. Total of 445pp. 5⅜ x 8½. Two-vol. set.
Vol. I: 0-486-65536-9 Vol. II: 0-486-65537-7

INTRODUCTION TO PARTIAL DIFFERENTIAL EQUATIONS WITH APPLICATIONS, E. C. Zachmanoglou and Dale W. Thoe. Essentials of partial differential equations applied to common problems in engineering and the physical sciences. Problems and answers. 416pp. 5⅜ x 8½. 0-486-65251-3

THE THEORY OF GROUPS, Hans J. Zassenhaus. Well-written graduate-level text acquaints reader with group-theoretic methods and demonstrates their usefulness in mathematics. Axioms, the calculus of complexes, homomorphic mapping, *p*-group theory, more. 276pp. 5⅜ x 8½. 0-486-40922-8

Math–Geometry and Topology

ELEMENTARY CONCEPTS OF TOPOLOGY, Paul Alexandroff. Elegant, intuitive approach to topology from set-theoretic topology to Betti groups; how concepts of topology are useful in math and physics. 25 figures. 57pp. 5⅜ x 8½. 0-486-60747-X

COMBINATORIAL TOPOLOGY, P. S. Alexandrov. Clearly written, well-organized, three-part text begins by dealing with certain classic problems without using the formal techniques of homology theory and advances to the central concept, the Betti groups. Numerous detailed examples. 654pp. 5⅜ x 8½. 0-486-40179-0

EXPERIMENTS IN TOPOLOGY, Stephen Barr. Classic, lively explanation of one of the byways of mathematics. Klein bottles, Moebius strips, projective planes, map coloring, problem of the Koenigsberg bridges, much more, described with clarity and wit. 43 figures. 210pp. 5⅜ x 8½. 0-486-25933-1

THE GEOMETRY OF RENÉ DESCARTES, René Descartes. The great work founded analytical geometry. Original French text, Descartes's own diagrams, together with definitive Smith-Latham translation. 244pp. 5⅜ x 8½. 0-486-60068-8

EUCLIDEAN GEOMETRY AND TRANSFORMATIONS, Clayton W. Dodge. This introduction to Euclidean geometry emphasizes transformations, particularly isometries and similarities. Suitable for undergraduate courses, it includes numerous examples, many with detailed answers. 1972 ed. viii+296pp. 6⅛ x 9¼. 0-486-43476-1

PRACTICAL CONIC SECTIONS: THE GEOMETRIC PROPERTIES OF ELLIPSES, PARABOLAS AND HYPERBOLAS, J. W. Downs. This text shows how to create ellipses, parabolas, and hyperbolas. It also presents historical background on their ancient origins and describes the reflective properties and roles of curves in design applications. 1993 ed. 98 figures. xii+100pp. 6½ x 9¼. 0-486-42876-1

THE THIRTEEN BOOKS OF EUCLID'S ELEMENTS, translated with introduction and commentary by Sir Thomas L. Heath. Definitive edition. Textual and linguistic notes, mathematical analysis. 2,500 years of critical commentary. Unabridged. 1,414pp. 5⅜ x 8½.. Three-vol. set.
Vol. I: 0-486-60088-2 Vol. II: 0-486-60089-0 Vol. III: 0-486-60090-4

SPACE AND GEOMETRY: IN THE LIGHT OF PHYSIOLOGICAL, PSYCHOLOGICAL AND PHYSICAL INQUIRY, Ernst Mach. Three essays by an eminent philosopher and scientist explore the nature, origin, and development of our concepts of space, with a distinctness and precision suitable for undergraduate students and other readers. 1906 ed. vi+148pp. 5⅜ x 8½. 0-486-43909-7

GEOMETRY OF COMPLEX NUMBERS, Hans Schwerdtfeger. Illuminating, widely praised book on analytic geometry of circles, the Moebius transformation, and two-dimensional non-Euclidean geometries. 200pp. 5⅝ x 8¼. 0-486-63830-8

DIFFERENTIAL GEOMETRY, Heinrich W. Guggenheimer. Local differential geometry as an application of advanced calculus and linear algebra. Curvature, transformation groups, surfaces, more. Exercises. 62 figures. 378pp. 5⅜ x 8½. 0-486-63433-7

History of Math

THE WORKS OF ARCHIMEDES, Archimedes (T. L. Heath, ed.). Topics include the famous problems of the ratio of the areas of a cylinder and an inscribed sphere; the measurement of a circle; the properties of conoids, spheroids, and spirals; and the quadrature of the parabola. Informative introduction. clxxxvi+326pp. 5⅜ x 8½. 0-486-42084-1

A SHORT ACCOUNT OF THE HISTORY OF MATHEMATICS, W. W. Rouse Ball. One of clearest, most authoritative surveys from the Egyptians and Phoenicians through 19th-century figures such as Grassman, Galois, Riemann. Fourth edition. 522pp. 5⅜ x 8½. 0-486-20630-0

THE HISTORY OF THE CALCULUS AND ITS CONCEPTUAL DEVELOPMENT, Carl B. Boyer. Origins in antiquity, medieval contributions, work of Newton, Leibniz, rigorous formulation. Treatment is verbal. 346pp. 5⅜ x 8½. 0-486-60509-4

THE HISTORICAL ROOTS OF ELEMENTARY MATHEMATICS, Lucas N. H. Bunt, Phillip S. Jones, and Jack D. Bedient. Fundamental underpinnings of modern arithmetic, algebra, geometry and number systems derived from ancient civilizations. 320pp. 5⅜ x 8½. 0-486-25563-8

A HISTORY OF MATHEMATICAL NOTATIONS, Florian Cajori. This classic study notes the first appearance of a mathematical symbol and its origin, the competition it encountered, its spread among writers in different countries, its rise to popularity, its eventual decline or ultimate survival. Original 1929 two-volume edition presented here in one volume. xxviii+820pp. 5⅜ x 8½. 0-486-67766-4

GAMES, GODS & GAMBLING: A HISTORY OF PROBABILITY AND STATISTICAL IDEAS, F. N. David. Episodes from the lives of Galileo, Fermat, Pascal, and others illustrate this fascinating account of the roots of mathematics. Features thought-provoking references to classics, archaeology, biography, poetry. 1962 edition. 304pp. 5⅜ x 8½. (Available in U.S. only.) 0-486-40023-9

OF MEN AND NUMBERS: THE STORY OF THE GREAT MATHEMATICIANS, Jane Muir. Fascinating accounts of the lives and accomplishments of history's greatest mathematical minds–Pythagoras, Descartes, Euler, Pascal, Cantor, many more. Anecdotal, illuminating. 30 diagrams. Bibliography. 256pp. 5⅜ x 8½. 0-486-28973-7

HISTORY OF MATHEMATICS, David E. Smith. Nontechnical survey from ancient Greece and Orient to late 19th century; evolution of arithmetic, geometry, trigonometry, calculating devices, algebra, the calculus. 362 illustrations. 1,355pp. 5⅜ x 8½. Two-vol. set. Vol. I: 0-486-20429-4 Vol. II: 0-486-20430-8

A CONCISE HISTORY OF MATHEMATICS, Dirk J. Struik. The best brief history of mathematics. Stresses origins and covers every major figure from ancient Near East to 19th century. 41 illustrations. 195pp. 5⅜ x 8½. 0-486-60255-9

Physics

OPTICAL RESONANCE AND TWO-LEVEL ATOMS, L. Allen and J. H. Eberly. Clear, comprehensive introduction to basic principles behind all quantum optical resonance phenomena. 53 illustrations. Preface. Index. 256pp. 5⅜ x 8½. 0-486-65533-4

QUANTUM THEORY, David Bohm. This advanced undergraduate-level text presents the quantum theory in terms of qualitative and imaginative concepts, followed by specific applications worked out in mathematical detail. Preface. Index. 655pp. 5⅜ x 8½. 0-486-65969-0

ATOMIC PHYSICS (8th EDITION), Max Born. Nobel laureate's lucid treatment of kinetic theory of gases, elementary particles, nuclear atom, wave-corpuscles, atomic structure and spectral lines, much more. Over 40 appendices, bibliography. 495pp. 5⅜ x 8½. 0-486-65984-4

A SOPHISTICATE'S PRIMER OF RELATIVITY, P. W. Bridgman. Geared toward readers already acquainted with special relativity, this book transcends the view of theory as a working tool to answer natural questions: What is a frame of reference? What is a "law of nature"? What is the role of the "observer"? Extensive treatment, written in terms accessible to those without a scientific background. 1983 ed. xlviii+172pp. 5⅜ x 8½. 0-486-42549-5

AN INTRODUCTION TO HAMILTONIAN OPTICS, H. A. Buchdahl. Detailed account of the Hamiltonian treatment of aberration theory in geometrical optics. Many classes of optical systems defined in terms of the symmetries they possess. Problems with detailed solutions. 1970 edition. xv + 360pp. 5⅜ x 8½. 0-486-67597-1

PRIMER OF QUANTUM MECHANICS, Marvin Chester. Introductory text examines the classical quantum bead on a track: its state and representations; operator eigenvalues; harmonic oscillator and bound bead in a symmetric force field; and bead in a spherical shell. Other topics include spin, matrices, and the structure of quantum mechanics; the simplest atom; indistinguishable particles; and stationary-state perturbation theory. 1992 ed. xiv+314pp. 6⅛ x 9¼. 0-486-42878-8

LECTURES ON QUANTUM MECHANICS, Paul A. M. Dirac. Four concise, brilliant lectures on mathematical methods in quantum mechanics from Nobel Prize-winning quantum pioneer build on idea of visualizing quantum theory through the use of classical mechanics. 96pp. 5⅜ x 8½. 0-486-41713-1

THIRTY YEARS THAT SHOOK PHYSICS: THE STORY OF QUANTUM THEORY, George Gamow. Lucid, accessible introduction to influential theory of energy and matter. Careful explanations of Dirac's anti-particles, Bohr's model of the atom, much more. 12 plates. Numerous drawings. 240pp. 5⅜ x 8½. 0-486-24895-X

ELECTRONIC STRUCTURE AND THE PROPERTIES OF SOLIDS: THE PHYSICS OF THE CHEMICAL BOND, Walter A. Harrison. Innovative text offers basic understanding of the electronic structure of covalent and ionic solids, simple metals, transition metals and their compounds. Problems. 1980 edition. 582pp. 6⅛ x 9¼. 0-486-66021-4